Textbooks in Electrical and Electronic Engineering

Series Editors

G. Lancaster J. K. Sykulski E. W. Williams

---

1. Introduction to fields and circuits
GORDON LANCASTER
2. The CD-ROM and optical recording systems
E. W. WILLIAMS
3. Engineering electromagnetism: physical processes and computation
P. HAMMOND and J. K. SYKULSKI
4. Integrated circuit engineering
L. J. HERBST
5. Electrical circuits and systems: an introduction for engineers and physical scientists
A.M. HOWATSON
6. Applied numerical modelling for engineers
D. DE COGAN and A. DE COGAN
7. Electromagnetism for engineers: an introductory course (Fourth edition)
P. HAMMOND
8. Instrumentation for engineers and scientists
JOHN TURNER and MARTYN HILL
9. Introduction to error control codes
SALVATORE GRAVANO

Introduction to Error Control Codes

Salvatore Gravano

University of Keele

OXFORD
UNIVERSITY PRESS

OXFORD
UNIVERSITY PRESS

Great Clarendon Street, Oxford OX2 6DP

Oxford University Press is a department of the University of Oxford.
It furthers the University's objective of excellence in research, scholarship,
and education by publishing worldwide in

Oxford New York

Athens Auckland Bangkok Bogotá Buenos Aires Cape Town
Chennai Dar es Salaam Delhi Florence Hong Kong Istanbul Karachi
Kolkata Kuala Lumpur Madrid Melbourne Mexico City Mumbai
Nairobi Paris São Paulo Singapore Taipei Tokyo Toronto Warsaw

with associated companies in Berlin Ibadan

Oxford is a registered trade mark of Oxford University Press
in the UK and in certain other countries

Published in the United States
by Oxford University Press Inc., New York

First published 2001

A catalogue record for this book is available from the British Library

Library of Congress Cataloging in Publication Data
(data applied for)

ISBN 0–19–856231–4

Typeset by Newgen Imaging Systems (P) Ltd., Chennai, India

Printed in Great Britain on acid-free paper by
Biddles Ltd., Guildford, Surrey

Dedicated to my wife Jo, and children Giuseppe, Rose and Antonio.

Preface

Error-control codes may not sound like the world's most exciting topic, but it is an important and interesting subject, that finds applications in data communications (e.g. satellite communications) and data storage systems (compact disc systems), and applications increase as the unstopable advance of digital technology continues.

This book has arisen out of notes put together whilst working on data communications in industry and later teaching error-control codes at both postgraduate and third year undergraduate level (and it is at the later level that the book is primarily aimed at). My main sources of reference have been *Error-Correcting Codes*, Peterson and Weldon, Mitt Press, 1972 (2nd Edition); *Error-Control Coding*, Lin and Costello, Prentice Hall, 1983; and *Theory and Practice of Error Control Codes*, Blahut, Addison-Wesley, 1984.

The basic principles of error detection and correction are first illustrated through the use of relatively simple codes such as single-parity-check codes and repetition codes. Here we address the question of what is the simplest way that we can detect the occurrence of errors in binary data, and then what is the simplest way that we can correct errors. Furthermore we see the price that has to be paid to enable error detection and correction, error control does not come free. Illustrative examples are given to lead the reader step by step from these basic error-control techniques through to linear codes, cyclic codes, BCH codes and convolutional codes. Vector spaces and Galois fields, required for linear codes and BCH codes respectively, are covered starting from a basic level to a level sufficient to understand the necessary codes. Where possible complex proofs have been omitted to keep the text concise and easy to follow.

Keele S.G.
March 2001

Contents

Tables

Block codes 1

Information transferred within an electronic communication channel is always liable to corruption by noise within the channel. Signals conveying information can be so contaminated by noise that the information becomes erroneous. It may be possible to reduce the level of noise but its complete elimination is not possible. The need therefore arises to be able to preserve the accuracy of information as it journeys through a noisy channel. Addressing this problem, Claude Shannon in 1948, showed that associated with every channel is an upper limit on the rate at which information can be transmitted reliably through the channel. This limitation on the capacity of a channel to transmit information is referred to as the *channel capacity*. Furthermore Shannon proved the existence of codes that enable information to be transmitted through a noisy channel such that the probability of errors is as small as required, providing that the transmission rate does not exceed the channel capacity. If information is transmitted at a rate greater than the channel capacity then it is not possible to achieve error-free transmission. Shannon's theoretical work on channel capacity and error-free transmission is now referred to as the *channel coding theorem*. The theorem does not say what the codes are or even how we go about finding them, it just proves their existence. It is a quite remarkable theorem as it tells us that there is no limit to the level of accuracy that can be achieved. It seems reasonable to expect a limit on the accuracy with which information can be reliably transmitted, however it is not accuracy that is limited but rather the rate at which information can be transmitted error free.

The codes referred to in the channel coding theorem do not prevent the occurrence of errors but rather allow their presence to be detected and corrected. As such the codes are known as *error-detecting* and *error-correcting codes* or for short *error-control codes*. Error-control codes fall into the categories of block codes and convolutional codes. We consider mainly block codes, convolutional codes are considered in Chapter 8. Before introducing block codes it is useful to consider the digital communication channel in general and then from the point of view of error-control codes.

1.1 The digital communication channel

The phrase *communication channel* is used here in a wide sense to describe any electronic system involving the transfer of information, and not just telecommunication systems. For example the transfer of data between the main memory of a computer and a data-storage device can be viewed as a communication system or subsystem. Applications of error-control codes tend to fall into the categories of digital telecommunication systems and data-storage systems. The main body of the theory of error-control codes, namely the construction of codes, encoding, decoding and performance evaluation, can be formulated without reference to the applications.

Figure 1.1 shows a block diagram of a communication system. The *information source* provides information, in either a digital or analogue form, to the system. The information can be a message or data from some other system or person. The *source encoder* generates a binary signal that gives an efficient representation of the source information. This may involve the use of codes, other than error-control codes, that minimize the number of bits needed to represent each message and allow the message to be uniquely reconstructed by the *source decoder*. If the output from the information source is in an analogue form, then source encoding needs to be preceded by analogue-to-digital conversion.

The *channel encoder* carries out error-control coding for the purpose of protecting information against errors incurred as it progresses through the noisy channel. This is achieved by including additional information such that the *channel decoder* is able to accurately recover the source information despite the presence of errors. The transmission of information into the *channel* is performed by the *modulator*. In a telecommunication system the channel could typically be a wire link, a microwave link, a satellite link or some other type of link. The channel output feeds into the *demodulator* which carries out the inverse operation of the modulator, so producing a stream of bits from the received signal. The source and the channel encoders along with the modulator form the *transmitter*, whilst the demodulator and the source and the channel decoders form the *receiver* within the system. In a data storage system the modulator, the channel and the demodulator can be thought of as the writing unit, the storage medium and the reading unit respectively. The reconstruction of the source information by the source decoder is the last stage of the communication process, beyond this lies the *information user* which hopefully

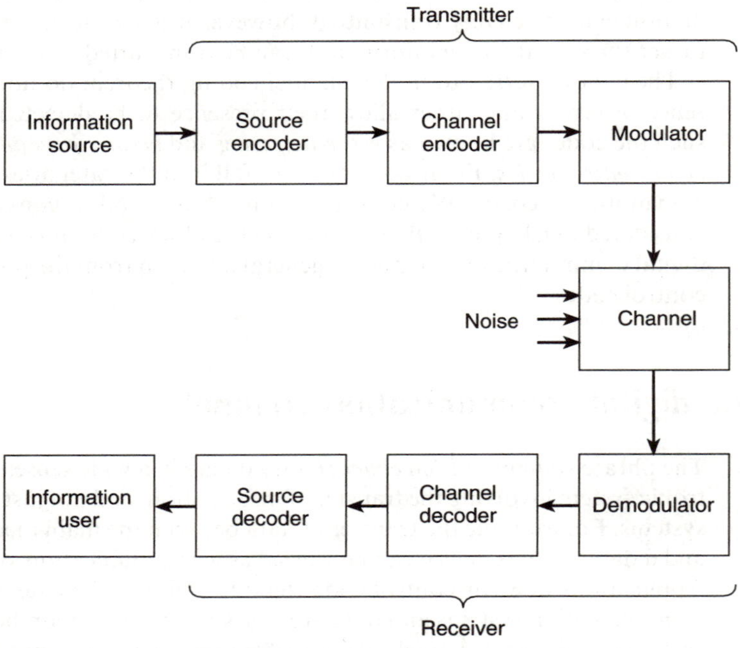

Fig. 1.1 Block diagram of a communication system.

accurately receives information generated at the source. Note that the nature and contents of the information is of no importance to the communication system, which serves solely to enable communication.

With regard to error-control coding the communication system shown in Fig. 1.1 can be simplified to that shown in Fig. 1.2. Here the *digital source* combines the information source and the source encoder, and the *digital sink* combines the source decoder and the information user. The channel now includes the modulator and demodulator, and the channel encoder and channel decoder are now referred to simply as the *encoder* and the *decoder* respectively. The channel input is a stream of binary bits with 'values' 0 or 1, and we no longer think of noise occurring within the channel but rather the occurrence of *bit errors*, so that 0 and 1 become 1 and 0 respectively. The probability of an error can be determined from the characteristics of the modulator and the demodulator along with the statistical nature of the noise.

Of particular importance is the *binary symmetric channel* in which the probabilities of bits 0 and 1 incurring errors are equal. Figure 1.3 shows the transitions that can occur in the binary symmetric channel, the probability of a bit incurring an error is p and is referred to as the *bit-error probability* or the *transition probability*. The probability of a bit being received error free is $1-p$. Errors occur randomly within the binary symmetric channel, so whether a bit incurs an error is independent of whether other bits incur errors. Such a channel with *random errors* is known as a *random-error channel*, codes developed for dealing specifically with random errors are called *random-error-control codes*. If errors have a tendency to occur in groups or bursts then the channel is called a burst-error channel. We assume the channel to be binary symmetric and we will be considering random-error-control codes.

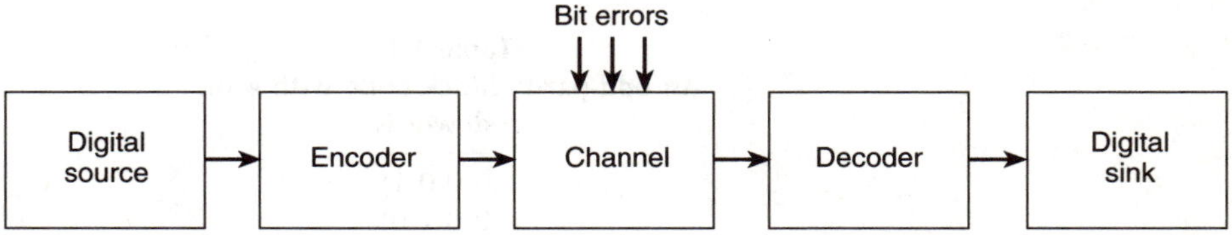

Fig. 1.2 A digital channel for error-control coding.

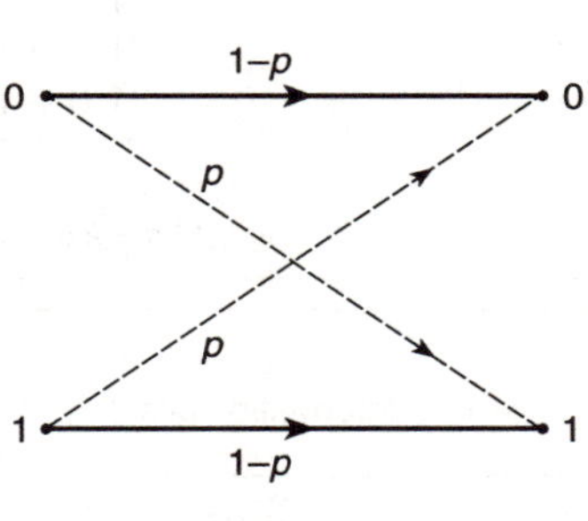

Fig. 1.3 Transitions in the binary symmetric channel.

1.2 Introduction to block codes

A *block code* is a set of words that has a well-defined mathematical property or structure, and where each word is a sequence of a fixed number of bits. The words belonging to a block code are called *codewords*. Table 1.1 shows an example of a simple block code with 4-bit codewords and where each codeword has *odd parity* (i.e. an odd number of ones). Note that a word with m bits is referred to as an m-bit word.

A codeword consists of *information bits* that carry information *per se*, and *parity-check bits* (also referred to as *parity bits* or *check bits*) that carry no information, in the sense of that carried by the information bits, but ensure that the codeword has the correct structure required by the block code. Blocks of information bits, referred to as *information words*, are encoded into codewords by an encoder for the code. The encoder determines the parity bits and appends them to the information word so giving a codeword. A code whose codewords have k information bits and r parity bits, has n-bit codewords where $n = k + r$. Such a code is referred to as an *(n, k) block code* where n and k are respectively the *blocklength* and *information length* of the code. The position of the parity bits within a codeword is quite arbitrary. They can be dispersed within the information bits or kept together and placed on either side of the information bits. Figure 1.4 shows a codeword whose parity bits are on the right-hand side of the information bits. A codeword whose information bits are kept together, so that they are readily identifiable, is said to be in a systematic form or to be *systematic*, otherwise the codeword is referred to as *nonsystematic*. The codeword in Fig. 1.4 is in a systematic form. A block code whose codewords are systematic is referred to as a *systematic code*. For clarity systematic codes are normally preferred to nonsystematic codes, however there are reasons for sometimes preferring nonsystematic codes (see Section 3.4).

Table 1.1
An odd-parity block code with 4-bit codewords

(0 0 0 1)
(0 0 1 0)
(0 1 0 0)
(1 0 0 0)
(0 1 1 1)
(1 1 1 0)
(1 1 0 1)
(1 0 1 1)

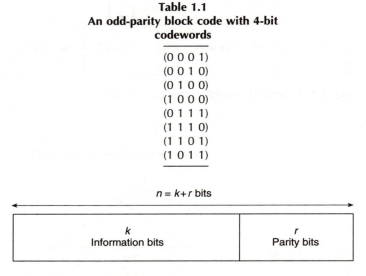

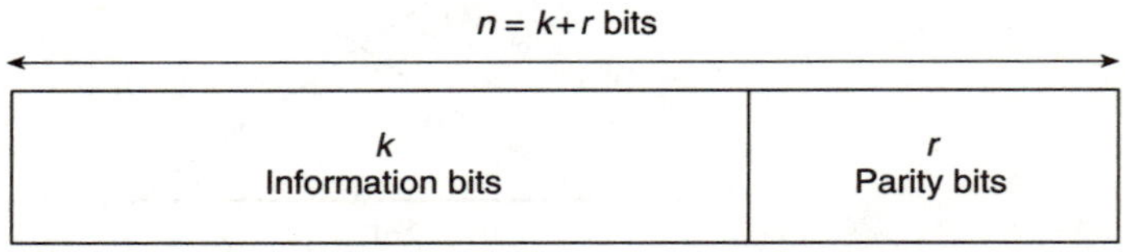

Fig. 1.4 An n-bit systematic codeword.

Table 1.2
The (5, 4) block code with odd parity

Codewords	*Redundant words*
(0 0 0 0 1)	(0 0 0 0 0)
(0 0 0 1 0)	(0 0 0 1 1)
(0 0 1 0 0)	(0 0 1 0 1)
(0 0 1 1 1)	(0 0 1 1 0)
(0 1 0 0 0)	(0 1 0 0 1)
(0 1 0 1 1)	(0 1 0 1 0)
(0 1 1 0 1)	(0 1 1 0 0)
(0 1 1 1 0)	(0 1 1 1 1)
(1 0 0 0 0)	(1 0 0 0 1)
(1 0 0 1 1)	(1 0 0 1 0)
(1 0 1 0 1)	(1 0 1 0 0)
(1 0 1 1 0)	(1 0 1 1 1)
(1 1 0 0 1)	(1 1 0 0 0)
(1 1 0 1 0)	(1 1 0 1 1)
(1 1 1 0 0)	(1 1 1 0 1)
(1 1 1 1 1)	(1 1 1 1 0)

In an (n, k) block code each different information word gives a codeword. As k bits give 2^k different words there are therefore 2^k codewords in an (n, k) code. Consider the $(5, 4)$ block code, with odd parity, shown in column 1 (the left-hand side column) of Table 1.2. The code has 4 information bits and therefore $2^4 = 16$ codewords. The parity bit is on the right-hand side of each codeword and the code is in a systematic form. A 5-bit word gives 32 different words of which 16 have odd parity and are therefore codewords of the $(5, 4)$ code. The remaining 16 words, shown in the second column, have even parity and are redundant to the code, these are referred to as *redundant words*. If information within a system is represented in the form of codewords then the presence of a redundant word can only be due to the occurrence of errors. It is by adding *redundancy* to information that error detection can be carried out. Furthermore it may be possible to carry out error correction if there is sufficient redundancy. The $(5, 4)$ block code enables some degree of error detection but no error correction. Section 1.4 gives the first example of an error-correcting code.

In an (n, k) block code we can think of the code's structure as partitioning or separating 2^n words into two sets:

(i) the 2^k codewords having the required structure;
(ii) the $2^n - 2^k$ redundant words not having the code's structure.

Consider the even-parity $(4, 3)$ block code shown in Table 1.3. Column 1 (on the left-hand side) shows the code's 8 information words and column 2 shows all the different words of a 4-bit word. The code's requirement that words have even parity separates the words in column 2 into codewords and redundant words, shown in columns 3 and 4 respectively. There are $2^k = 2^3 = 8$ codewords and $2^n - 2^k = 2^4 - 2^3 = 8$ redundant words. The 'space' between some of the codewords, in column

Table 1.3
The (4, 3) block code with even parity

3-bit words	*4-bit words*	*Codewords*	*Redundant words*
(0 0 0)	(0 0 0 0)	(0 0 0 0)	
	(0 0 0 1)		(0 0 0 1)
(0 0 1)	(0 0 1 0)		(0 0 1 0)
	(0 0 1 1)	(0 0 1 1)	
(0 1 0)	(0 1 0 0)		(0 1 0 0)
	(0 1 0 1)	(0 1 0 1)	
(0 1 1)	(0 1 1 0)	(0 1 1 0)	
	(0 1 1 1)		(0 1 1 1)
(1 0 0)	(1 0 0 0)		(1 0 0 0)
	(1 0 0 1)	(1 0 0 1)	
(1 0 1)	(1 0 1 0)	(1 0 1 0)	
	(1 0 1 1)		(1 0 1 1)
(1 1 0)	(1 1 0 0)	(1 1 0 0)	
	(1 1 0 1)		(1 1 0 1)
(1 1 1)	(1 1 1 0)		(1 1 1 0)
	(1 1 1 1)	(1 1 1 1)	

3, reflects the redundancy added by the parity-check bits. Increasing the number of parity-check bits has the effect of increasing the redundancy and so 'pushing' the codewords further apart. The further apart the codewords, the more likely it is that errors will be detected and corrected. The idea of codewords being separated by redundancy leads to the notion of distance, this as we shall see is fundamental to the theory of error-control codes.

A useful measure of the redundancy within a block code is given by the ratio of the number of information bits to the blocklength

$$R = \frac{k}{n}$$

and is known as the *code rate*. For a fixed number of information bits the code rate R tends to 0 as the number of parity bits r increases. Low code rates reflect high levels of redundancy. In the case of no coding there are no parity bits and so $n = k$ and the code rate $R = 1$ (whilst this is of no practical interest it does nevertheless set an upper limit to the code rate). We can see that the code rate is therefore bounded by

$$0 \leq R \leq 1.$$

A word with n bits can be represented by a vector with n components. For now we associate words with vectors in a rather loose way, purely for the purpose of representation. In Section 5.6 a formal definition of codewords in terms of vectors is given.

For an (n, k) block code the input to the encoder is the information word

$$\boldsymbol{i} = (i_1, i_2, \ldots, i_k)$$

where $i_j = 0$ or 1 and j is an integer $1 \leq j \leq k$. The encoder determines $r = n - k$ parity-check bits $p_1, p_2, \ldots, p_r$ according to the encoding rule of the code and appends them to the information bits so giving the codeword

$$c = (i_1, i_2, \ldots, i_k, p_1, p_2, \ldots, p_r).$$

The encoding rule of a code is such that the combination of the parity bits and the information bits (i.e. the codeword) has the mathematical property required by the code. It is usual to represent codewords as

$$c = (c_1, c_2, \ldots, c_n)$$

where

$$\begin{aligned} c_j &= i_j \quad \text{for } 1 \leq j \leq k \\ &= p_{j-k} \quad \text{for } k < j \leq n. \end{aligned}$$

Errors can also be represented by vectors. An n-bit codeword is liable to a maximum of n errors which can be represented by the *error vector* or *error pattern*

$$e = (e_1, e_2, \ldots, e_n)$$

where

$$\begin{aligned} e_j &= 1 \quad \text{if there is an error in the } j\text{th position} \\ &= 0 \quad \text{if the position is error free.} \end{aligned}$$

A codeword c that incurs an error e results in the word

$$v = c + e \tag{1.1}$$

where the components of v are given by the components of c and e added pairwise

$$\begin{aligned} v &= (c_1, c_2, \ldots, c_n) + (e_1, e_2, \ldots, e_n) \\ &= (c_1 + e_1, c_2 + e_2, \ldots, c_n + e_n) \\ &= (v_1, v_2, \ldots, v_n) \end{aligned}$$

where $v_j = c_j + e_j$ for $1 \leq j \leq n$ and where modulo-2 addition (see Table 1.4) is used when adding c_j and e_j together. From Table 1.4 we see that $1 + 1 = 0$, which gives $-1 = 1$ and so subtraction is equivalent to addition when using modulo-2 addition. To emphasize the use of modulo-2 addition we sometimes use the notation $1 + 1 = 0$ modulo-2. Note that for two bits b_1 and b_2 we get

$$\begin{aligned} b_1 + b_2 &= 0 \text{ modulo-2} \quad \text{if } b_1 = b_2 \\ b_1 + b_2 &= 1 \text{ modulo-2} \quad \text{if } b_1 \neq b_2. \end{aligned}$$

Table 1.4
Modulo-2 addition

$0+0=0$
$0+1=1$
$1+0=1$
$1+1=0$

The word $\boldsymbol{v}$ represents the codeword after it has been subjected to the error $\boldsymbol{e}$. If all the components of $\boldsymbol{e}$ are zero, then

$$\begin{aligned} v_1 &= c_1 \\ v_2 &= c_2 \\ &\vdots \\ v_n &= c_n \end{aligned}$$

and therefore $\boldsymbol{v} = \boldsymbol{c}$, otherwise $\boldsymbol{v} \neq \boldsymbol{c}$. Equation 1.1 is central to the decoding process. A decoder has no *a priori* knowledge of $\boldsymbol{c}$, the only information that it has is the word $\boldsymbol{v}$ that it receives, which is referred to as the *received word* or the *decoder input*. For an error-detecting code the task of the decoder is to establish whether $\boldsymbol{v}$ is a codeword. This can be achieved by checking $\boldsymbol{v}$ against a table of codewords or by checking whether $\boldsymbol{v}$ has the mathematical property required by the code (e.g. if $\boldsymbol{v}$ has the correct parity). For an error-correcting code the decoder has to estimate or guess the codeword from $\boldsymbol{v}$. If the decoder's estimate of the error pattern is $\hat{\boldsymbol{e}}$ then, from eqn 1.1, its estimate of the codeword is $\hat{\boldsymbol{c}} = \boldsymbol{v} - \hat{\boldsymbol{e}}$, and given that modulo-2 addition is used then

$$\hat{\boldsymbol{c}} = \boldsymbol{v} + \hat{\boldsymbol{e}} \tag{1.2}$$

is the decoder's estimate of $\boldsymbol{c}$. Whether a decoder can determine the correct codeword from $\boldsymbol{v}$ depends upon the code, the errors incurred and the decoding algorithm.

1.3 Single-parity-check codes

The (4, 3) even-parity and the (5, 4) odd-parity block codes considered in the previous section are just two examples of the *single-parity-check codes*. These are a class of error-detecting block codes that give the simplest form of error control. The codes use a single parity bit to generate codewords with even or odd parity (i.e. an even or odd number of ones respectively). For all integer values $k \geq 1$ there exists an (n, k) single-parity-check code with $n = k + 1$.

A good example of a single-parity-check code is the (8, 7) block code used in the *American Standard Code for Information Interchange* (*ASCII*). The ASCII format uses 10-bit words to represent alphanumeric characters (see Fig. 1.5). The first and last bits are the start and stop bits respectively, these are not used for error

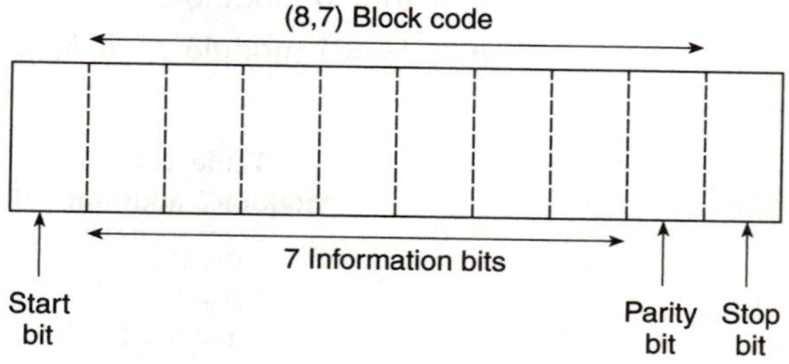

Fig. 1.5 The ASCII format.

control but help to achieve word synchronization. The remaining 8 bits consist of 7 information bits, used to identify a particular character, and a parity bit to provide some degree of error control. The parity bit is set to 0 or 1 depending upon the information bits and the parity required.

For an (n, k) even-parity code with information bits $i_1, i_2, \ldots, i_k$ the parity bit is given by

$$p = i_1 + i_2 + \cdots + i_k \tag{1.3}$$

where modulo-2 addition is again used. If the information bits contain an even number of ones then $p = 0$, otherwise $p = 1$. If odd parity is required then $p = (i_1 + i_2 + \cdots + i_k) + 1$ where now $p = 1$ if the information bits contain an even number of ones. Appending p to the information bits gives the codeword

$$c = (i_1, i_2, \ldots, i_k, p). \tag{1.4}$$

Example 1.1

Given the $(5, 4)$ even-parity block code find the codewords corresponding to $\boldsymbol{i}_1 = (1\ 0\ 1\ 1)$ and $\boldsymbol{i}_2 = (1\ 0\ 1\ 0)$.

Consider first the parity bit for $\boldsymbol{i}_1 = (1\ 0\ 1\ 1)$. Using eqn 1.3 the parity bit is given by

$$p_1 = 1 + 0 + 1 + 1.$$

Recall that modulo-2 addition is used and so

$$0 + 0 = 0$$
$$0 + 1 = 1$$
$$1 + 0 = 1$$
$$1 + 1 = 0.$$

The bits in p_1 can be added together two at a time, therefore p_1 can be written as

$$p_1 = (1 + 0) + (1 + 1)$$

which gives

$$p_1 = 1 + 0$$

and so

$$p_1 = 1.$$

Using eqn 1.4 the codeword for $\boldsymbol{i}_1 = (1\ 0\ 1\ 1)$ is therefore

$$\boldsymbol{c}_1 = (1\ 0\ 1\ 1\ p_1) = (1\ 0\ 1\ 1\ 1).$$

Likewise for $\boldsymbol{i}_2 = (1\ 0\ 1\ 0)$ the parity-check bit is

$$\begin{aligned} p_2 &= 1 + 0 + 1 + 0 \\ &= (1 + 0) + (1 + 0) \\ &= 1 + 1 \\ &= 0 \end{aligned}$$

which gives the codeword

$$\boldsymbol{c}_2 = (1\ 0\ 1\ 0\ p_2) = (1\ 0\ 1\ 0\ 0).$$ □

Equations 1.3 and 1.4 describe the encoding stage of a single-parity-check code. At the decoding stage the input to the decoder is the received word

$$\boldsymbol{v} = (v_1, v_2, \ldots, v_n)$$

which may or may not be a codeword, depending on whether any errors have occurred. To check whether $\boldsymbol{v}$ is a codeword, the decoder determines the *parity-check sum*

$$s = v_1 + v_2 + \cdots + v_n \tag{1.5}$$

where $s = 0$ and $s = 1$ when $\boldsymbol{v}$ has even and odd parity respectively. Consider now an even-parity code. If $s = 0$ then $\boldsymbol{v}$ has even parity and is therefore a codeword. If $s = 1$ then $\boldsymbol{v}$ has odd parity and is therefore not a codeword, this is known as a *parity-check failure*. If there are no errors then $s = 0$ and $\boldsymbol{v}$ is a codeword, however the occurrence of a single error changes the parity of a codeword to odd, resulting in a parity-check failure. The decoder is therefore able to detect all single-bit errors irrespective of the position of the error within the codeword (including an error in the parity bit). Two errors, however, do not alter the parity of $\boldsymbol{v}$ and so $s = 0$ as obtained when no errors occur. The decoder is therefore unable to distinguish between the occurrence of two errors and no errors.

Example 1.2
Given the (8, 7) even-parity code determine whether $\boldsymbol{v}_1 = (1\ 0\ 1\ 1\ 0\ 1\ 1\ 0)$ and $\boldsymbol{v}_2 = (0\ 1\ 1\ 0\ 1\ 0\ 0\ 1)$ give parity-check failures.

The parity-check sum for $\boldsymbol{v}_1$ is

$$s_1 = 1 + 0 + 1 + 1 + 0 + 1 + 1 + 0 = 1$$

and so $\boldsymbol{v}_1$ gives a parity-check failure. However for $\boldsymbol{v}_2$ we get

$$s_2 = 0 + 1 + 1 + 0 + 1 + 0 + 0 + 1 = 0$$

which is the correct parity-check sum for an even-parity code. □

It is easy to see that the occurrence of an odd number of errors always give $s = 1$, whilst an even number of errors gives $s = 0$. When $s = 1$ the decoder knows that at least 1 error has occurred. However, when $s = 0$ the decoder has no *a priori* way of knowing whether this is due to no errors or the occurrence of an even number of errors. The decoder cannot distinguish between the events that can give rise to $s=0$. In a binary symmetric channel the probability of obtaining r errors decreases as r increases, and so a zero parity-check sum is more likely to be due to 0 errors than 2 errors, or any other even number of errors. The best decision that the decoder can make is to assume that if $s=0$ then no errors have occurred. Such an approach, where 'the most probable reason' is used to make a decision is referred to as a *maximum-likelihood decision* and the decoder known as a *maximum-likelihood decoder*. The decoder is basically making a guess, but it is more likely to be right than wrong.

Decoding an odd-parity code is the same as decoding an even-parity code, except that $s=1$ is taken to indicate correct parity. A word of caution is appropriate here, the use of the words 'odd' and 'even' when describing the type of parity should not be confused with their use when describing the number of errors. Given that even-parity codes detect odd numbers of errors, then odd-parity codes do not detect even numbers of errors, they too detect odd numbers of errors.

The error-control capabilities of the single-parity-check codes are quite poor, a measure of the codes' performance can be obtained by considering all the possible outcomes at the decoder. Consider a codeword $\boldsymbol{c}$ which at the decoding stage gives rise to the received word $\boldsymbol{v}$ and let:

$p_c =$ the probability that the decoder gives the correct codeword $\boldsymbol{c}$;

$p_e =$ the probability that the decoder gives an incorrect codeword $\boldsymbol{c}' \neq \boldsymbol{c}$;

$p_f =$ the probability that the decoder fails to give a codeword.

These give the three possible outcomes at the decoder, namely *correct decoding*, a *decoding error* and a *decoding failure* for p_c, p_e, and p_f respectively. Note that correct decoding is also referred to as successful decoding. A decoding failure occurs whenever an uncorrectable error pattern is detected, the decoder does not return a codeword and in this sense decoding is said to have failed (i.e. failed to return a codeword). Note also that the decoder never knows whether decoding has been correct or incorrect, it is only aware of decoding failures. There are no other possible outcomes at the decoder and therefore

$$p_c + p_e + p_f = 1.$$

Expressions for p_c, p_e, and p_f can be found by first considering the probability of obtaining j errors in an n-bit word. The number of ways (combinations) of having j bits in error, out of n bits, is ${}^n\mathrm{C}_j$ where

$${}^n\mathrm{C}_j = \frac{n!}{(n-j)!j!}$$

and if p is the bit-error probability then

$$P_j = {}^n\mathrm{C}_j p^j (1-p)^{n-j} \tag{1.6}$$

is the probability of obtaining j errors in an n-bit word. If $\boldsymbol{v}$ contains no errors then $\boldsymbol{v}$ will have the correct parity and the decoder will correctly assume that $\boldsymbol{v}$ is the required codeword. Hence the probability of correct decoding p_c is given by the probability of no errors occurring, i.e. by taking $j=0$, and so

$$p_c = P_0. \tag{1.7}$$

The decoder gives an incorrect codeword whenever an even number of errors occur, as this preserves the correct parity so making the errors undetectable. Hence taking $j=2, 4, 6, \ldots$ gives

$$p_e = P_2 + P_4 + P_6 + \cdots + P_{n'} \tag{1.8}$$

where $n' = n$ if n is even, otherwise $n' = n-1$. In the event of an odd number of errors the parity of $\boldsymbol{v}$ is incorrect and the errors are therefore detected. The decoder has no way of establishing the number of errors or their positions and therefore the decoder is unable to return a codeword. Hence the occurrence of odd numbers of errors causes a decoding failure, and so taking $j=1, 3, 5, \ldots$ gives

$$p_f = P_1 + P_3 + P_5 + \cdots + P_{n'} \tag{1.9}$$

where this time $n' = n$ if n is odd, otherwise $n' = n-1$.

Example 1.3
Given the even parity $(5, 4)$ block code find p_c, p_e and p_f for (i) $p=0.1$ and (ii) $p=0.001$.

(i) For the even parity $(5, 4)$ block code $n=5$ and we substitute $n'=4$ and $n'=5$ into eqns 1.8 and 1.9 respectively. This, along with eqn 1.7, gives

$$\begin{aligned}
p_c &= P_0 \\
p_e &= P_2 + P_4 \\
p_f &= P_1 + P_3 + P_5.
\end{aligned}$$

Using eqn 1.6 and taking $p=0.1$ gives

$$\begin{aligned}
P_0 &= {}^5C_0 p^0 (1-p)^5 = (1-p)^5 = 0.5905 \\
P_1 &= {}^5C_1 p^1 (1-p)^4 = 5p(1-p)^4 = 0.3281 \\
P_2 &= {}^5C_2 p^2 (1-p)^3 = 10p^2(1-p)^3 = 0.0729 \\
P_3 &= {}^5C_3 p^3 (1-p)^2 = 10p^3(1-p)^2 = 0.0081 \\
P_4 &= {}^5C_4 p^4 (1-p)^1 = 5p^4(1-p) = 0.0005 \\
P_5 &= {}^5C_5 p^5 (1-p)^0 = p^5 = 0.00001.
\end{aligned}$$

Substituting P_0 to P_5 into p_c, p_f and p_e gives

$$\begin{aligned}
p_c &= 0.5905 \\
p_e &= 0.0734 \\
p_f &= 0.3361
\end{aligned}$$

note that $p_c + p_e + p_f = 1$ as required.

(ii) Repeating the above with $p = 0.001$ gives

$$P_0 = 0.9950$$
$$P_1 = 4.98 \times 10^{-3}$$
$$P_2 = 9.97 \times 10^{-6}$$
$$P_3 = 9.98 \times 10^{-9}$$
$$P_4 = 5 \times 10^{-12}$$
$$P_5 = 1 \times 10^{-15}$$

and so

$$p_c = 0.9950$$
$$p_e = 9.97 \times 10^{-6}$$
$$p_f = 4.98 \times 10^{-3}$$

which again add to 1 as required. Note that when $p = 0.1$ only 59% of codewords will be received error free. Whereas when $p = 0.001$ over 99% of codewords are error-free. □

The expressions for p_c, p_e, and p_f can be written in terms of the bit-error probability p by substituting eqn 1.6 into eqns 1.7, 1.8, and 1.9. The probability of correct decoding is then given by

$$p_c = P_0 = {}^nC_0 p^0 (1-p)^{n-0}$$

which gives

$$p_c = (1-p)^n. \tag{1.10}$$

The probability of a decoding error becomes

$$p_e = \sum_{j=2(\text{even})}^{n'} {}^nC_j p^j (1-p)^{n-j} \tag{1.11}$$

where the series summation is over even values of j and where $n' = n$ if n is even, otherwise $n' = n - 1$. The probability of a decoding failure is

$$p_f = \sum_{j=1(\text{odd})}^{n'} {}^nC_j p^j (1-p)^{n-j} \tag{1.12}$$

where now $n' = n$ if n is odd, otherwise $n' = n - 1$ and the summation is now over odd values of j. As a check it can be easily shown that the sum of p_c, p_e, and p_f gives 1.

Adding eqns 1.10, 1.11, and 1.12 together gives

$$
\begin{aligned}
p_c + p_e + p_f &= (1-p)^n + \sum_{j=2(\text{even})}^{n'} {}^nC_j p^j (1-p)^{n-j} + \sum_{j=1(\text{odd})}^{n'} {}^nC_j p^j (1-p)^{n-j} \\
&= (1-p)^n + \sum_{j=1}^{n} {}^nC_j p^j (1-p)^{n-j} \\
&= \sum_{j=0}^{n} {}^nC_j p^j (1-p)^{n-j} \\
&= (p + (1-p))^n = 1
\end{aligned}
$$

where we have made use of the binomial expansion theorem

$$
(a+b)^n = \sum_{j=0}^{n} {}^nC_j a^j b^{n-j}.
$$

Example 1.4
Given the (8, 7) single-parity-check code, with even parity, find the probability of correct decoding, a decoding error and a decoding failure when $p = 0.01$.

Substituting $p = 0.01$ and $n = 8$ into eqn 1.10 gives the probability of correct decoding as:

$$
p_c = (1-p)^n = (1-0.01)^8 = 0.9227.
$$

The probability of a decoding error is obtained by letting $n' = n = 8$ in eqn 1.11:

$$
\begin{aligned}
p_e &= {}^8C_2 p^2 (1-p)^6 + {}^8C_4 p^4 (1-p)^4 + {}^8C_6 p^6 (1-p)^2 + {}^8C_8 p^8 \\
&= 28p^2(1-p)^6 + 70p^4(1-p)^4 + 28p^6(1-p)^2 + p^8
\end{aligned}
$$

and substituting $p = 0.01$ gives $p_e = 0.0026$.

We could find p_f using $p_f = 1 - (p_c + p_e)$, however we will use eqn 1.12 to obtain p_f and then check that $p_c + p_e + p_f = 1$. Letting $n = 8$ and $n' = 7$ in eqn 1.12 we get

$$
\begin{aligned}
p_f &= {}^8C_1 p(1-p)^7 + {}^8C_3 p^3 (1-p)^5 + {}^8C_5 p^5 (1-p)^3 + {}^8C_7 p^7 (1-p) \\
&= 8p(1-p)^7 + 56p^3(1-p)^5 + 56p^5(1-p)^3 + 8p^7(1-p)
\end{aligned}
$$

which gives $p_f = 0.0746$ when $p = 0.01$.

Adding together p_c, p_e and p_f gives 1, so the results are consistent. □

In eqns 1.7 to 1.12, p_c, p_e, and p_f are probabilities with values lying between 0 and 1. Multiplying each term by 100 gives values of 0 to 100%, in which case p_c, p_e, and p_f can be referred to as the *success rate*, the *error rate* and the *failure rate* respectively.

If $p \ll 1$ we can obtain approximations to p_e and p_f by neglecting the occurrence of 3 or more errors. This then gives

$$p_e \approx P_2$$
$$p_f \approx P_1.$$

There is no need to approximate p_c as this can be easily determined from eqn 1.10.

Example 1.5
Repeat Example 1.3 using $p_e \approx P_2$ and $p_f \approx P_1$.

For $p = 0.1$ we get $p_f \approx 3.28 \times 10^{-1}$ and $p_e \approx 7.29 \times 10^{-2}$, compared to $p_f \approx 3.36 \times 10^{-1}$ and $p_e = 7.34 \times 10^{-2}$ obtained in Example 1.3. For $p = 0.001$ we get $p_f \approx 4.98 \times 10^{-3}$ and $p_e \approx 9.97 \times 10^{-6}$, compared to $p_f = 4.98 \times 10^{-3}$ and $p_e = 9.97 \times 10^{-6}$ obtained in Example 1.3. Note that there is greater agreement between the values obtained using eqns 1.8 and 1.9, and their corresponding approximations when $p = 0.001$ than when $p = 0.1$. This is clearly to be expected. □

1.4 Product codes

The single-parity-check codes have no error-correction capability, they can only detect the occurrence of odd numbers of errors. However, error correction can be achieved by combining two single-parity-check codes in the form of a rectangular array as shown in Fig. 1.6. Each row in Fig. 1.6 shows a 4-bit word encoded into 5 bits using a (5, 4) single-parity-check code. Even parity has been used and the parity bit is on the right-hand side of each row. After the sixth word a (7, 6) code is

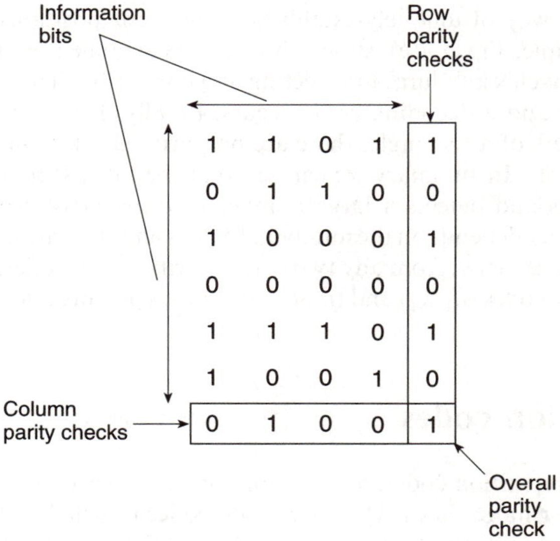

Fig. 1.6 A (35, 24) product code.

used to encode bits within the same column. Again even parity has been used, and the parity bits are now at the bottom of each column. There are therefore two sets of parity bits, *row* and *column parity-check bits*. At the bottom right-hand corner of the array is an *overall parity-check bit* which can be generated from either code as they give the same parity check. There are a total of 35 bits in the array, of which 24 are information bits and 11 are parity-check bits. The two codes can therefore be thought of as a single (35, 24) block code. It is referred to as a *product code* because n and k are given by the product of the blocklengths and of the information lengths of the individual codes, we can think of this as a $(5,4) \times (7,6) = (5 \times 7, 4 \times 6) = (35,24)$ code. Codewords now take the form of arrays and as such are referred to as *code arrays*. Product codes are also referred to as *rectangular codes* because of the rectangular appearance of codewords.

When decoding a product code, parity-check sums are calculated for rows and columns. In the event of the array containing no errors there will be no parity-check failures, and the array will be taken as the required code array. The occurrence of a single error gives parity-check failures in the row and column containing the error, and the decoder can determine the error position from the intersection of the row and column parity-check failures. Figure 1.7(a) shows an example of a single error in a code array belonging to the (35, 24) product code. The erroneous bit is enclosed in a circle and the resulting parity-check failures are shown by the highlighted row and column. The error position is uniquely determined by the intersection of the highlighted row and column, and the error can therefore be corrected. Figure 1.7(b) likewise shows a single error, in a different code array to that shown in Fig. 1.7(a), which again can clearly be located and corrected. Figure 1.7(c) shows two errors in the same column but different rows. Here the parity check in the middle column is correct, but there are two row parity-check failures and there is no way of establishing the position of the errors within the two rows. The same problem arises when there are two errors in the same row but different columns. Two errors in different columns and different rows give four points of intersection and again there is no way of uniquely establishing the error positions (see Fig. 1.7(d)). The next example, Fig. 1.7(e), shows three errors arranged in an 'L' shape, resulting in the parity-check failures intersecting at a correct bit. This therefore appears as a single error and a decoding error occurs. Finally, Fig. 1.7(f) shows four errors on the corners of a rectangle, there are no parity-check failures and the errors are undetectable. In summary we can say that the (35, 24) product code corrects all single errors and detects a large number of other error patterns. The outcome of the decoder depends on the number of errors and their arrangement. A product code can be constructed from any two single-parity-check codes, furthermore given any two block codes (n_1, k_1) and (n_2, k_2) an $(n_1 n_2, k_1 k_2)$ product code can be constructed.

1.5 Repetition codes

The repetition codes, as the name suggests, are codes that repeat information bits two or more times. They are block codes in which the parity bits are set equal to a single information bit and if the number of parity bits is $n-1$ then the code is referred to as an $(n, 1)$ *repetition code*. As an example we consider the (3, 1)

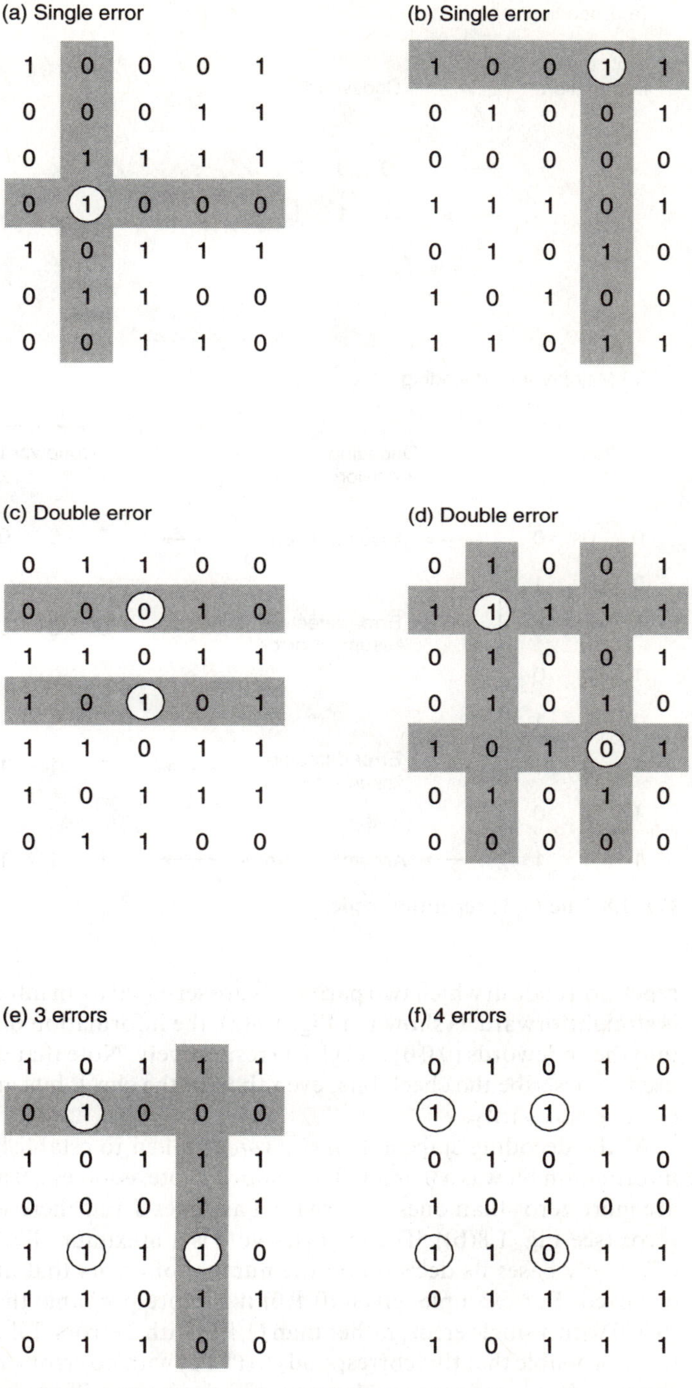

Fig. 1.7 Error patterns in the (35, 24) product code.

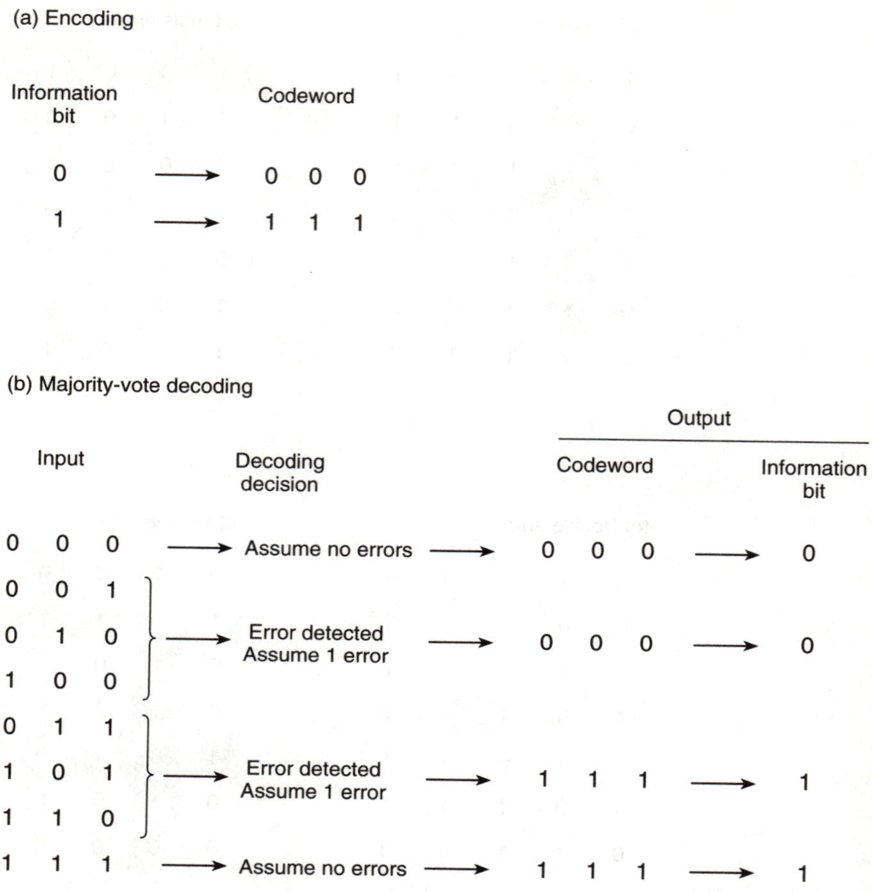

Fig. 1.8 The (3, 1) repetition code.

repetition code in which two parity bits are set equal to an information bit. Encoding is straightforward as shown in Fig. 1.8(a), the information bits 0 and 1 are encoded into the codewords (0 0 0) and (1 1 1) respectively. Note that the word 'parity' is still used to describe the check bits, even though the check bits are not used for setting even or odd parity.

At the decoding stage a *majority vote* is taken to establish whether the original information bit was a 0 or a 1. The majority vote assumes a 0 information bit if there are more zeros than ones, otherwise it assumes a 1 as there will be more ones than zeros (see Fig. 1.8(b)). The majority vote is a maximum-likelihood rule because it effectively bases its decision on the number of errors that are most likely to have occurred. For example, given (0 1 0) it is more probable that this corresponds to (0 0 0) with a single error, rather than (1 1 1) with 2 errors. Likewise, given (1 1 1) it is more probable that this corresponds to (1 1 1) with no errors, rather than (0 0 0) with 3 errors. If no errors occur then the decoder input will be (0 0 0) or (1 1 1) and the majority vote gives 0 or 1 respectively, both decisions being correct. If one of the codewords incurs a single error then the majority vote will always give the correct codeword and information bit. So (0 0 0) with 1 error gives (0 0 1), (0 1 0) or (1 0 0) all of which are correctly decoded by the majority vote. However, if (0 0 0) incurs 2 errors then the majority vote is taken over (0 1 1), (1 1 0) or (1 0 1) and so giving the incorrect codeword (1 1 1). Triple errors obviously give an incorrect codeword and furthermore the decoder is unaware of the errors. Note that for the (3, 1) code a majority vote over (v_1, v_2, v_3) gives

$$i = v_1 v_2 + v_2 v_3 + v_3 v_1$$

as the required information bit. For example given (0 1 1) then

$$i = 0 \cdot 1 + 1 \cdot 1 + 1 \cdot 0 = 1$$

and for (0 1 0) we get

$$i = 0 \cdot 1 + 1 \cdot 0 + 0 \cdot 0 = 0$$

which are clearly correct.

In Section 1.3 expressions for the probability of correct decoding p_c, a decoding error p_e and decoding failure p_f were determined for the single-parity-check codes. Similar expressions can easily be obtained for the repetition codes. With the single-parity-check codes correct decoding occurs only when there are no errors and so, as we saw in Section 1.3, $p_c = P_0$ where P_0 is the probability of obtaining no errors in an n-bit word (see eqn 1.6). However, with the (3, 1) repetition code single errors are corrected and so decoding is successful when 0 errors or 1 error occur. Therefore

$$p_c = P_0 + P_1 = (1-p)^3 + 3p(1-p)^2 \tag{1.13}$$

where P_0 and P_1 give the probability of getting 0 and 1 error in a 3-bit word respectively and again p is the bit-error probability. The occurrence of 2 or 3 errors gives a decoding error and so

$$p_e = P_2 + P_3 = 3p^2(1-p) + p^3 \tag{1.14}$$

where P_2 and P_3 are the probability of getting 2 and 3 errors in a 3-bit word respectively. The (3, 1) repetition code always gives correct decoding or a decoding error, it never gives a decoding failure because a majority vote can always be made irrespective of the number of errors. Majority voting, over an odd number of bits (as considered here) is an example of *complete decoding*, it never fails to give a codeword (even though it may be the wrong codeword). Hence the decoding failure rate is zero, and so

$$p_f = 0.$$

A decoder in which the failure rate is nonzero is referred to as an *incomplete decoder*, this can return codewords as erroneous and uncorrectable (i.e. decoding failures can occur). A zero failure rate may seem like a good characteristic of a decoder but this is not necessarily the case. Whilst a complete decoder gives a zero failure rate, the probability of a decoding error is larger than that of an incomplete decoder for the same code (note that complete or incomplete decoders can be used for most codes). Complete decoding is normally used when it is better to have the best guess of a codeword as opposed to nothing at all, however in doing so more decoding errors will occur. If a low probability of decoding errors is required then an incomplete decoder is better suited.

Example 1.6

Given that an information bit is encoded using the (3, 1) repetition code, find the probability p_c that the bit is correct after decoding when the bit-error probability $p = 0.01$ and when $p = 0.001$.

Substituting $p = 0.01$ into eqn 1.13 gives $p_c = 0.999702$ and $p = 0.001$ gives $p_c = 0.999997$. Furthermore the probability of the bit being incorrect after decoding is 2.98×10^{-4} and 3.00×10^{-6} for $p = 0.01$ and $p = 0.001$ respectively. □

The (3, 1) code is just one example of a repetition code. For any positive integer n there is an $(n,1)$ repetition code that has $n - 1$ parity-check bits all equal to a single information bit. If the blocklength n is even, decoding failures occur in the event of $n/2$ errors, because a majority vote over an equal number of zeros and ones does not give a unique answer, decoding is therefore incomplete. Hence the blocklength n is usually taken to be odd as this avoids the problem of an even number of zeros and ones, so giving complete decoding. An $(n, 1)$ repetition code has two codewords—the words with n zeros and n ones. As the blocklength n increases, the number of errors that can be corrected also increases. Consider the (5, 1) repetition code with codewords (0 0 0 0 0) and (1 1 1 1 1). Assuming that 2 errors occur in (0 0 0 0 0) so giving, for example, (0 1 0 0 1) then a maority vote taken over (0 1 0 0 1) gives the correct codeword (0 0 0 0 0). The occurrence of 3 errors, to give say (1 0 1 1 0), will result in the decoder giving the wrong codeword (1 1 1 1 1). The (5, 1) repetition code is able to correct up to 2 errors and is therefore a double-error-correcting code. Clearly an $(n, 1)$ repetition code can correct up to $(n - 1)/2$ errors. If the blocklength n is even then $(n - 1)/2$ is rounded down to the nearest integer.

Expressions for the probabilities of obtaining correct decoding and a decoding error for an $(n, 1)$ repetition code are readily obtained by extending the expressions for p_c and p_e already given for the (3, 1) repetition code (eqns 1.13 and 1.14 respectively). Assuming n to be odd then

$$p_c = P_0 + P_1 + \cdots + P_{(n-1)/2}$$

and

$$p_e = P_{(n+1)/2} + P_{(n+3)/2} + \cdots + P_n = 1 - p_c$$

where P_j is the probability of obtaining j errors in an n-bit codeword (see eqn 1.6) and $0 \le j \le n$. In terms of the bit-error probability p these give

$$p_c = \sum_{j=0}^{(n-1)/2} {}^nC_j p^j (1 - p)^{n-j} \tag{1.15}$$

and

$$p_e = \sum_{j=(n+1)/2}^{n} {}^{n}C_j p^j (1-p)^{n-j}. \tag{1.16}$$

The probability of a decoding failure $p_f = 0$, as we have already seen for the (3, 1) code, as so $p_c + p_e = 1$.

Example 1.7
A (7, 1) repetition code is used to encode information sent through a channel with a bit-error probability of 0.01. Find the probability that an information bit is erroneous after decoding.

A decoding error occurs whenever 4, 5, 6, or 7 errors occur, and the probability of this is given by substituting $n=7$ into eqn 1.16, so giving

$$p_e = {}^{7}C_4 p^4(1-p)^3 + {}^{7}C_5 p^5(1-p)^2 + {}^{7}C_6 p^6(1-p)^2 + {}^{7}C_7 p^7$$

Substituting $p=0.01$ into p_e gives $p_e = 3.4 \times 10^{-7}$. Hence approximately 1 bit in every 3×10^6 bits are erroneous, compared to 1 in every 100 if information bits are sent through the channel without the use of the (7, 1) code. □

Examples 1.6 and 1.7 show that for $p=0.01$ the probability of a decoding error is 3.0×10^{-4} and 3.4×10^{-7} for the (3, 1) and (7, 1) repetition codes respectively. Table 1.5 shows values of p_e for repetition codes with odd blocklengths up to $n=11$ and a bit-error probability of $p=0.01$. As the blocklength n increases, the probability of obtaining an incorrect bit decreases significantly. However this is at the expense of the high levels of redundancy required by the codes, as indicated by the low code rates R (also shown in Table 1.5).

As already mentioned, a repetition code with an even blocklength gives incomplete decoding and therefore a nonzero decoding failure. This is considered in the next example.

Example 1.8
Determine the probability of a decoding failure for the (6, 1) repetition code, assuming a bit-error probability of 0.01.

A decoding failure occurs whenever a codeword incurs 3 errors because the errors are detectable and a majority vote will fail (as there will be an even number of zeros

Table 1.5
Probability of decoding errors for repetition codes

n	p_e	R
3	3.0×10^{-4}	0.33
5	9.9×10^{-6}	0.20
7	3.4×10^{-7}	0.14
9	1.2×10^{-8}	0.11
11	4.4×10^{-10}	0.09

and ones). The probability of obtaining 3 errors in a 6-bit word is given by letting $n=6$ and $j=3$ in eqn 1.6, which gives

$$P_3 = {}^6C_3 p^3 (1-p)^3$$

and substituting $p=0.01$ gives $P_3 = 1.94 \times 10^{-5}$. The probability of a decoding failure is therefore $p_f = P_3 = 1.94 \times 10^{-5}$. □

The repetition codes provide us with a very simple way of carrying out error correction and may be suitable in a system where a high level of redundancy is acceptable. But if a channel is to be used efficiently so that high levels of redundancy are not acceptable, then the repetition codes are quite inadequate and error-correcting codes are required that make better use of redundancy.

1.6 Hamming codes

The single-parity-check codes, product codes and repetition codes considered in the previous sections can be thought of as 'first steps' towards achieving error control. The codes allow a limited degree of error detection and correction, and are simple to implement. Moving on towards somewhat more interesting codes are the *Hamming codes* which were the first class of linear codes devised for error control and as we shall see, in Chapter 2, linearity is a good property for a code to possess. The Hamming codes occupy an important position in the history of error-control codes and we will refer to them repeatedly throughout the book. Here the Hamming codes are introduced and later they are reconsidered in terms of their linear properties.

The first in the class of Hamming codes is the (7, 4) code that takes 4-bit information words and encodes them into 7-bit codewords. Three parity-check bits are required, these are determined from the information bits using eqns 1.17 shown below. Given an information word $\boldsymbol{i} = (i_1, i_2, i_3, i_4)$ then the parity-check bits are

$$\begin{aligned} p_1 &= i_1 + i_2 + i_3 \\ p_2 &= i_2 + i_3 + i_4 \\ p_3 &= i_1 + i_2 + i_4 \end{aligned} \qquad (1.17)$$

where the information bits are added together using modulo-2 addition (see Table 1.4). Appending the parity bits to the information word gives the codeword

$$\boldsymbol{c} = (i_1, i_2, i_3, i_4, p_1, p_2, p_3). \qquad (1.18)$$

Table 1.6 shows the set of codewords for the (7, 4) code. There are 16 codewords, one for each information word. For reference purposes, the codewords and information words are labelled $\boldsymbol{c}_0$ to $\boldsymbol{c}_{15}$, and $\boldsymbol{i}_0$ to $\boldsymbol{i}_{15}$ respectively. The subscript i of the codeword $\boldsymbol{c}_i$ gives the numerical value of the corresponding information word, for example $\boldsymbol{c}_5$ is the codeword corresponding to the information word $\boldsymbol{i}_5 = (1\,0\,0\,1)$. Note that here, as with the repetition codes, the word 'parity' does not refer to whether there are an even or odd number of ones in a word, but rather refers to the code's check bits irrespective of the code's property or structure.

Table 1.6
The (7, 4) Hamming code

Information words $i=(i_1, i_2, i_3, i_4)$	*Codewords* $c=(i_1, i_2, i_3, i_4, p_1, p_2, p_3)$
$i_0=(0\ 0\ 0\ 0)$	$c_0=(0\ 0\ 0\ 0\ 0\ 0\ 0)$
$i_1=(0\ 0\ 0\ 1)$	$c_1=(0\ 0\ 0\ 1\ 0\ 1\ 1)$
$i_2=(0\ 0\ 1\ 0)$	$c_2=(0\ 0\ 1\ 0\ 1\ 1\ 0)$
$i_3=(0\ 0\ 1\ 1)$	$c_3=(0\ 0\ 1\ 1\ 1\ 0\ 1)$
$i_4=(0\ 1\ 0\ 0)$	$c_4=(0\ 1\ 0\ 0\ 1\ 1\ 1)$
$i_5=(0\ 1\ 0\ 1)$	$c_5=(0\ 1\ 0\ 1\ 1\ 0\ 0)$
$i_6=(0\ 1\ 1\ 0)$	$c_6=(0\ 1\ 1\ 0\ 0\ 0\ 1)$
$i_7=(0\ 1\ 1\ 1)$	$c_7=(0\ 1\ 1\ 1\ 0\ 1\ 0)$
$i_8=(1\ 0\ 0\ 0)$	$c_8=(1\ 0\ 0\ 0\ 1\ 0\ 1)$
$i_9=(1\ 0\ 0\ 1)$	$c_9=(1\ 0\ 0\ 1\ 1\ 1\ 0)$
$i_{10}=(1\ 0\ 1\ 0)$	$c_{10}=(1\ 0\ 1\ 0\ 0\ 1\ 1)$
$i_{11}=(1\ 0\ 1\ 1)$	$c_{11}=(1\ 0\ 1\ 1\ 0\ 0\ 0)$
$i_{12}=(1\ 1\ 0\ 0)$	$c_{12}=(1\ 1\ 0\ 0\ 0\ 1\ 0)$
$i_{13}=(1\ 1\ 0\ 1)$	$c_{13}=(1\ 1\ 0\ 1\ 0\ 0\ 1)$
$i_{14}=(1\ 1\ 1\ 0)$	$c_{14}=(1\ 1\ 1\ 0\ 1\ 0\ 0)$
$i_{15}=(1\ 1\ 1\ 1)$	$c_{15}=(1\ 1\ 1\ 1\ 1\ 1\ 1)$

At the decoding stage the received word is

$$\boldsymbol{v} = (v_1, v_2, v_3, v_4, v_5, v_6, v_7)$$

where

$$\begin{aligned} v_1 &= i_1 \\ v_2 &= i_2 \\ v_3 &= i_3 \\ v_4 &= i_4 \\ v_5 &= p_1 \\ v_6 &= p_2 \\ v_7 &= p_3 \end{aligned}$$

if no errors occur. The decoder determines 3 parity-check sums

$$\begin{aligned} s_1 &= (v_1 + v_2 + v_3) + v_5 \\ s_2 &= (v_2 + v_3 + v_4) + v_6 \\ s_3 &= (v_1 + v_2 + v_4) + v_7. \end{aligned} \tag{1.19}$$

The first 3 bits in each parity-check sum correspond to the same combination of information bits as that used in the construction of the parity bits (see eqns 1.17), they are enclosed in parenthesis to emphasize this correspondence. From the parity-check sums we can define

$$\boldsymbol{s} = (s_1, s_2, s_3) \tag{1.20}$$

which is known as the *error syndrome* or *syndrome* of $\boldsymbol{v}$. The parity-check sums are constructed such that they are zero if no errors occur. For example, if there are no errors then

$$s_1 = (v_1 + v_2 + v_3) + v_5 = (i_1 + i_2 + i_3) + p_1$$

and using eqns 1.17 gives

$$s_1 = p_1 + p_1 = 0.$$

Note that $p_1 + p_1 = 0$ irrespective of whether p_1 is 0 or 1 (because we are using modulo-2 addition). Likewise we can show that $s_2 = s_3 = 0$ when there are no errors. Therefore the error syndrome $\boldsymbol{s} = (0\,0\,0)$ when there are no errors. Consider now the codeword $\boldsymbol{c}_{11} = (1\,0\,1\,1\,0\,0\,0)$, if it incurs no errors then it will give $\boldsymbol{v} = (1\,0\,1\,1\,0\,0\,0)$ as the decoder input and the resulting parity-check sums will be

$$\begin{aligned} s_1 &= (1 + 0 + 1) + 0 = 0 \\ s_2 &= (0 + 1 + 1) + 0 = 0 \\ s_3 &= (1 + 0 + 1) + 0 = 0 \end{aligned}$$

which again given $\boldsymbol{s} = (0\ 0\ 0)$. Likewise if we take any codeword from Table 1.6 the error syndrome $\boldsymbol{s} = (0\ 0\ 0)$ and so the error syndrome of a codeword is always zero. The construction of the parity-check bits and parity-check sums is such that the error syndrome of any codeword is zero.

Example 1.9
Let $\boldsymbol{c}_7 = (0\ 1\ 1\ 1\ 0\ 1\ 0)$ be a codeword sent over a channel. If no errors occur then the received word is $\boldsymbol{v} = \boldsymbol{c}_7 = (0\ 1\ 1\ 1\ 0\ 1\ 0)$ and using eqns 1.19 gives the parity-check sums

$$\begin{aligned} s_1 &= (0 + 1 + 1) + 0 = 0 \\ s_2 &= (1 + 1 + 1) + 1 = 0 \\ s_3 &= (0 + 1 + 1) + 0 = 0 \end{aligned}$$

giving an error syndrome of $\boldsymbol{s} = (s_1, s_2, s_3) = (0\,0\,0)$. □

A zero error syndrome is always obtained when there are no errors and a nonzero error syndrome can only arise if errors have occurred. However, the occurrence of errors does not necessarily give a nonzero error syndrome, as a zero error syndrome is also obtained whenever an error pattern changes a codeword into a different codeword. For example, if the codeword $\boldsymbol{c}_5 = (0\,1\,0\,1\,1\,0\,0)$ incurs the error $\boldsymbol{e} = (0\,1\,1\,0\,0\,0\,1)$ then, using eqn 1.1, the decoder input is

$$\begin{aligned} \boldsymbol{v} &= \boldsymbol{c} + \boldsymbol{e} \\ &= (0\ 1\ 0\ 1\ 1\ 0\ 0) + (0\ 1\ 1\ 0\ 0\ 0\ 1) \\ &= (0\ 0\ 1\ 1\ 1\ 0\ 1) \end{aligned}$$

which is the codeword c_3. The resulting parity-check sums are

$$s_1 = (0 + 0 + 1) + 1 = 0$$
$$s_2 = (0 + 1 + 1) + 0 = 0$$
$$s_3 = (0 + 0 + 1) + 1 = 0$$

giving $s = (0\ 0\ 0)$ and therefore the error pattern is undetectable. The occurrence of a single error gives an error syndrome s that depends uniquely on the position of the error within the codeword. Furthermore s is independent of the codeword incurring the error because the error syndrome of a codeword is always zero. For example consider the codeword $c_{12} = (1\ 1\ 0\ 0\ 0\ 1\ 0)$ incurring the error pattern $e = (0\ 0\ 0\ 1\ 0\ 0\ 0)$ this gives

$$\begin{aligned} v &= c + e \\ &= (1\ 1\ 0\ 0\ 0\ 1\ 0) + (0\ 0\ 0\ 1\ 0\ 0\ 0) \\ &= (1\ 1\ 0\ 1\ 0\ 1\ 0) \end{aligned}$$

and the parity-check sums are

$$s_1 = (1 + 1 + 0) + 0 = 0$$
$$s_2 = (1 + 0 + 1) + 1 = 1$$
$$s_3 = (1 + 1 + 1) + 0 = 1$$

giving $s = (0\ 1\ 1)$. Consider now some other codeword, say, $c_3 = (0\ 0\ 1\ 1\ 1\ 0\ 1)$ incurring the same error pattern $e = (0\ 0\ 0\ 1\ 0\ 0\ 0)$, here $v = (0\ 0\ 1\ 0\ 1\ 0\ 1)$ and again we get $s = (0\ 1\ 1)$. Any of the 16 codewords, belonging to the (7, 4) code, incurring the error $e = (0\ 0\ 0\ 1\ 0\ 0\ 0)$ will give the error syndrome $s = (0\ 1\ 1)$. Table 1.7 shows the 7 different single errors e that can occur in a 7-bit word, along with their corresponding error syndromes s. The error syndrome $s = (0\ 0\ 0)$, obtained when no errors occur, is included in the table. The table can be used for single-error correction by 'looking up' the error pattern e corresponding to a given error syndrome s. The position of the nonzero bit in e gives the position of the error in v, and on locating the error the erroneous bit is corrected by inverting it. Table 1.7 is referred to as a *syndrome table*. Note that the error pattern obtained from the syndrome table and

Table 1.7
Syndrome table for the (7, 4) Hamming code

Error pattern e $(e_1, e_2, e_3, e_4, e_5, e_6, e_7)$	*Error syndrome* s (s_1, s_2, s_3)
(0 0 0 0 0 0 0)	(0 0 0)
(0 0 0 0 0 0 1)	(0 0 1)
(0 0 0 0 0 1 0)	(0 1 0)
(0 0 0 0 1 0 0)	(1 0 0)
(0 0 0 1 0 0 0)	(0 1 1)
(0 0 1 0 0 0 0)	(1 1 0)
(0 1 0 0 0 0 0)	(1 1 1)
(1 0 0 0 0 0 0)	(1 0 1)

the resulting codeword are the decoder's guess or estimate of the error pattern $\boldsymbol{e}$ and codeword $\boldsymbol{c}$ and are denoted by $\hat{\boldsymbol{e}}$ and $\hat{\boldsymbol{c}}$ respectively. Decoding can be summarized in the three steps:

1. Calculate $\boldsymbol{s}$ from the decoder input $\boldsymbol{v}$.
2. From the syndrome table obtain the error pattern $\hat{\boldsymbol{e}}$ that corresponds to $\boldsymbol{s}$.
3. The required codeword is then given by $\hat{\boldsymbol{c}} = \boldsymbol{v} + \hat{\boldsymbol{e}}$, this has the effect of inverting the bit in $\boldsymbol{v}$ given by the position of the nonzero bit in $\hat{\boldsymbol{e}}$.

In the event of a single error occurring, the resulting error syndrome gives the correct error pattern and the correct codeword is obtained. All single errors can be corrected and therefore the (7, 4) Hamming code is a single-error-correcting code.

Example 1.10

Given that a codeword $\boldsymbol{c}$, of the (7, 4) Hamming code, incurs a single error so giving $\boldsymbol{v} = (1\ 0\ 1\ 1\ 0\ 0\ 1)$, find $\boldsymbol{c}$.

Using eqns 1.19 gives the parity-check sums

$$\begin{aligned} s_1 &= (1+0+1)+0 = 0 \\ s_2 &= (0+1+1)+0 = 0 \\ s_3 &= (1+0+1)+1 = 1 \end{aligned}$$

and so the error syndrome $\boldsymbol{s} = (s_1, s_2, s_3) = (0\ 0\ 1)$. From Table 1.7, $\boldsymbol{s} = (0\ 0\ 1)$ gives the error pattern $(0\ 0\ 0\ 0\ 0\ 0\ 1)$. Inverting the right-hand bit of $\boldsymbol{v}$ gives $(1\ 0\ 1\ 1\ 0\ 0\ 0)$ which is the codeword $\boldsymbol{c}_{11}$.

□

The syndrome table for the (7, 4) code contains all possible values of $\boldsymbol{s} = (s_1, s_2, s_3)$, so whenever two or more errors occur the resulting error syndrome $\boldsymbol{s}$ will always be one that corresponds to no errors ($\boldsymbol{s} = 0$) or to a single error ($\boldsymbol{s} \neq 0$), and on both occasions a decoding error occurs. There are no decoding failures and therefore decoding is complete. For example consider $\boldsymbol{c}_9 = (1\ 0\ 0\ 1\ 1\ 1\ 0)$, along with the double-error pattern $\boldsymbol{e} = (0\ 0\ 1\ 0\ 0\ 1\ 0)$, so giving $\boldsymbol{v} = \boldsymbol{c}_9 + \boldsymbol{e} = (1\ 0\ 1\ 1\ 1\ 0\ 0)$. The error syndrome of $\boldsymbol{v}$ is $\boldsymbol{s} = (1\ 0\ 0)$, and referring to Table 1.7 we get $\hat{\boldsymbol{e}} = (0\ 0\ 0\ 0\ 1\ 0\ 0)$. Hence $\hat{\boldsymbol{c}} = \boldsymbol{v} + \hat{\boldsymbol{e}} = (1\ 0\ 1\ 1\ 0\ 0\ 0)$ which is $\boldsymbol{c}_{11}$ and not $\boldsymbol{c}_9$. Whilst the two errors have been detected (because $\boldsymbol{s} \neq 0$), a decoding error has ultimately occurred. All double errors give a nonzero syndrome and are therefore detectable. We have already seen that the code can detect and correct single errors, and therefore the (7, 4) Hamming code can detect single and double errors or can correct single errors. Note that although the code can detect up to 2 errors, this is only interpreted as error detection if error correction is not implemented. If error correction is carried out then decoding is viewed as a correction, and not a detection, process. Hence it is in this sense that we think of the code as being able to detect up to two errors or correct single errors. The code cannot be used for jointly carrying out error detection and correction, for this requires codes with greater error-control capability (see Section 1.7).

We have seen that the (7, 4) code is guaranteed to detect all single and double errors, however other errors patterns are also detectable. For example some triple errors and 4-bit errors are detectable as shown in the example below.

Example 1.11
Given that the $(7, 4)$ code is used for error detection only, determine the outcome when the codeword $\boldsymbol{c}_6 = (0\ 1\ 1\ 0\ 0\ 0\ 1)$ incurs:
(a) the triple error $\boldsymbol{e}_1 = (0\ 1\ 0\ 1\ 1\ 0\ 0)$;
(b) the triple error $\boldsymbol{e}_2 = (1\ 0\ 0\ 1\ 0\ 1\ 0)$;
(c) the 4-bit error $\boldsymbol{e}_3 = (1\ 1\ 1\ 0\ 1\ 0\ 0)$;
(d) the 4-bit error $\boldsymbol{e}_4 = (0\ 1\ 0\ 1\ 1\ 0\ 1)$.

(a) If $\boldsymbol{c}_6 = (0\ 1\ 1\ 0\ 0\ 0\ 1)$ incurs the error $\boldsymbol{e}_1 = (0\ 1\ 0\ 1\ 1\ 0\ 0)$ then $\boldsymbol{v} = \boldsymbol{c}_6 + \boldsymbol{e}_1 = (0\ 0\ 1\ 1\ 1\ 0\ 1)$. The parity-check sums are $s_1 = 0$, $s_2 = 0$, and $s_3 = 0$, and so $\boldsymbol{s} = (s_1, s_2, s_3) = (0\ 0\ 0)$. The error is therefore not detected and so a decoding error has occurred.

(b) When $\boldsymbol{c}_6$ incurs the error $\boldsymbol{e}_2 = (1\ 0\ 0\ 1\ 0\ 1\ 0)$ we get $\boldsymbol{v} = (1\ 1\ 1\ 1\ 0\ 1\ 1)$. The error syndrome is now $\boldsymbol{s} = (1\ 0\ 0)$ and so the error has been detected.

(c) For $\boldsymbol{c}_6$ and $\boldsymbol{e}_3$, $\boldsymbol{v} = (1\ 0\ 0\ 0\ 1\ 0\ 1)$ and the error syndrome $\boldsymbol{s} = (0\ 0\ 0)$. The four errors are not detected.

(d) Here $\boldsymbol{c}_6$ and $\boldsymbol{e}_4$ give $\boldsymbol{v} = (0\ 0\ 1\ 1\ 1\ 0\ 0)$ and error syndrome $\boldsymbol{s} = (0\ 0\ 1)$. The four errors are therefore detected.

Note that the triple-error pattern $\boldsymbol{e}_1$ and the 4-bit error pattern $\boldsymbol{e}_3$ are undetected because they are identical to the codewords $\boldsymbol{c}_5$ and $\boldsymbol{c}_{14}$ respectively. The other two error patterns do not resemble any of the codewords and are therefore detectable. □

The $(7, 4)$ Hamming code is the first code in the class of single-error-correcting codes whose blocklengths n and information lengths k satisfy

$$\begin{aligned} n &= 2^r - 1 \\ k &= 2^r - 1 - r \end{aligned} \tag{1.21}$$

for any integer $r \geq 3$, and where $r = n - k$ gives the number of parity-check bits. Taking $r = 3$ gives the $(7, 4)$ code already considered. For $r = 4$ we get the $(15, 11)$ Hamming code which has 11-bit information words, 15-bit codewords and 4 parity-check bits. Given the information word $\boldsymbol{i} = (i_1, i_2, \ldots, i_{11})$ the parity bits are

$$\begin{aligned} p_1 &= i_1 + i_2 + i_3 + i_4 + i_6 + i_8 + i_9 \\ p_2 &= i_2 + i_3 + i_4 + i_5 + i_7 + i_9 + i_{10} \\ p_3 &= i_3 + i_4 + i_5 + i_6 + i_8 + i_{10} + i_{11} \\ p_4 &= i_1 + i_2 + i_3 + i_5 + i_7 + i_8 + i_{11} \end{aligned} \tag{1.22}$$

so giving the codeword $\boldsymbol{c} = (i_1, i_2, \ldots, i_{11}, p_1, p_2, p_3, p_4)$. The parity-check sums and syndrome table are constructed in the same way as those for the $(7, 4)$ code. Table 1.8 shows the number of codewords and error syndromes for the $(2^r - 1, 2^r - 1 - r)$ Hamming codes for values of $r = 3$, 4, 5 and 6. Note that the number of error syndromes rises much less rapidly with r than the number of codewords. Error detection and correction can be achieved through the use of tables of codewords, but this becomes impractical for large values of n and k. Decoding based on a syndrome

Table 1.8
The Hamming codes for $r = 3$ to 6

r	(n, k)	*Number of codewords*	*Number of syndromes*
3	(7,4)	16	8
4	(15,11)	2,048	16
5	(31, 26)	$\approx 67 \times 10^6$	32
6	(63, 57)	$\approx 10^{17}$	64
.	.	.	.
.	.	.	.
.	.	.	.
r	$(2^r - 1, 2^r - 1 - r)$	2^k	2^r

Where $k = 2^r - 1 - r$.

table is usually a practical alternative, due to the number of error syndromes being significantly less than the number of codewords.

An additional parity-check bit can be added to the (7, 4) code to give 8-bit codewords with even parity. The resulting code is known as the (8, 4) *extended Hamming code* and is capable of jointly correcting single errors and detecting double errors (see Section 2.7).

The use of a syndrome table for error correction is not restricted to the Hamming codes but can be applied to any block code. The syndrome table consists of all correctable error patterns along with their corresponding error syndromes. An (n, k) single-error-correcting code has ${}^n\mathrm{C}_1 = n$ single-error patterns and corresponding error syndromes in its syndrome table. A double-error-correcting code has ${}^n\mathrm{C}_1$ single-error patterns and ${}^n\mathrm{C}_2$ double-error patterns and error syndromes, in its syndrome table. A code capable of correcting t errors requires ${}^n\mathrm{C}_1, {}^n\mathrm{C}_2, \ldots, {}^n\mathrm{C}_t$ error patterns and error syndromes.

1.7 Minimum distance of block codes

In Section 1.2 we introduced the idea of parity-check bits having the effect of adding redundancy and so producing codewords that are separated from each other. Increasing the parity-check bits increases the redundancy and the separation or distance between codewords. Here we extend these ideas further and consider how distance between codewords determines the error-control capability of a code. As we shall see the concept of distance between codewords, and in particular the minimum distance within a code, is fundamental to error-control codes.

The *Hamming weight* or *weight* of a word $\boldsymbol{v}$ is defined as the number of nonzero components of $\boldsymbol{v}$ and is denoted by $w(\boldsymbol{v})$. The *Hamming distance* or *distance* between two words $\boldsymbol{v}_1$ and $\boldsymbol{v}_2$, having the same number of bits, is defined as the number of places in which they differ and is denoted by $d(\boldsymbol{v}_1, \boldsymbol{v}_2)$. For example the words $\boldsymbol{v}_1 = (0\,1\,1\,0\,1\,0)$ and $\boldsymbol{v}_2 = (1\,0\,1\,0\,0\,0)$ have weights 3 and 2 respectively and are separated by a distance of 3, therefore

$$w(0\ 1\ 1\ 0\ 1\ 0) = 3$$
$$w(1\ 0\ 1\ 0\ 0\ 0) = 2$$
$$d(0\ 1\ 1\ 0\ 1\ 0,\ 1\ 0\ 1\ 0\ 0\ 0) = 3.$$

The *minimum distance* d_{min} of a block code is the smallest distance between codewords. Hence codewords differ by d_{min} or more bits. The minimum distance is found by taking a pair of codewords, determining the distance between them and then repeating this for all pairs of different codewords. The smallest value obtained is the minimum distance of the code.

Example 1.12
Determine the minimum distance of the even-parity (3, 2) block code.

Here the codewords are (0 0 0) , (0 1 1), (1 1 0) and (1 0 1). Taking codewords pairwise gives

$$d(0\ 0\ 0,\ 0\ 1\ 1) = 2$$
$$d(0\ 0\ 0,\ 1\ 1\ 0) = 2$$
$$d(0\ 0\ 0,\ 1\ 0\ 1) = 2$$
$$d(0\ 1\ 1,\ 1\ 1\ 0) = 2$$
$$d(0\ 1\ 1,\ 1\ 0\ 1) = 2$$
$$d(1\ 1\ 0,\ 1\ 0\ 1) = 2$$

and the minimum distance of the code is therefore 2. □

Consider the (7, 4) Hamming code whose 16 codewords are shown in Table 1.6. This has 120 pairs of different codewords, and it can be shown that any pair of codewords has its 2 codewords separated by a distance of 3, 4 or 7 and therefore the minimum distance of the (7, 4) Hamming code is 3. The code has 8 pairs of codewords where the 2 codewords in each pair are separated by a distance of 7, 56 pairs have their 2 codewords separated by a distance of 4 and the remaining 56 pairs have codewords separated by a distance of 3 (see Table 1.9).

It is not usually practical to determine the minimum distance of a code by considering the distance between all pairs of different codewords. An (n, k) block code has $m = 2^k$ codewords and therefore ${}^{m}C_2$ different pairs of codewords, a term that rises very rapidly with increasing k. The (7, 4) Hamming code has 120 pairs of different codewords, and the (15, 11) Hamming code (see Table 1.8, $r = 4$) has 2^{11} codewords which gives over 2×10^6 pairs of codewords. An arbitrary block code could require a considerable degree of computation to determine its minimum distance. However, the codes that are important are not arbitrary but have a linear property (already referred to at the start of Section 1.6) that allows the minimum distance to be determined easily, this is considered in Section 2.1.

It is interesting to consider a block code from a geometric point of view, as this helps to illustrate the concept of distance between words. Codewords belonging to an (n, k) block code can be thought of as lying within an n-dimensional space

Table 1.9
Distance between codewords in the (7, 4) code

	c_0	c_1	c_2	c_3	c_4	c_5	c_6	c_7	c_8	c_9	c_{10}	c_{11}	c_{12}	c_{13}	c_{14}	c_{15}
c_0	0	3	3	4	4	3	3	4	3	4	4	3	3	4	4	7
c_1	3	0	4	3	3	4	4	3	4	3	3	4	4	3	7	4
c_2	3	4	0	3	3	4	4	3	4	3	3	4	4	7	3	4
c_3	4	3	3	0	4	3	3	4	3	4	4	3	7	4	4	3
c_4	4	3	3	4	0	3	3	4	3	4	4	7	3	4	4	3
c_5	3	4	4	3	3	0	4	3	4	3	7	4	4	3	3	4
c_6	3	4	4	3	3	4	0	3	4	7	3	4	4	3	3	4
c_7	4	3	3	4	4	3	3	0	7	4	4	3	3	4	4	3
c_8	3	4	4	3	3	4	4	7	0	3	3	4	4	3	3	4
c_9	4	3	3	4	4	3	7	4	3	0	4	3	3	4	4	3
c_{10}	4	3	3	4	4	7	3	4	3	4	0	3	3	4	4	3
c_{11}	3	4	4	3	7	4	4	3	4	3	3	0	4	3	3	4
c_{12}	3	4	4	7	3	4	4	3	4	3	3	4	0	3	3	4
c_{13}	4	3	7	4	4	3	3	4	3	4	4	3	3	0	4	3
c_{14}	4	7	3	4	4	3	3	4	3	4	4	3	3	4	0	3
c_{15}	7	4	4	3	3	4	4	3	4	3	3	4	4	3	3	0

(1 dimension for each bit). For example Fig. 1.9 shows the (3, 2) even-parity code in a 3-dimensional space. Each codeword is of the form $\boldsymbol{c} = (c_x, c_y, c_z)$ where c_x, c_y, and c_z are 0 or 1 and are the coordinates of $\boldsymbol{c}$ along the X, Y, and Z axes respectively. The shaded circles are codewords and the open circles show words that are redundant to the code. Each codeword has 3 redundant words a distance 1 away from the codeword and is at a distance of 2 away from the other 3 codewords (the dotted lines represent a distance of 1 within the space). A codeword incurring a single error will result in a 'transition', along a dotted line, to a redundant word. A second error will give another transition, this time to 1 of 2 codewords. For example, say (1 0 1) incurs a single error to give (0 0 1), then a second error will then give the codeword (0 1 1) or (0 0 0). A minimum of 2 errors/transitions are required to change any one codeword into another codeword, which is consistent with the code's minimum distance of 2.

The distance within a code can be illustrated without reference to a coordinate system, as shown in Fig. 1.10(a). This shows the main features of the (3, 2) code seen in Fig. 1.9, namely that each codeword is connected to 3 redundant words and a minimum of 2 transitions are required to change any one codeword into another. Figure 1.10(b) shows the corresponding diagram for the (3, 1) repetition code. Here there are only 2 codewords (0 0 0) and (1 1 1). To get from either of the codewords to the other requires 3 transitions and the minimum distance of the code is therefore 3. This is a rather obvious result because comparing (0 0 0) to (1 1 1) gives $d_{\min} = 3$ immediately, nevertheless it is always interesting to view a code in this manner.

If we attempt to produce a similar diagram for the (7, 4) code we find that it soon becomes rather cluttered. Whilst there are only 16 codewords, the total number of words is 128 each of which has to be linked to 7 other words (one link for each transition). A further simplification to Fig. 1.9 is to show words along one dimension only, as shown in Fig. 1.11. Each site no longer represents a specific word, but

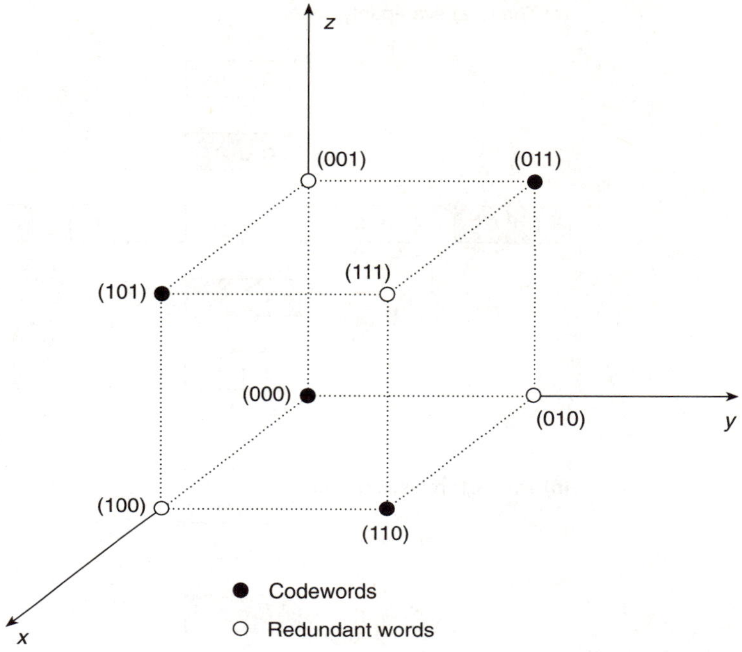

Fig. 1.9 The (3, 2) code in a 3-dimensional space.

only whether or not it is a codeword (shaded and open circles represent codewords and redundant words respectively). Moving from one circle to an adjacent circle represents a change of 1 bit (i.e. a distance of 1). The codewords $\boldsymbol{c}_1$ and $\boldsymbol{c}_2$ are separated by a distance 3, and $\boldsymbol{c}_2$ and $\boldsymbol{c}_3$ by a distance of 2, five redundant words $\boldsymbol{r}_1$ to $\boldsymbol{r}_5$ are shown.

Consider now the arrangement of codewords shown in Fig. 1.12(a), this typifies the separation of words in a code with minimum distance 3. Here A and B indicate examples of transitions that can occur if $\boldsymbol{c}_1$ incurs 1 or 2 errors respectively. Examples C and D show triple errors occurring at $\boldsymbol{c}_2$ and $\boldsymbol{c}_4$ respectively. To determine the decoding decisions for the errors A, B, C, and D we consider a maximum-likelihood decoder with input $\boldsymbol{v}$. For a binary-symmetric channel, maximum-likelihood decoding is equivalent to selecting a codeword that is closest to $\boldsymbol{v}$ than any other codeword and is referred to as *minimum-distance decoding* or *nearest-neighbour decoding*. The error patterns in Fig. 1.12(a) will therefore be decoded as follows:

A: c_1 incurs a single error

The error is detected because it gives a redundant word $\boldsymbol{r}_2$. Furthermore, minimum-distance decoding is able to correct the error as $\boldsymbol{r}_2$ lies closer to $\boldsymbol{c}_1$ than $\boldsymbol{c}_2$.

B: c_1 incurs a double error

The error is again detected, however this time the redundant word $\boldsymbol{r}_3$ lies closer to $\boldsymbol{c}_2$ than $\boldsymbol{c}_1$ and minimum-distance decoding estimates $\boldsymbol{c}_2$ to be the required codeword. This will give a decoding error.

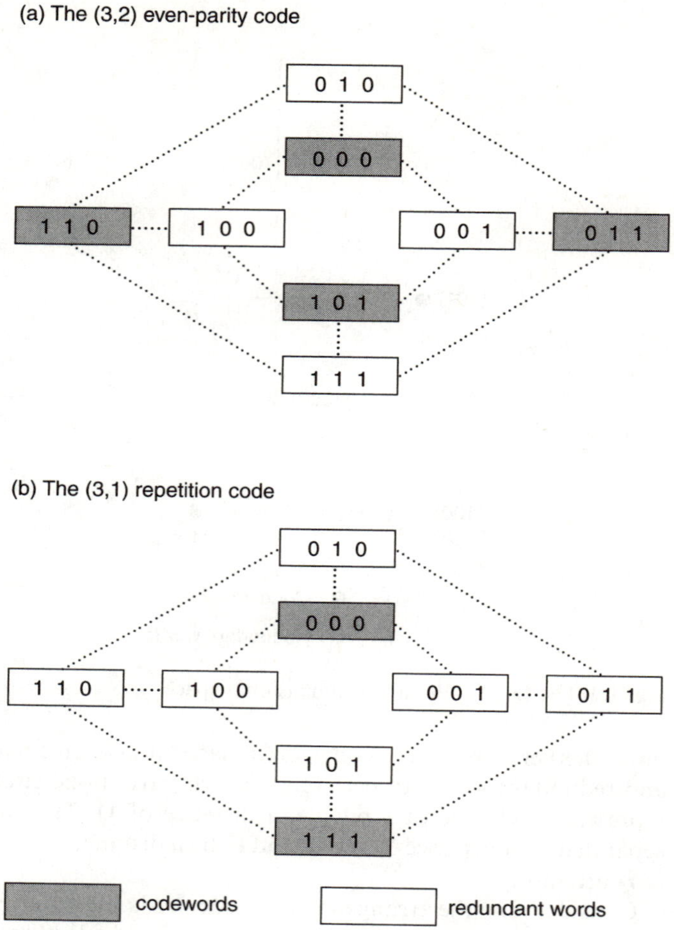

Fig. 1.10 Illustrating distance.

C: c_2 incurs a triple error

The error changes c_2 into c_3 and cannot be detected or corrected. Again this will give a decoding error.

D: c_4 incurs a triple error

Here the number of errors equals the minimum distance of the code but a redundant word r_6 is still obtained as c_3 and c_4 are separated by a distance of 4. The error is therefore detected but a decoding error occurs as c_3 is the nearest codeword to r_6.

In a code with $d_{\min} = 3$ the occurrence of single or double errors within a codeword always gives a redundant word (as in *A* and *B* above) and the detection of the errors is guaranteed. The occurrence of 3 or more errors may be detected (e.g. *D* above) but not all such error patterns are detectable (e.g. *C* above). Hence a block code with minimum distance 3 can detect all single and double errors. A code with minimum distance 5 can detect 4 or fewer errors, as illustrated by *A*, *B*, *C* and *D* in Fig. 1.12(b), *E* shows an undetectable 5-bit error. A code with minimum distance 7 can detect 6 or

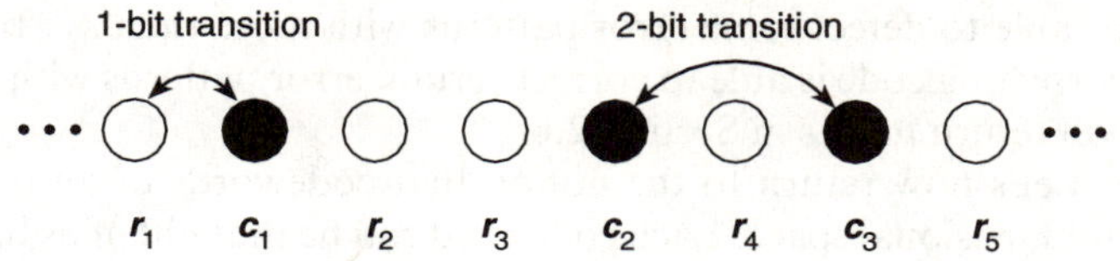

Fig. 1.11 A simplified way of illustrating distance.

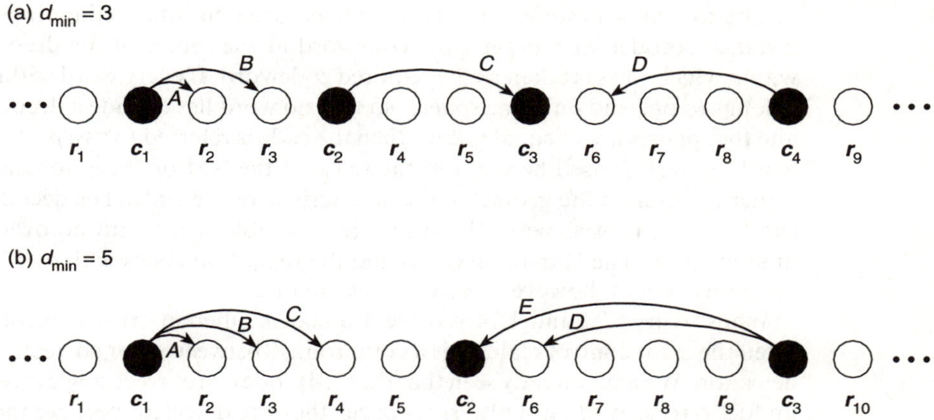

Fig. 1.12 Examples of errors.

fewer errors. It follows that a block code with minimum distance $d_{\min}$ can detect all error patterns with

$$\ell = d_{\min} - 1 \tag{1.23}$$

or fewer errors.

An error pattern incurred by a codeword $\boldsymbol{c}$ is correctable only if the resulting redundant word is closer to $\boldsymbol{c}$ than any other codeword. For a code with $d_{\min} = 3$ it is only single errors that satisfy this requirement Hence a block code with minimum distance 3 can correct all single errors. A code with minimum distance 5 can correct all single and double errors, a codeword $\boldsymbol{c}$ incurring a single or double error will give a redundant word that is closer to $\boldsymbol{c}$ than any other codeword. A code with minimum distance 7 can correct 3 or fewer errors and it follows that a block code with minimum distance $d_{\min}$ can correct all error patterns with

$$t = \tfrac{1}{2}(d_{\min} - 1) \tag{1.24}$$

or fewer errors.

We refer to t and ℓ as the *error-correction* and *error-detection limits* respectively and they give the *error-control limits* of a code. Codes with error-correction limit t and error-detection limit ℓ are referred to as *t-error-correcting codes* and *ℓ-error-detecting codes* respectively. Note that whilst a code with error-detection limit ℓ is guaranteed to detect all error patterns with ℓ errors or less, the code will also be able to detect some error patterns with more than ℓ errors. Likewise a t-error-correcting code is able to correct certain error patterns with more than t errors. We will return to this in Section 2.6.

Let's now return to the notion that codewords of an (n, k) code lie within an n-dimensional space. Each codeword can be thought of as having a *decoding sphere* of radius t around it. Each decoding sphere contains redundant words that are at a distance of t or less away from the codeword at the centre of the sphere. There will usually be redundant words lying outside the decoding spheres but no word will belong to two or more spheres as the spheres are non-intersecting. In a minimum-distance decoder with input $\boldsymbol{v}$ the codeword at the centre of the decoding sphere within which $\boldsymbol{v}$ lies is taken as the required codeword. If every word within the space belongs to one and only one sphere, so that no word lies outside a decoding sphere, and the spheres are of equal radius, then the code is referred to as a *perfect code*. The word 'perfect' is used here not in the sense of the best or exceptionally good, but rather to describe the geometrical characteristic of the code. The decoding spheres can be thought of as perfectly fitting the available space with no overlap and no unused space. The Hamming codes and the repetition codes with odd blocklength are perfect codes, however perfect codes are rare.

Whilst eqns 1.23 and 1.24 represent a code's inherent error-control capability, often the error control realized is a compromise between error correction and error detection. We have already seen that the (7, 4) code can correct single errors or detect up to 2 errors. When double errors occur they are detected, because the error syndrome is nonzero, but subsequently 'corrected'. The decoding process is not so much one of double-error detection, which would result in a decoding failure, but rather error correction resulting in a decoding error. When carrying out single-error correction the double-error detection capability of the code is not used. However, if the decoder does not carry out single-error correction then double errors give a decoding failure and are said to be detected. The (7, 4) code, or any other code with $d_{\min} = 3$, cannot carry out double-error detection and single-error correction jointly. To do so requires a larger minimum distance and it can be shown that a block code with minimum distance $d_{\min}$ can jointly correct t' or fewer errors and detect ℓ' or fewer errors providing

$$t' + \ell' \leq d_{\min} - 1 \tag{1.25}$$

where $\ell' > t'$. Table 1.10 shows values of t' and ℓ' that satisfy eqn 1.25 for minimum distances of 1 to 7. Note that for each value of t' the value of ℓ' shown gives the maximum number of errors that can be detected excluding error patterns with t' or fewer errors. For example for $d_{\min} = 5$ and $t' = 1$ we get $\ell' = 3$ which means that all double and triple errors can be detected. Likewise for $d_{\min} = 7$ and $t' = 2$, which give $\ell' = 4$, all triple and 4-bit errors can be detected. Note also that for odd values of $d_{\min}$ error detection is not possible when the maximum number of errors are corrected. Whereas when $d_{\min}$ is even then $d_{\min}/2$ errors can be detected when the maximum number of errors, now given by $(d_{\min} - 2)/2$, are corrected.

The four ways of using the error-control capability of a code with $d_{\min} = 7$ are illustrated in Fig. 1.13. Two codewords $\boldsymbol{c}_1$ and $\boldsymbol{c}_2$, separated by a distance of 7, are shown along with six redundant words $\boldsymbol{r}_1$ to $\boldsymbol{r}_6$ and we consider $\boldsymbol{c}_1$ incurring 6 or fewer errors. In Fig. 1.13(a) the decoder is correcting the maximum number of errors 3 and so decoding spheres of radius $t' = t = 3$ are shown around $\boldsymbol{c}_1$ and $\boldsymbol{c}_2$. Single,

Table 1.10
Joint error correction and detection

d_{min}	Number of errors corrected t'	Number of errors detected ℓ'
1	0	0
2	0	1
3	1	0
	or 0	2
4	1	2
	or 0	3
5	2	0
	or 1	3
	or 0	4
6	2	3
	or 1	4
	or 0	5
7	3	0
	or 2	4
	or 1	5
	or 0	6

double, and triple errors incurred by $\boldsymbol{c}_1$ result in redundant words lying within $\boldsymbol{c}_1$'s decoding sphere and the errors are correctable. There is no error detection capability because error patterns with 4, 5, or 6 errors give words lying within $\boldsymbol{c}_2$'s decoding sphere and will therefore give decoding errors. Figure 1.13(b) illustrates decoding when only 2 errors are corrected ($t' = 2$). Each decoding sphere now has a radius of 2 and the redundant words $\boldsymbol{r}_3$ and $\boldsymbol{r}_4$ are excluded from both spheres. Single and double errors can still be corrected, however 3- and 4-bit errors lie outside $\boldsymbol{c}_1$'s and $\boldsymbol{c}_2$'s decoding spheres and cannot be corrected. This is therefore an example of 1- and 2-bit error correction, jointly with 3- and 4-bit error detection. Reducing t' to 1 gives $\boldsymbol{r}_2$, $\boldsymbol{r}_3$, $\boldsymbol{r}_4$, and $\boldsymbol{r}_5$ lying outside the decoding spheres (Fig. 1.13(c)). This allows single-error-correction and the detection of 2, 3, 4, and 5 errors to take place jointly. If no error correction is implemented ($t' = 0$), then there are no decoding spheres and 1 to 6 errors can be detected (Fig. 1.13(d)).

1.8 Soft-decision decoding

In the preceding sections it has been assumed that the output of the demodulator (see Fig. 1.1) is always a 0 or a 1. Such a demodulator is said to make *hard decisions*. A demodulator that is not constrained to return 0 or 1 but is allowed to return a third symbol, say X, is said to make *soft decisions*. A soft decision is made whenever the demodulator input is so noisy that a 0 and 1 are equiprobable. The symbol X is referred to as an *erasure* and Fig. 1.14 shows a channel that includes erasures. An erasure is treated as an error whose location is known but with unknown magnitude and is thought of as being at a distance of 1/2 away from 0 and from 1 (i.e. equidistant from 0 and 1). A demodulator that makes hard decisions gives no

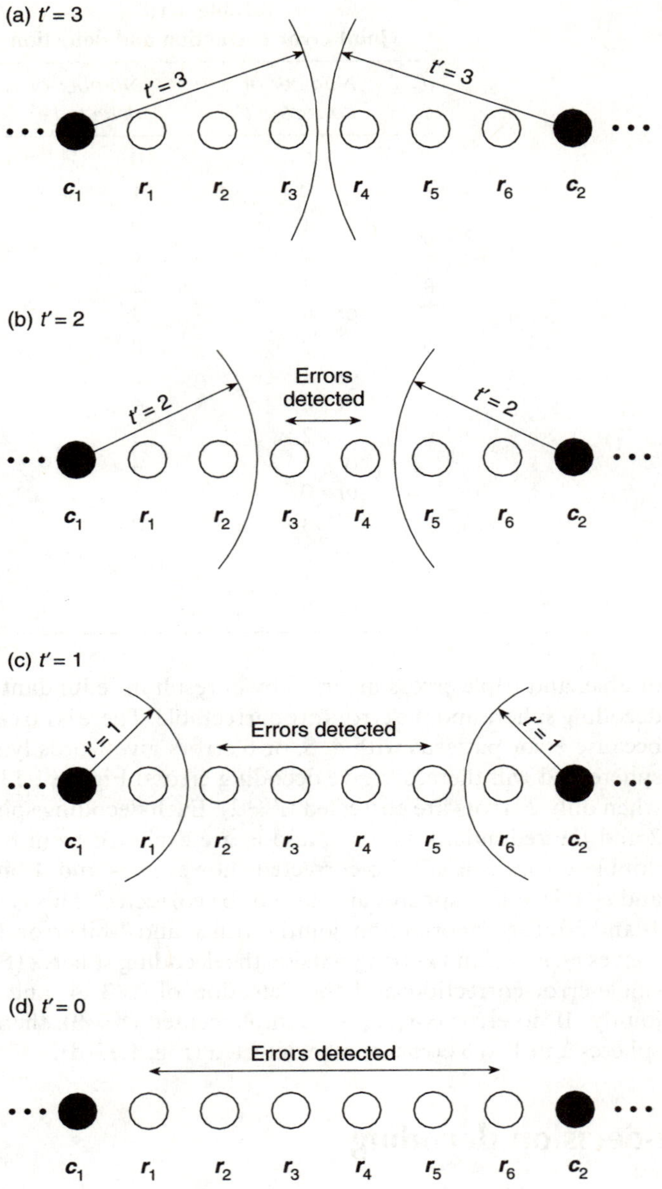

Fig. 1.13 Joint error correction and detection for $d_{\min} = 7$.

indication of the quality of the 0s and 1s entering the decoder. The decoder makes decisions on the presence or absence of errors according to the error-control code being used. A soft-decision demodulator, however, provides the decoder with additional information that can be used to improve error control. In the event of the decoder detecting errors, any erasures in the word being decoded will be the bits that are most likely to be in error.

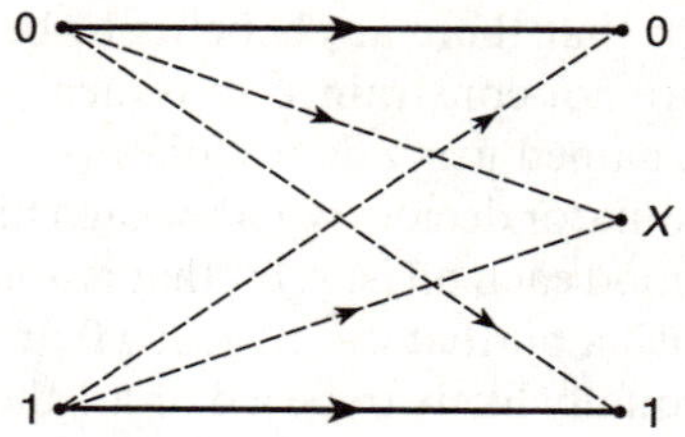

Fig. 1.14 An erasure channel.

Table 1.11
Error and erasure correction for $d_{\min} = 10$

t'	s'
0	9
1	7
2	5
3	3
4	1

Consider the (8, 7) even-parity code and let's assume that the input to the decoder is $\boldsymbol{v} = (1\,0\,0\,1\,0\,1\,0\,0)$. The parity of $\boldsymbol{v}$ is odd and therefore the decoder knows that at least 1 error has occurred. Based on the parity of $\boldsymbol{v}$ alone the decoder has no way of establishing the position of the error, or errors, in $\boldsymbol{v}$. Consider now a decoder whose input is taken from a soft-decision demodulator and let $\boldsymbol{v} = (0\,1\,1\,1\,0\,X\,1\,1)$ be the decoder input. The parity of $\boldsymbol{v}$ is incorrect but here it is reasonable to assume that the position of the erasure gives the bit that is most likely to be in error. If the erasure is assumed to have a 0 value then $\boldsymbol{v}$ still has the wrong parity. However setting $X = 1$ gives $\boldsymbol{v} = (0\,1\,1\,1\,0\,1\,1\,1)$ which has the correct parity and can be taken as the most likely even-parity codeword that $\boldsymbol{v}$ corresponds to. If $\boldsymbol{v}$ has the correct parity and contains an erasure, then the erasure is set to 0 or 1 as so to preserve the correct parity, for example given $\boldsymbol{v} = (1\,X\,0\,0\,0\,1\,0\,0)$ we would set $X = 0$. If $\boldsymbol{v}$ contains a single erasure, and no other errors, then single-error correction is guaranteed. The (8, 7) even-parity code is an error-detecting code, it has no error-correcting capability. However, here we see that in conjunction with a soft-decision demodulator single-error correction can be achieved. The combination of error-control coding with soft-decision demodulation is referred to as *soft-decision decoding*.

It can be shown that for a code with minimum distance $d_{\min}$ any pattern of t' errors and s' erasures can be corrected providing

$$2t' + s' \leq d_{\min} - 1. \tag{1.26}$$

Table 1.11 shows values of t' and s' for $d_{\min} = 10$. Note that, because erasures and errors are respectively at a distance of $1/2$ and 1 from the correct value, for every extra bit corrected the number of erasures that can be corrected is reduced by 2. The (8, 7) even-parity code has $d_{\min} = 2$ and $t = 0$ therefore only 1 erasure can be corrected. The (7, 4) code, with $d_{\min} = 3$, cannot correct erasures when used for error correction, but can correct up to 2 erasures if error correction is not carried out.

We have seen that there are benefits to be gained by using soft-decision demodulators that are not constrained to return 0 or 1 but can return erasures. Further benefits can be gained using demodulators that can assign a *bit quality* to each bit. Here the demodulator decides whether each bit is a 0 or 1 and assign a bit quality that indicates how good each bit is. A bit that is a clear-cut 0 or 1 would be assigned a high bit quality, whilst a bit that is only just a 0 or 1 is given a low bit quality. In the case where a bit is equally likely to be a 0 or a 1 then an erasure is returned. The decoder then makes its decision not just on the basis of the bit values, 0, 1, or X, but also on the quality associated with each bit. The use of bit-quality information can give considerable coding gains, but is at the expense of increased complexity for both the demodulator and the decoder.

1.9 Automatic-repeat-request schemes

The communication channel that we have considered so far is one in which information transfer takes place in one direction only, namely from the point at which information is generated to the point at which the information is used (see Fig. 1.1). Error-control encoding takes place prior to transmission, and on reception of the information, decoding takes place with a view to detecting, and if possible correcting, any errors incurred during the transmission. The direction of information transfer from the source to the user is referred to as the *forward path* and the error-correction techniques previously considered are known as *forward-error-correction* schemes. A channel within which transmission is possible from the user to the source is said to have a *return path*. The existence of a return path allows requests to be made for retransmission of information in the event of a decoding failure. Strategies of error control based on requests for retransmission are referred to as *Automatic-Repeat-Request* (ARQ) schemes.

Figure 1.15(a) shows one of the simplest ARQ schemes, namely a *stop-and-wait* scheme. Here the transmitter sends a word $\boldsymbol{w}_1$, on the forward path, and waits for an *acknowledgement* (*ACK*) on the return path before sending the next word $\boldsymbol{w}_2$. If the decoder at the receiver detects no errors in $\boldsymbol{w}_1$ then the receiver sends an *ACK* to the transmitter. The transmitter, upon receipt of the *ACK* transmits the next word $\boldsymbol{w}_2$. However, if $\boldsymbol{w}_1$ is found to contain errors then the receiver sends a *negative acknowledgement* (*NACK*) reply, in which case the transmitter will retransmit $\boldsymbol{w}_1$ instead of transmitting $\boldsymbol{w}_2$. Communication continues in this way with the transmitter waiting for a reply to each word sent, and sending a new word whenever the reply is an *ACK* and retransmitting the previous word if the reply is a *NACK*. Such a stop-and-wait scheme is simple to implement but quite inefficient in terms of the usage of the communication channel, as the forward path lies idle whilst the transmitter waits for the *ACK*/*NACK* replies.

The *go-back-N* ARQ scheme, shown in Fig. 1.15(b) for $N=4$, allows continuous transmission on the forward path, therefore avoiding idle transmission time. Here the transmitter does not wait for an *ACK* to each word sent, but transmits continuously until it receives a *NACK*. We assume, that because of delays within the system, that if the ith word sent by the transmitter is erroneous then the *NACK* is received before the transmitter sends the $(i+N)$th word. The receiver does not send an *ACK* upon receipt of each error-free frame, but only sends a *NACK* whenever it

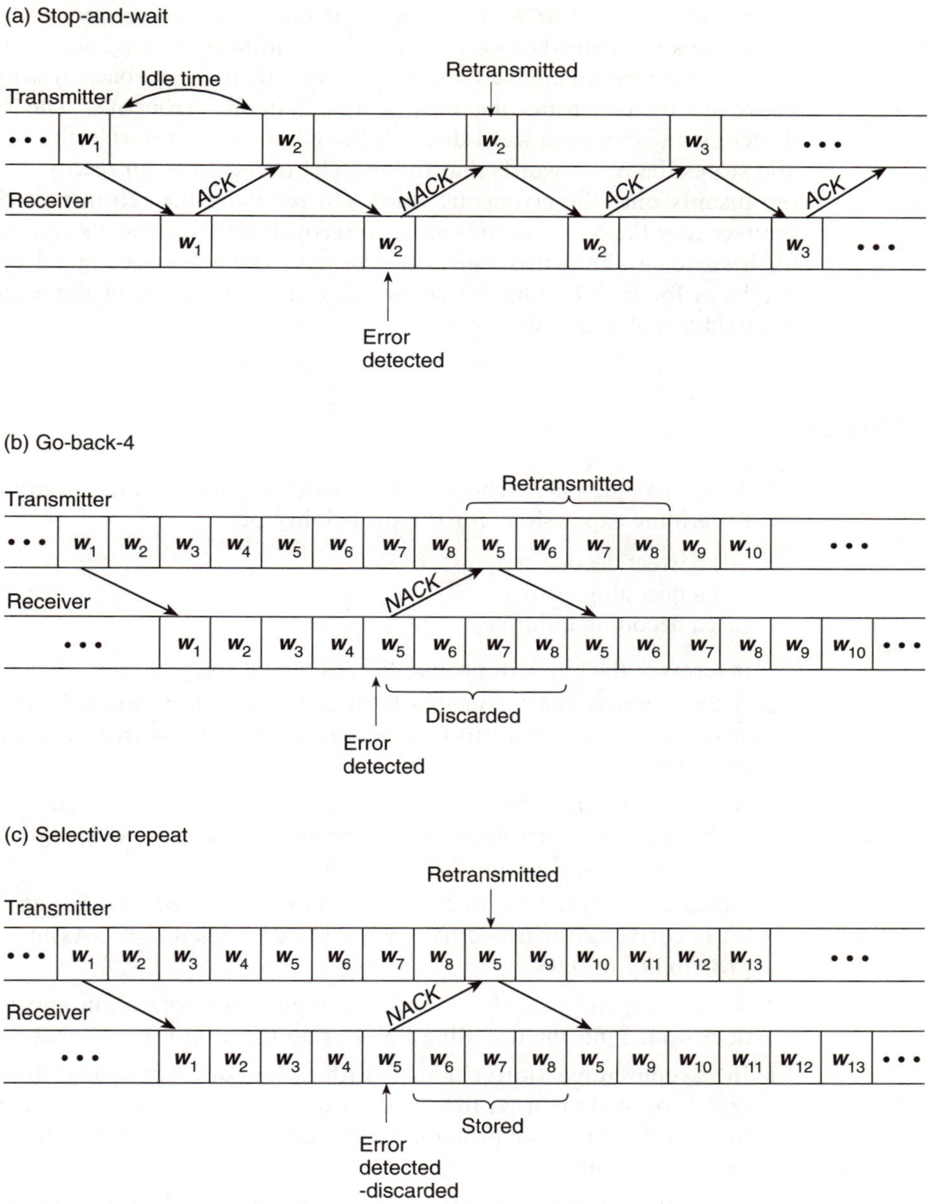

Fig. 1.15 Automatic-repeat-request schemes.

detects an error. Furthermore the receiver discards the erroneous word and the $N-1$ words that follow. The transmitter, upon receipt of a *NACK* goes back N words and retransmits the ith word along with the $N-1$ words that followed. By respectively discarding and retransmitting N words the receiver and the transmitter ensure that the correct sequence of words is preserved, without the receiver having to store words. In Fig. 1.15(b) $\boldsymbol{w}_5$ is in error and so the receiver replies with a *NACK*.

On receipt of the *NACK* the transmitter interrupts the sequence of words and retransmits $\boldsymbol{w}_5$, instead of $\boldsymbol{w}_9$, after which it continues with $\boldsymbol{w}_6$, $\boldsymbol{w}_7, \ldots$ and so on.

If the receiver is capable of storing words, then the go-back-N scheme can be improved by retransmitting only words that are erroneous. The receiver, on detecting an erroneous word discards the word, sends a *NACK* to the transmitter and stores the $N-1$ words that follow. The transmitter, on receipt of the *NACK*, retransmits only the erroneous word. On receiving the retransmitted word, the receiver uses the $N-1$ words stored to reconstruct the correct sequence of words. Such a scheme is known as a *selective-repeat* ARQ scheme (see Fig. 1.15(c)). Here the emphasis for maintaining the correct sequence of words, in the event of errors occurring, is placed at the receiver.

Problems

1.1 A $(4, 3)$ single-parity-check code is used to generate even-parity codewords. Determine expressions for the probability of:

(i) correct decoding p_c
(ii) a decoding error p_e
(iii) a decoding failure p_f

in terms of the bit-error probability p. Evaluate p_c, p_e, and p_f when $p = 5 \times 10^{-2}$.

1.2 A single-parity check code has 8-bit codewords. Determine the maximum bit-error probability that can be tolerated so that codewords have a success rate of 99.9%.

1.3 An $(n, n-1)$ single-parity-check code is used for error detection in a channel with bit-error probability 10^{-3}. Find the maximum blocklength n such that the success rate does not fall below 99%.

1.4 Given an $(n, 1)$ repetition code determine the probability that an information bit is correct after decoding when $n = 3$ and when $n = 5$. Assume a bit-error probability of 0.05.

1.5 A $(4, 1)$ repetition code is used for single-error correction and double-error detection. Find the decoding failure rate for a bit-error probability of 0.01.

1.6 In a communication channel with bit-error probability 10^{-2} an $(n, 1)$ repetition code is used for error correction. Find the minimum blocklength that gives a bit-error probability less than 10^{-6} after decoding. Assume odd values of n only.

1.7 A product code is constructed from the $(4, 3)$ and $(5, 4)$ single-parity-check codes. The information bits in the code arrays are denoted by i_{j1}, i_{j2}, i_{j3} and i_{j4} where $j = 1$, 2, and 3. Show that the overall parity-check bit p is the same whether it is constructed from the row parity-checks or from the column parity-checks. Assume even parity.

1.8 A $(32, 21)$ product code is constructed from the $(8, 7)$ and $(4, 3)$ single-parity-check codes. Even parity is used to construct both the column and row parity-check bits. Figure 1.16 shows 4 arrays each of which represents a code array. Determine the decisions that a decoder is likely to make.

(a)

1	0	0	1
1	1	1	0
0	0	0	0
1	0	1	0
0	1	1	0
0	1	0	1
0	1	1	0
1	1	0	0

(b)

0	1	0	1
1	0	1	0
1	1	1	1
1	0	1	0
1	1	0	0
0	0	0	0
0	1	0	1
0	0	1	1

(c)

0	1	1	0
1	0	0	1
1	1	0	0
0	1	1	0
1	1	1	1
1	1	0	0
1	1	1	1
0	0	1	1

(d)

0	0	0	1
0	1	0	1
0	1	0	1
1	0	0	1
1	1	1	0
1	0	1	0
0	1	1	0
0	0	1	1

Fig. 1.16 Arrays for Problem 1.8.

1.9 Given that $\boldsymbol{c} = (1\ 0\ 0\ 1\ 1\ 1\ 0)$ is a codeword of the (7, 4) Hamming code determine the received word $\boldsymbol{v}$, error syndrome $\boldsymbol{s}$, the decoder's estimate of the error pattern and estimate of the codeword when $\boldsymbol{c}$ incurs the error patterns

(i) $\boldsymbol{e}_1 = (0\ 0\ 1\ 0\ 0\ 0\ 0)$
(ii) $\boldsymbol{e}_2 = (0\ 1\ 0\ 1\ 0\ 0\ 0)$
(iii) $\boldsymbol{e}_3 = (1\ 0\ 0\ 0\ 1\ 0\ 1)$.

1.10 Using the parity-check equations (eqns 1.22) for the (15, 11) Hamming code determine expressions for the parity-check sums s_1, s_2, s_3, and s_4 given $\boldsymbol{v} = (v_1, v_2, \ldots, v_{15})$. Find the error syndrome $\boldsymbol{s} = (s_1, s_2, s_3, s_4)$ for each different single-bit error in a 15-bit word. Hence construct the syndrome table for the (15, 11) Hamming code.

1.11 Using the syndrome table for the (15, 11) Hamming code, constructed in Problem 1.10, find the codeword that each of the following words is most likely to represent

(i) $\boldsymbol{v}_1 = (1\ 0\ 0\ 0\ 1\ 0\ 0\ 1\ 1\ 0\ 0\ 1\ 0\ 0\ 1)$
(ii) $\boldsymbol{v}_2 = (0\ 0\ 1\ 0\ 0\ 1\ 1\ 1\ 0\ 1\ 0\ 0\ 1\ 1\ 0)$.

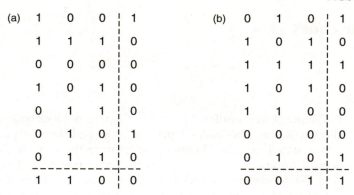

Linear codes

2

In Chapter 1 we considered some basic aspects of block codes, and the single-parity check codes, product codes, repetition codes and Hamming codes were introduced. It was stated that the Hamming codes were the first class of linear codes to be devised for error control, and it is the linear property of codes that we consider here. Whilst nonlinear codes do exist, by far most codes, and indeed the best codes, are linear. Linearity is an important structural property of codes, allowing a concise representation of codes and the accompanying encoding and decoding rules. Of particular importance is the interpretation of linear codes in terms of vector spaces, but this is deferred until we meet vector spaces in Chapter 5.

2.1 Definition of linear codes

A blockcode is said to be a *linear code* if its codewords satisfy the condition that the sum of any two codewords gives another codeword. Hence, given the codewords $\boldsymbol{c}_i$ and $\boldsymbol{c}_j$ of a linear code, then

$$\boldsymbol{c}_k = \boldsymbol{c}_i + \boldsymbol{c}_j \tag{2.1}$$

is also a codeword, where the components of $\boldsymbol{c}_i$ and $\boldsymbol{c}_j$ are added together pairwise using modulo-2 addition. A linear code has the properties that

1. The all-zero word (0 0 0 . . . 0) is always a codeword.
2. Given any three codewords $\boldsymbol{c}_i$, $\boldsymbol{c}_j$, and $\boldsymbol{c}_k$ such that

   $$\boldsymbol{c}_k = \boldsymbol{c}_i + \boldsymbol{c}_j$$

   then

   $$d(\boldsymbol{c}_i, \boldsymbol{c}_j) = w(\boldsymbol{c}_k)$$

   that is, the distance between two codewords equals the weight of the sum of the codewords.
3. The minimum distance of the code

   $$d_{\min} = w_{\min}$$

   where $w_{\min}$ is the weight of any nonzero codeword with the smallest weight.

The third property is of particular importance because it enables the minimum distance to be found quite easily. For an arbitrary block code, the minimum distance is found by considering the distance between all codewords. However, with a linear code we only need to evaluate the weight of every nonzero codeword. The minimum distance of the code is then given by the smallest weight obtained. This is much quicker than considering the distance between all codewords.

The $(7, 4)$ Hamming code considered in Section 1.6 is a linear code. Taking any pair of codewords (see Table 1.6) we find that the sum modulo-2 is always another codeword. For example consider say $\boldsymbol{c}_1$ and $\boldsymbol{c}_{10}$ then

$$\begin{aligned}\boldsymbol{c}_1 + \boldsymbol{c}_{10} &= (0\ 0\ 0\ 1\ 0\ 1\ 1) + (1\ 0\ 1\ 0\ 0\ 1\ 1)\\ &= (1\ 0\ 1\ 1\ 0\ 0\ 0)\\ &= \boldsymbol{c}_{11}.\end{aligned}$$

Likewise taking say $\boldsymbol{c}_{14}$ and $\boldsymbol{c}_3$ gives

$$\begin{aligned}\boldsymbol{c}_{14} + \boldsymbol{c}_3 &= (1\ 1\ 1\ 0\ 1\ 0\ 0) + (0\ 0\ 1\ 1\ 1\ 0\ 1)\\ &= (1\ 1\ 0\ 1\ 0\ 0\ 1)\\ &= \boldsymbol{c}_{13}.\end{aligned}$$

Table 2.1 shows the sum of codewords, taken pairwise, belonging to the $(7, 4)$ code. As a linear code, the $(7, 4)$ code must have an all-zero codeword, and from Table 1.6 we can see that this is so because $\boldsymbol{c}_0 = (0\,0\,0\,0\,0\,0\,0)$ is a codeword. If we take the sum of any two codewords, then the weight of the resulting codeword should equal the distance between the two codewords. Take for example $\boldsymbol{c}_1$ and $\boldsymbol{c}_{10}$, as shown above their sum is $\boldsymbol{c}_{11}$. The weight of $\boldsymbol{c}_{11}$ is

$$w(\boldsymbol{c}_{11}) = w(1\ 0\ 1\ 1\ 0\ 0\ 0) = 3$$

and the distance between $\boldsymbol{c}_1$ and $\boldsymbol{c}_{10}$ is

$$d(\boldsymbol{c}_1, \boldsymbol{c}_{10}) = d(0\ 0\ 0\ 1\ 0\ 1\ 1,\ 1\ 0\ 1\ 0\ 0\ 1\ 1) = 3$$

as required. The minimum distance of a linear code is given by the weight of a nonzero codeword (or codewords) with smallest weight. Referring to Table 1.6

Table 2.1
Addition of codewords in the (7, 4) linear code

$+$	c_1	c_2	c_3	c_4	c_5	c_6	c_7	c_8	c_9	c_{10}	c_{11}	c_{12}	c_{13}	c_{14}	c_{15}
c_1	c_0	c_3	c_2	c_5	c_4	c_7	c_6	c_9	c_8	c_{11}	c_{10}	c_{13}	c_{12}	c_{15}	c_{14}
c_2	c_3	c_0	c_1	c_6	c_7	c_4	c_5	c_{10}	c_{11}	c_8	c_9	c_{14}	c_{15}	c_{12}	c_{13}
c_3	c_2	c_1	c_0	c_7	c_6	c_5	c_4	c_{11}	c_{10}	c_9	c_8	c_{15}	c_{14}	c_{13}	c_{12}
c_4	c_5	c_6	c_7	c_0	c_1	c_2	c_3	c_{12}	c_{13}	c_{14}	c_{15}	c_8	c_9	c_{10}	c_{11}
c_5	c_4	c_7	c_6	c_1	c_0	c_3	c_2	c_{13}	c_{12}	c_{15}	c_{14}	c_9	c_8	c_{11}	c_{10}
c_6	c_7	c_4	c_5	c_2	c_3	c_0	c_1	c_{14}	c_{15}	c_{12}	c_{13}	c_{10}	c_{11}	c_8	c_9
c_7	c_6	c_5	c_4	c_3	c_2	c_1	c_0	c_{15}	c_{14}	c_{13}	c_{12}	c_{11}	c_{10}	c_9	c_8
c_8	c_9	c_{10}	c_{11}	c_{12}	c_{13}	c_{14}	c_{15}	c_0	c_1	c_2	c_3	c_4	c_5	c_6	c_7
c_9	c_8	c_{11}	c_{10}	c_{13}	c_{12}	c_{15}	c_{14}	c_1	c_0	c_3	c_2	c_5	c_4	c_7	c_6
c_{10}	c_{11}	c_8	c_9	c_{14}	c_{15}	c_{12}	c_{13}	c_2	c_3	c_0	c_1	c_6	c_7	c_4	c_5
c_{11}	c_{10}	c_9	c_8	c_{15}	c_{14}	c_{13}	c_{12}	c_3	c_2	c_1	c_0	c_7	c_6	c_5	c_4
c_{12}	c_{13}	c_{14}	c_{15}	c_8	c_9	c_{10}	c_{11}	c_4	c_5	c_6	c_7	c_0	c_1	c_2	c_3
c_{13}	c_{12}	c_{15}	c_{14}	c_9	c_8	c_{11}	c_{10}	c_5	c_4	c_7	c_6	c_1	c_0	c_3	c_2
c_{14}	c_{15}	c_{12}	c_{13}	c_{10}	c_{11}	c_8	c_9	c_6	c_7	c_4	c_5	c_2	c_3	c_0	c_1
c_{15}	c_{14}	c_{13}	c_{12}	c_{11}	c_{10}	c_9	c_8	c_7	c_6	c_5	c_4	c_3	c_2	c_1	c_0

we can see that all the nonzero codewords, apart from $c_{15} = (1\ 1\ 1\ 1\ 1\ 1\ 1)$, have weight 3 or 4. There are no nonzero codewords with weight less than 3 and therefore the minimum distance of the (7, 4) code is 3 (as we have already seen in Section 1.7).

The single-parity-check codes with even parity are also linear. Consider, for example, the codewords $c_1 = (0\ 1\ 1\ 0\ 1\ 0\ 1\ 0)$, $c_2 = (1\ 0\ 1\ 1\ 0\ 1\ 0\ 0)$ and $c_3 = (0\ 0\ 0\ 0\ 1\ 1\ 0\ 0)$ belonging to the (8, 7) single-parity-check code with even parity. The sum $c_1 + c_2 = (1\ 1\ 0\ 1\ 1\ 1\ 1\ 0)$ has even parity and is therefore a codeword. Likewise $c_1 + c_3 = (0\ 1\ 1\ 0\ 0\ 1\ 1\ 0)$ and $c_2 + c_3 = (1\ 0\ 1\ 1\ 1\ 0\ 0\ 0)$, which again have even parity and are therefore codewords. Taking the sum of any 2 codewords belonging to the (8, 7) even-parity code always gives an even parity word and therefore a codeword. The (8, 7) even-parity code is therefore linear.

Example 2.1

Consider the (4, 3) even-parity code with the 8 codewords

$$(0\ 0\ 0\ 0), (0\ 0\ 1\ 1), (0\ 1\ 1\ 0), (1\ 1\ 0\ 0)$$
$$(0\ 1\ 0\ 1), (1\ 0\ 1\ 0), (1\ 0\ 0\ 1), (1\ 1\ 1\ 1).$$

Show that the code is linear.

Taking the sum of the first nonzero codeword (0 0 1 1) with the other 7 codewords gives

$$\begin{aligned}
(0\ 0\ 1\ 1) + (0\ 0\ 0\ 0) &= (0\ 0\ 1\ 1)\\
(0\ 0\ 1\ 1) + (0\ 1\ 1\ 0) &= (0\ 1\ 0\ 1)\\
(0\ 0\ 1\ 1) + (1\ 1\ 0\ 0) &= (1\ 1\ 1\ 1)\\
(0\ 0\ 1\ 1) + (0\ 1\ 0\ 1) &= (0\ 1\ 1\ 0)\\
(0\ 0\ 1\ 1) + (1\ 0\ 1\ 0) &= (1\ 0\ 0\ 1)\\
(0\ 0\ 1\ 1) + (1\ 0\ 0\ 1) &= (1\ 0\ 1\ 0)\\
(0\ 0\ 1\ 1) + (1\ 1\ 1\ 1) &= (1\ 1\ 0\ 0)
\end{aligned}$$

all of which are even-parity words. The reader can verify that the remaining combinations also give even-parity words. The (4, 3) even-parity code is therefore linear. □

Given that even-parity codes are linear, we might expect that odd-parity codes are also linear. However, this is not so because the sum of any two odd-parity words gives an even-parity word. Consider for example the codewords (0 0 0 0 0 1 0 0) and (1 0 1 0 0 0 0 1) belonging to the (8, 7) odd-parity code. The sum of the two codewords is (1 0 1 0 0 1 0 1) which has even parity and therefore is not a codeword. Furthermore note that the all-zero word has even parity and therefore odd-parity codes cannot be linear as the all-zero word is not a codeword.

Example 2.2

Show that the sum of any 2 codewords from the (4, 3) odd-parity code fails to give a codeword.

The codewords are

$$(0\ 0\ 0\ 1), (0\ 0\ 1\ 0), (0\ 1\ 0\ 0), (1\ 0\ 0\ 0)$$
$$(0\ 1\ 1\ 1), (1\ 1\ 1\ 0), (1\ 0\ 1\ 1), (1\ 1\ 0\ 1)$$

and taking the sum of, say, (0 0 0 1) with the other 7 codewords gives

$$\begin{aligned}
(0\ 0\ 0\ 1) + (0\ 0\ 1\ 0) &= (0\ 0\ 1\ 1)\\
(0\ 0\ 0\ 1) + (0\ 1\ 0\ 0) &= (0\ 1\ 0\ 1)\\
(0\ 0\ 0\ 1) + (1\ 0\ 0\ 0) &= (1\ 0\ 0\ 1)\\
(0\ 0\ 0\ 1) + (1\ 1\ 1\ 0) &= (1\ 1\ 1\ 1)\\
(0\ 0\ 0\ 1) + (1\ 1\ 0\ 1) &= (1\ 1\ 0\ 0)\\
(0\ 0\ 0\ 1) + (1\ 0\ 1\ 1) &= (1\ 0\ 1\ 0)\\
(0\ 0\ 0\ 1) + (0\ 1\ 1\ 1) &= (0\ 1\ 1\ 0)
\end{aligned}$$

none of which have odd parity. The reader can verify that the remaining combinations also give even-parity words. The (4, 3) odd-parity code is therefore not linear, unlike the (4, 3) even-parity code. □

2.2 Generator matrices

In Section 1.6 parity-check equations were used to generate codewords for the (7, 4) Hamming code. Given an information word $\boldsymbol{i} = (i_1, i_2, i_3, i_4)$ the parity-check bits p_1, p_2, and p_3 are obtained using eqns 1.17 and then added to $\boldsymbol{i}$ to give the codeword $\boldsymbol{c} = (i_1, i_2, i_3, i_4, p_1, p_2, p_3)$. A better approach to encoding is through the use of matrices and as we shall see, linear codes lend themselves naturally to a matrix representation. There is a unique correspondence between information words $\boldsymbol{i}$ and codewords $\boldsymbol{c}$, which can be expressed as

$$\boldsymbol{c} = \boldsymbol{i}\boldsymbol{G} \tag{2.2}$$

where $\boldsymbol{G}$ is a matrix and is referred to as the *generator matrix* of the linear code. The generator matrix of an (n, k) linear code has k rows and n columns (note that a matrix with k rows and n columns is referred to as a 'k by n' or $k \times n$ matrix and is known as the order of the matrix). The generator matrix for the (7, 4) code is a 4 by 7 matrix given by

$$\boldsymbol{G} = \begin{bmatrix} 1 & 0 & 0 & 0 & 1 & 0 & 1 \\ 0 & 1 & 0 & 0 & 1 & 1 & 1 \\ 0 & 0 & 1 & 0 & 1 & 1 & 0 \\ 0 & 0 & 0 & 1 & 0 & 1 & 1 \end{bmatrix} \tag{2.3}$$

and later we shall see the relationship between the generator matrix and the code's parity-check equations. The product $\boldsymbol{i}\boldsymbol{G}$ is determined by taking the product of $\boldsymbol{i}$ with every column in $\boldsymbol{G}$. Each element in $\boldsymbol{i}$ is multiplied by the corresponding element in the column and then summed, using modulo-2 addition, over all elements. Consider $\boldsymbol{i} = (1\ 1\ 1\ 0)$, then using eqns 2.2 and 2.3 the corresponding codeword $c = (c_1, c_2, c_3, c_4, c_5, c_6, c_7)$ is given by

$$c = (1\ 1\ 1\ 0)\begin{bmatrix} 1 & 0 & 0 & 0 & 1 & 0 & 1 \\ 0 & 1 & 0 & 0 & 1 & 1 & 1 \\ 0 & 0 & 1 & 0 & 1 & 1 & 0 \\ 0 & 0 & 0 & 1 & 0 & 1 & 1 \end{bmatrix}.$$

Multiplying $\boldsymbol{i}$ by the first (left-hand side) column of $\boldsymbol{G}$ gives the first bit c_1

$$(1\ 1\ 1\ 0)\begin{bmatrix} 1 \\ 0 \\ 0 \\ 0 \end{bmatrix} = 1 \cdot 1 + 1 \cdot 0 + 1 \cdot 0 + 0 \cdot 0 = 1$$

and the second column gives c_2

$$(1\ 1\ 1\ 0)\begin{bmatrix} 0 \\ 1 \\ 0 \\ 0 \end{bmatrix} = 1 \cdot 0 + 1 \cdot 1 + 1 \cdot 0 + 0 \cdot 0 = 1.$$

Likewise columns 3, 4, 5, and 6 give 1, 0, 1, and 0 for c_3, c_4, c_5, and c_6 respectively. The last column gives c_7

$$(1\ 1\ 1\ 0)\begin{bmatrix} 1 \\ 1 \\ 0 \\ 1 \end{bmatrix} = 1 \cdot 1 + 1 \cdot 1 + 1 \cdot 0 + 0 \cdot 1 = 1 + 1 = 0$$

and so $\boldsymbol{c} = (c_1, c_2, c_3, c_4, c_5, c_6, c_7) = (1\ 1\ 1\ 0\ 1\ 0\ 0)$, which is the correct codeword for $\boldsymbol{i} = (1\ 1\ 1\ 0)$. Equation 2.2 can be used to generate all the 16 codewords belonging to the (7, 4) code. Note that when $\boldsymbol{i} = (0\,0\,0\,0)$, eqn 2.2 gives $\boldsymbol{c} = (0\,0\,0\,0\,0\,0\,0)$. Clearly the generator matrix provides a concise and convenient way of constructing codewords.

Example 2.3

Determine the set of codewords for the (6, 3) code with generator matrix

$$\boldsymbol{G} = \begin{bmatrix} 1 & 0 & 0 & 0 & 1 & 1 \\ 0 & 1 & 0 & 1 & 0 & 1 \\ 0 & 0 & 1 & 1 & 1 & 0 \end{bmatrix}. \tag{2.4}$$

We need to consider the information words (0 0 1), (0 1 0), (1 0 0), ..., (1 1 1). Substituting $\boldsymbol{G}$ and $\boldsymbol{i} = (0\ 0\ 1)$ into eqn 2.2 gives the codeword

$$c = (0\ 0\ 1)\begin{bmatrix} 1 & 0 & 0 & 0 & 1 & 1 \\ 0 & 1 & 0 & 1 & 0 & 1 \\ 0 & 0 & 1 & 1 & 1 & 0 \end{bmatrix} = (0\ 0\ 1\ 1\ 1\ 0)$$

Table 2.2
The (6, 3) linear code

i	c
(0 0 0)	(0 0 0 0 0 0)
(0 0 1)	(0 0 1 1 1 0)
(0 1 0)	(0 1 0 1 0 1)
(0 1 1)	(0 1 1 0 1 1)
(1 0 0)	(1 0 0 0 1 1)
(1 0 1)	(1 0 1 1 0 1)
(1 1 0)	(1 1 0 1 1 0)
(1 1 1)	(1 1 1 0 0 0)

and $\boldsymbol{i} = (0\ 1\ 0)$ gives the codeword

$$\boldsymbol{c} = (0\ 1\ 0)\begin{bmatrix} 1 & 0 & 0 & 0 & 1 & 1 \\ 0 & 1 & 0 & 1 & 0 & 1 \\ 0 & 0 & 1 & 1 & 1 & 0 \end{bmatrix} = (0\ 1\ 0\ 1\ 0\ 1).$$

Substituting the other 5 nonzero information words into eqn 2.2 gives the remaining 5 nonzero codewords (the codeword for (0 0 0) is (0 0 0 0 0 0)). The resulting 8 codewords are shown in Table 2.2. □

Note that the codewords given by the generator matrices of the (7, 4) and (6, 3) codes are in a systematic form, that is the information bits are placed together and are readily identifiable in each codeword. The reason for this lies in the way in which the generator matrices are constructed. The generator matrix for the (7, 4) code can be expressed as

$$\boldsymbol{G} = \left[\begin{array}{cccc|ccc} 1 & 0 & 0 & 0 & 1 & 0 & 1 \\ 0 & 1 & 0 & 0 & 1 & 1 & 1 \\ 0 & 0 & 1 & 0 & 1 & 1 & 0 \\ 0 & 0 & 0 & 1 & 0 & 1 & 1 \end{array}\right]$$

that is a 4 by 4 *identity matrix*

$$\boldsymbol{I}_4 = \begin{bmatrix} 1 & 0 & 0 & 0 \\ 0 & 1 & 0 & 0 \\ 0 & 0 & 1 & 0 \\ 0 & 0 & 0 & 1 \end{bmatrix}$$

augmented with a 4 by 3 matrix

$$\boldsymbol{P} = \begin{bmatrix} 1 & 0 & 1 \\ 1 & 1 & 1 \\ 1 & 1 & 0 \\ 0 & 1 & 1 \end{bmatrix}$$

known as the *parity matrix*. In terms of $\boldsymbol{I}_4$ and $\boldsymbol{P}$, the generator matrix $\boldsymbol{G}$ can be expressed as

$$\boldsymbol{G} = [\boldsymbol{I}_4 | \boldsymbol{P}].$$

The role of $\boldsymbol{I}_4$ is to keep the 4 information bits together and to position them at one end of the codeword, so that the resulting codeword is systematic. The parity matrix $\boldsymbol{P}$ is responsible for the way in which the parity-check bits are constructed from the information bits. To illustrate this consider the codeword $\boldsymbol{c}$ for $\boldsymbol{i}=(i_1, i_2, i_3, i_4)$

$$\boldsymbol{c} = \boldsymbol{i}\boldsymbol{G}$$

$$= (i_1, i_2, i_3, i_4)\left[\begin{array}{cccc|ccc} 1 & 0 & 0 & 0 & 1 & 0 & 1 \\ 0 & 1 & 0 & 0 & 1 & 1 & 1 \\ 0 & 0 & 1 & 0 & 1 & 1 & 0 \\ 0 & 0 & 0 & 1 & 0 & 1 & 1 \end{array}\right]$$

which gives

$$\boldsymbol{c} = (i_1, i_2, i_3, i_4, p_1, p_2, p_3)$$

where

$$\begin{aligned} p_1 &= 1 \cdot i_1 + 1 \cdot i_2 + 1 \cdot i_3 + 0 \cdot i_4 = i_1 + i_2 + i_3 \\ p_2 &= 0 \cdot i_1 + 1 \cdot i_2 + 1 \cdot i_3 + 1 \cdot i_4 = i_2 + i_3 + i_4 \\ p_3 &= 1 \cdot i_1 + 1 \cdot i_2 + 0 \cdot i_3 + 1 \cdot i_4 = i_1 + i_2 + i_4 \end{aligned}$$

are the parity-check bits. The codeword $\boldsymbol{c}$ is in a systematic form and the parity-check bits are the same as those given previously by the parity-check equations for the (7, 4) Hamming code (see eqns 1.17). Each column in the parity matrix contributes one parity bit and determines how the parity bit is derived from the information bits.

A generator matrix in the form

$$\boldsymbol{G} = [\boldsymbol{I}_k | \boldsymbol{P}] \tag{2.5}$$

where $\boldsymbol{I}_k$ is a k by k identity matrix and $\boldsymbol{P}$ is a k by $n - k$ matrix, is said to be in a *systematic form* as it generates systematic codewords. However, a generator matrix $\boldsymbol{G}$ may not necessarily be in a systematic form, in which case it generates a non-systematic code. If two generator matrices differ only by elementary row operations, that is by swapping any two rows or by adding any row to another row, then the matrices generate the same set of codewords and therefore the same code (there is a third row operation which is trivial here, see Section 5.5). If however two matrices differ by column permutations then the codewords generated by the matrices will differ by the order in which bits occur. Two codes that differ only by a permutation of bits are said to be *equivalent*. Hence two generator matrices generate equivalent codes if the matrices are related by

- column permutations
- elementary row operations.

The generator matrix of an (n, k) code is a k by n matrix (k rows and n columns) which in a systematic form is a k by k identity matrix augmented with a k by $n - k$ parity matrix. The k by k identity matrix ensures that codewords are systematic, whilst each of the $n - k$ columns of the parity matrix gives a parity-check bit. It can be shown that every linear code is equivalent to a systematic linear code and therefore every nonsystematic generator matrix can be put into a systematic form by column permutations and elementary row operations.

Example 2.4
The (5, 3) linear code has generator matrix

$$\boldsymbol{G}=\begin{bmatrix}1&0&1&0&0\\0&1&0&0&1\\0&1&1&1&0\end{bmatrix}.$$

Determine the systematic form of $\boldsymbol{G}$.

In its systematic form $\boldsymbol{G}$ will be of the form $\boldsymbol{G}=[\boldsymbol{I}|\boldsymbol{P}]$, where $\boldsymbol{I}$ is a 3 by 3 identity matrix and $\boldsymbol{P}$ is a 3 by 2 matrix. It is helpful to include a line in $\boldsymbol{G}$ showing where the matrix is augmented. To reduce $\boldsymbol{G}$ to its systematic form, we first add row 2 to row 3, so giving

$$\boldsymbol{G}=\left[\begin{array}{ccc|cc}1&0&1&0&0\\0&1&0&0&1\\0&0&1&1&1\end{array}\right]$$

and then add row 3 to row 1 to give

$$\boldsymbol{G}=\left[\begin{array}{ccc|cc}1&0&0&1&1\\0&1&0&0&1\\0&0&1&1&1\end{array}\right] \tag{2.6}$$

which is the required systematic form. Table 2.3 shows the resulting systematic and nonsystematic codewords. □

The codewords generated by a systematic and a nonsystematic generator matrix of a linear code differ only in the correspondence or mapping between information words and codewords. Consider, for example, the (5, 3) linear code with systematic and nonsystematic codewords shown in Table 2.3(a) and (b) respectively (see Example 2.4). Tables 2.3(a) and (b) have the same set of codewords but arranged in a different order. The tables differ only in the correspondence between information words $\boldsymbol{i}$ and codewords $\boldsymbol{c}$. In both tables the correspondence between

Table 2.3
The (5, 3) linear code

(a) *Systematic codewords*		(b) *Nonsystematic codewords*	
$\boldsymbol{i}$	$\boldsymbol{c}$	$\boldsymbol{i}$	$\boldsymbol{c}$
(0 0 0)	(0 0 0 0 0)	(0 0 0)	(0 0 0 0 0)
(0 0 1)	(0 0 1 1 1)	(0 0 1)	(0 1 1 1 0)
(0 1 0)	(0 1 0 0 1)	(0 1 0)	(0 1 0 0 1)
(1 0 0)	(1 0 0 1 1)	(1 0 0)	(1 0 1 0 0)
(0 1 1)	(0 1 1 1 0)	(0 1 1)	(0 0 1 1 1)
(1 0 1)	(1 0 1 0 0)	(1 0 1)	(1 1 0 1 0)
(1 1 0)	(1 1 0 1 0)	(1 1 0)	(1 1 1 0 1)
(1 1 1)	(1 1 1 0 1)	(1 1 1)	(1 0 0 1 1)

information words and codewords is unique and therefore either set of codewords can be used to represent the information words. We can clearly see now that a linear code does not have just one generator matrix, but many matrices all of which are equivalent and generate equivalent codes. The reader will find that for a given code different generator matrices may be given in different text books.

Example 2.5

Given $\boldsymbol{i}=(i_1, i_2, i_3, i_4)$ determine the codewords for the (7, 4) code when (a) rows 2 and 3 of its generator matrix (eqn 2.3) are interchanged and (b) columns 5 and 6 are interchanged.

(a) The generator matrix of the code is

$$\boldsymbol{G}=\begin{bmatrix} 1 & 0 & 0 & 0 & 1 & 0 & 1 \\ 0 & 1 & 0 & 0 & 1 & 1 & 1 \\ 0 & 0 & 1 & 0 & 1 & 1 & 0 \\ 0 & 0 & 0 & 1 & 0 & 1 & 1 \end{bmatrix}$$

and interchanging rows 2 and 3 gives

$$\boldsymbol{G}=\begin{bmatrix} 1 & 0 & 0 & 0 & 1 & 0 & 1 \\ 0 & 0 & 1 & 0 & 1 & 1 & 0 \\ 0 & 1 & 0 & 0 & 1 & 1 & 1 \\ 0 & 0 & 0 & 1 & 0 & 1 & 1 \end{bmatrix}.$$

Taking $\boldsymbol{i}=(i_1, i_2, i_3, i_4)$ gives the codeword

$$\boldsymbol{c}=\boldsymbol{iG}=(i_1, i_3, i_2, i_4,\ i_1+i_2+i_3,\ i_2+i_3+i_4,\ i_1+i_3+i_4).$$

In Section 1.6 codewords of the (7, 4) code were expressed as $(i_1, i_2, i_3, i_4, i_1+i_2+i_3, i_2+i_3+i_4, i_1+i_2+i_4)$. We see therefore that i_2 and i_3 have interchanged positions and the third parity bit of $\boldsymbol{c}$ is now given by $i_1+i_3+i_4$ whereas previously it was $i_1+i_2+i_4$. Nevertheless the set of codewords generated by $\boldsymbol{G}$ is the same as that previously obtained, except that there is a different correspondence between the information words and codewords. Table 2.4 shows the codewords generated by $\boldsymbol{G}$, and we can see that, for example, the codeword for $\boldsymbol{i}_5=(0\ 1\ 0\ 1)$ is now $\boldsymbol{c}_3=(0\ 0\ 1\ 1\ 1\ 0\ 1)$ instead of $\boldsymbol{c}_5$. Likewise the codeword for, say, $\boldsymbol{i}_{12}=(1\ 1\ 0\ 0)$ is now $\boldsymbol{c}_{10}=(1\ 0\ 1\ 0\ 0\ 1\ 1)$ instead of $\boldsymbol{c}_{12}$.

(b) Interchanging columns 5 and 6 in $\boldsymbol{G}$ gives codewords as

$$\boldsymbol{c}=(i_1, i_2, i_3, i_4,\ i_2+i_3+i_4,\ i_1+i_2+i_3,\ i_1+i_2+i_4)$$

and so this has had the effect of swapping the first and second parity bits.

□

The repetition codes are linear and therefore encoding can be achieved using a generator matrix. An $(n, 1)$ repetition code has generator matrix.

$$\boldsymbol{G}=(1\ 1\ 1\ldots 1\ 1). \tag{2.7}$$

For example the (3, 1) code has

$$\boldsymbol{G}=(1\ 1\ 1)$$

Table 2.4
Nonsystematic codewords of the (7, 4) code

Information words $\boldsymbol{i}$	*Codewords* $\boldsymbol{c}$
$i_0 = (0\ 0\ 0\ 0)$	$(0\ 0\ 0\ 0\ 0\ 0\ 0) = \boldsymbol{c}_0$
$i_1 = (0\ 0\ 0\ 1)$	$(0\ 0\ 0\ 1\ 0\ 1\ 1) = \boldsymbol{c}_1$
$i_2 = (0\ 0\ 1\ 0)$	$(0\ 1\ 0\ 0\ 1\ 1\ 1) = \boldsymbol{c}_4$
$i_3 = (0\ 0\ 1\ 1)$	$(0\ 1\ 0\ 1\ 1\ 0\ 0) = \boldsymbol{c}_5$
$i_4 = (0\ 1\ 0\ 0)$	$(0\ 0\ 1\ 0\ 1\ 1\ 0) = \boldsymbol{c}_2$
$i_5 = (0\ 1\ 0\ 1)$	$(0\ 0\ 1\ 1\ 1\ 0\ 1) = \boldsymbol{c}_3$
$i_6 = (0\ 1\ 1\ 0)$	$(0\ 1\ 1\ 0\ 0\ 0\ 1) = \boldsymbol{c}_6$
$i_7 = (0\ 1\ 1\ 1)$	$(0\ 1\ 1\ 1\ 0\ 1\ 0) = \boldsymbol{c}_7$
$i_8 = (1\ 0\ 0\ 0)$	$(1\ 0\ 0\ 0\ 1\ 0\ 1) = \boldsymbol{c}_8$
$i_9 = (1\ 0\ 0\ 1)$	$(1\ 0\ 0\ 1\ 1\ 1\ 0) = \boldsymbol{c}_9$
$i_{10} = (1\ 0\ 1\ 0)$	$(1\ 1\ 0\ 0\ 0\ 1\ 0) = \boldsymbol{c}_{12}$
$i_{11} = (1\ 0\ 1\ 1)$	$(1\ 1\ 0\ 1\ 0\ 0\ 1) = \boldsymbol{c}_{13}$
$i_{12} = (1\ 1\ 0\ 0)$	$(1\ 0\ 1\ 0\ 0\ 1\ 1) = \boldsymbol{c}_{10}$
$i_{13} = (1\ 1\ 0\ 1)$	$(1\ 0\ 1\ 1\ 0\ 0\ 0) = \boldsymbol{c}_{11}$
$i_{14} = (1\ 1\ 1\ 0)$	$(1\ 1\ 1\ 0\ 1\ 0\ 0) = \boldsymbol{c}_{14}$
$i_{15} = (1\ 1\ 1\ 1)$	$(1\ 1\ 1\ 1\ 1\ 1\ 1) = \boldsymbol{c}_{15}$

and given $\boldsymbol{i} = i_1$ the corresponding codeword is

$$\boldsymbol{c} = \boldsymbol{iG} = (i_1)(1\ 1\ 1) = (i_1, i_1, i_1)$$

as required. The single-parity check codes, with even parity, are also linear and for an (n, k) single-parity-check code the systematic generator matrix consists of an $\boldsymbol{I}_k$ identity matrix augmented with a single column with k rows all equal to 1, so that

$$\boldsymbol{G} = \left[\begin{array}{ccccc|c} 1 & 0 & 0 & \ldots & 0 & 1 \\ 0 & 1 & 0 & \ldots & 0 & 1 \\ 0 & 0 & 1 & \ldots & 0 & 1 \\ \vdots & & & & & \vdots \\ 0 & 0 & 0 & \ldots & 1 & 1 \end{array}\right]. \tag{2.8}$$

For example the (8, 7) even-parity code has

$$\boldsymbol{G} = \left[\begin{array}{ccccccc|c} 1 & 0 & 0 & 0 & 0 & 0 & 0 & 1 \\ 0 & 1 & 0 & 0 & 0 & 0 & 0 & 1 \\ 0 & 0 & 1 & 0 & 0 & 0 & 0 & 1 \\ 0 & 0 & 0 & 1 & 0 & 0 & 0 & 1 \\ 0 & 0 & 0 & 0 & 1 & 0 & 0 & 1 \\ 0 & 0 & 0 & 0 & 0 & 1 & 0 & 1 \\ 0 & 0 & 0 & 0 & 0 & 0 & 1 & 1 \end{array}\right]$$

and so given $\boldsymbol{i} = (i_1, i_2, i_3, i_4, i_5, i_6, i_7)$ we get

$$\boldsymbol{c} = \boldsymbol{iG} = (i_1, i_2, i_3, i_4, i_5, i_6, i_7, i_1 + i_2 + i_3 + i_4 + i_5 + i_6 + i_7)$$

as required.

When constructing codewords, using $\boldsymbol{c}=\boldsymbol{i}\boldsymbol{G}$, the information word $\boldsymbol{i}$ selects different rows of $\boldsymbol{G}$ to form the codeword $\boldsymbol{c}$. Each 1 in $\boldsymbol{i}$ contributes a row to $\boldsymbol{c}$ and if $\boldsymbol{i}$ consists of m 1s then $\boldsymbol{c}$ will be made up of m rows of $\boldsymbol{G}$ added together using modulo-2 addition. The first (i.e. the left-hand side) bit of $\boldsymbol{i}$ determines whether the top row of $\boldsymbol{G}$ contributes to $\boldsymbol{c}$, the second bit in $\boldsymbol{i}$ determines if the second row contributes and so forth. Consider for example the (7, 4) code and let $\boldsymbol{i}=(0\ 0\ 0\ 1)$, the resulting codeword is

$$\boldsymbol{C}=(0\,0\,0\,1)\begin{bmatrix} 1 & 0 & 0 & 0 & 1 & 0 & 1 \\ 0 & 1 & 0 & 0 & 1 & 1 & 1 \\ 0 & 0 & 1 & 0 & 1 & 1 & 0 \\ 0 & 0 & 0 & 1 & 0 & 1 & 1 \end{bmatrix}=(0\,0\,0\,1\,0\,1\,1)$$

which is simply the bottom row of $\boldsymbol{G}$. Likewise the codeword for $\boldsymbol{i}=(0\,1\,0\,0)$ is simply the second row $(0\ 1\ 0\ 0\ 1\ 1\ 1)$. If we now take $\boldsymbol{i}=(0\ 1\ 0\ 1)$ then the resulting codeword is constructed from the second and bottom row, so that $\boldsymbol{c}=(0\ 1\ 0\ 0\ 1\ 1\ 1)+(0\,0\,0\,1\,0\,1\,1)=(0\,1\,0\,1\,1\,0\,0)$. Taking every possible combination of the 4 rows gives the 16 codewords belonging to the (7, 4) code. If we imagine the rows of $\boldsymbol{G}$ that are not required as being 'crossed out' then the required codeword can be obtained by going down each column of $\boldsymbol{G}$ and adding together the 1s in the rows that are not crossed out. A note of the resulting sum is then made at the bottom of each column, and forms the required codeword. For example given $\boldsymbol{i}=(1\ 0\ 1\ 0)$ we can visualize the construction of $\boldsymbol{c}$ as

$$\boldsymbol{c}=\begin{array}{c} (1\,0\,1\,0) \\ \rightarrow \\ \\ \rightarrow \\ \\ \\ \end{array}\begin{array}{c} \begin{bmatrix} 1 & 0 & 0 & 0 & 1 & 0 & 1 \\ \cancel{0} & \cancel{1} & \cancel{0} & \cancel{0} & \cancel{1} & \cancel{1} & \cancel{1} \\ 0 & 0 & 1 & 0 & 1 & 1 & 0 \\ \cancel{0} & \cancel{0} & \cancel{0} & \cancel{1} & \cancel{0} & \cancel{1} & \cancel{1} \end{bmatrix} \\ \begin{array}{ccccccc} \downarrow & \downarrow & \downarrow & \downarrow & \downarrow & \downarrow & \downarrow \\ 1 & 0 & 1 & 0 & 0 & 1 & 1 \end{array} \end{array}$$

which gives $\boldsymbol{c}=(1\,0\,1\,0\,0\,1\,1)$. Likewise, say for $\boldsymbol{i}=(1\,1\,0\,1)$ we can cross out the third row to get

$$\boldsymbol{c}=\begin{array}{c} (1\,1\,0\,1) \\ \\ \rightarrow \\ \\ \\ \\ \end{array}\begin{array}{c} \begin{bmatrix} 1 & 0 & 0 & 0 & 1 & 0 & 1 \\ 0 & 1 & 0 & 0 & 1 & 1 & 1 \\ \cancel{0} & \cancel{0} & \cancel{1} & \cancel{0} & \cancel{1} & \cancel{1} & \cancel{0} \\ 0 & 0 & 0 & 1 & 0 & 1 & 1 \end{bmatrix} \\ \begin{array}{ccccccc} \downarrow & \downarrow & \downarrow & \downarrow & \downarrow & \downarrow & \downarrow \\ 1 & 1 & 0 & 1 & 0 & 0 & 1 \end{array} \end{array}$$

which gives $\boldsymbol{c}=(1\ 1\ 0\ 1\ 0\ 0\ 1)$. The construction of codewords by taking combinations of rows of $\boldsymbol{G}$ is reconsidered in Section 5.6 but this time in the context of vector spaces.

2.3 The standard array

A scheme for error correction in any linear code can be easily developed using a *standard array* of the code. For an (n, k) code the array consists of all 2^n n-bit words, and the array is constructed such that for an n-bit word $\boldsymbol{v}$ the array gives the codeword that is the least distance away from $\boldsymbol{v}$. Hence, the array can be used for minimum-distance decoding. A standard array for an (n, k) linear code is constructed as follows:

1. Place all the codewords in the first row, starting with the all-zero codeword.
2. Take one of the remaining words $\boldsymbol{w}$ that has least weight and place it under the column containing the all-zero codeword.
3. Fill in the row containing $\boldsymbol{w}$ by adding $\boldsymbol{w}$ to the codeword at the top of each column.
4. Continue steps 2 and 3 until all the 2^n words appear within the array.

The completed array (see Table 2.5) has 2^k columns (one column for each codeword) and 2^{n-k} rows, so the total number of words in the array is $2^k \times 2^{n-k} = 2^n$. Each word appears once in the array. The rows of the standard array are called *cosets* and the first word in each coset is called the *coset leader*. For a t-error-correcting code a horizontal line is drawn in the array to separate rows whose coset leaders have weights less than or equal to t, from the rows whose coset leaders have weights greater than t.

The standard array for the (6, 3) code, whose codewords are given in Table 2.2, is shown in Table 2.6. The first row contains the 8 codewords, starting with (0 0 0 0 0 0) on the left-hand side of the row and ending with (1 1 1 0 0 0). The order in which the codewords appear in the first row of the array is arbitrary. The second row starts with (0 0 0 0 0 1) as this is one of the words with least weight that is not in the first row. We could have used (0 0 0 0 1 0), (0 0 0 1 0 0), (0 0 1 0 0 0), (0 1 0 0 0 0), or (1 0 0 0 0 0) as they all have the same weight as (0 0 0 0 0 1), but it is usual to start the second row with the (0 0 . . . 0 1) word. The second row is completed by adding (0 0 0 0 0 1) to the codeword heading the column. On completing the second row, the word (0 0 0 0 1 0) has been used to start the third row and the row is completed by adding (0 0 0 0 1 0) to the codeword at the top of each column. Rows 4, 5, 6, and 7 are likewise completed using words of weight 1 as the coset leaders. However, row 8 requires a coset leader of weight 2 or more as all the words of weight 1 are already in the array. Rows 1 to 7 show that (0 0 1 0 0 1), (0 1 0 0 1 0), and (1 0 0 1 0 0) are words of weight 2 that are not yet in the array and therefore any of these 3 words can be used as the coset leader of row 8. Taking say (0 0 1 0 0 1) as the coset leader, row 8 is completed by adding (0 0 1 0 0 1) to the codeword heading each column. As the code is a single-error-correcting code a horizontal line is drawn below the row containing the coset

Table 2.5
Standard array for an (n, k) linear code

c_0	c_1	c_2	$\cdots$	c_{2^k-1}
w_1	$c_1 + w_1$	$c_2 + w_1$	$\cdots$	$c_{2^k-1} + w_1$
w_2	$c_1 + w_2$	$c_2 + w_2$	$\cdots$	$c_{2^k-1} + w_2$
w_3	$c_1 + w_3$	$c_2 + w_3$	$\cdots$	$c_{2^k-1} + w_3$
$\vdots$				$\vdots$
$w_{2^{n-k}-1}$	$c_1 + w_{2^{n-k}-1}$	$c_2 + w_{2^{n-k}-1}$	$\cdots$	$c_{2^k-1} + w_{2^{n-k}-1}$

Table 2.6
Standard array for the (6, 3) code

0 0 0 0 0 0	0 0 1 1 1 0	0 1 0 1 0 1	0 1 1 0 1 1	1 0 0 0 1 1	1 0 1 1 0 1	1 1 0 1 1 0	1 1 1 0 0 0
0 0 0 0 0 1	0 0 1 1 1 1	0 1 0 1 0 0	0 1 1 0 1 0	1 0 0 0 1 0	1 0 1 1 0 0	1 1 0 1 1 1	1 1 1 0 0 1
0 0 0 0 1 0	0 0 1 1 0 0	0 1 0 1 1 1	0 1 1 0 0 1	1 0 0 0 0 1	1 0 1 1 1 1	1 1 0 1 0 0	1 1 1 0 1 0
0 0 0 1 0 0	0 0 1 0 1 0	0 1 0 0 0 1	0 1 1 1 1 1	1 0 0 1 1 1	1 0 1 0 0 1	1 1 0 0 1 0	1 1 1 1 0 0
0 0 1 0 0 0	0 0 0 1 1 0	0 1 1 1 0 1	0 1 0 0 1 1	1 0 1 0 1 1	1 0 0 1 0 1	1 1 1 1 1 0	1 1 0 0 0 0
0 1 0 0 0 0	0 1 1 1 1 0	0 0 0 1 0 1	0 0 1 0 1 1	1 1 0 0 1 1	1 1 1 1 0 1	1 0 0 1 1 0	1 0 1 0 0 0
1 0 0 0 0 0	1 0 1 1 1 0	1 1 0 1 0 1	1 1 1 0 1 1	0 0 0 0 1 1	0 0 1 1 0 1	0 1 0 1 1 0	0 1 1 0 0 0
0 0 1 0 0 1	0 0 0 1 1 1	0 1 1 1 0 0	0 1 0 0 1 0	1 0 1 0 1 0	1 0 0 1 0 0	1 1 1 1 1 1	1 1 0 0 0 1

leader (1 0 0 0 0 0). The order in which words appear in the last row is arbitrary, had (0 1 0 0 1 0) or (1 0 0 1 0 0) been taken as the coset leader, then the words would be permuted.

Example 2.6
Construct a standard array for the (5, 2) single-error-correcting code with generator matrix

$$G = \left[\begin{array}{cc|ccc} 1 & 0 & 1 & 1 & 1 \\ 0 & 1 & 1 & 0 & 1 \end{array}\right]. \tag{2.9}$$

The code has 4 codewords (0 0 0 0 0), (0 1 1 0 1), (1 0 1 1 1), and (1 1 0 1 0). The codewords form the top row of the standard array, starting with (0 0 0 0 0) (see Table 2.7). The second row is formed by taking (0 0 0 0 1) as the coset leader and completing the row by adding (0 0 0 0 1) to each codeword. Rows 3, 4, 5, and 6 are constructed using (0 0 0 1 0), (0 0 1 0 0), (0 1 0 0 0), and (1 0 0 0 0) as the coset leaders. As the code is a single-error-correcting code a horizontal line is drawn below row 6. Four of the remaining words have weight 2, namely (1 0 0 0 1), (0 0 1 1 0), (1 0 1 0 0), and (0 0 0 1 1). Any of these can be used to construct the 7th row, we have used (1 0 0 0 1). Finally (1 0 1 0 0) has been used to construct the last row. □

In each column of the standard array the words above the horizontal line fall within the decoding sphere of the codeword heading the column. For a t-error-correcting code the coset leaders above the horizontal line represent error patterns with t or less errors. Coset leaders below the horizontal line represent error patterns with more than t errors, but nevertheless are still correctable (recall that a t-error-correcting code can correct some error patterns with more than t errors). A complete decoder for a standard array locates the decoder input $\boldsymbol{v}$ in the array and, irrespective of whether $\boldsymbol{v}$ is above or below the horizontal line, the codeword at the top of the column that $\boldsymbol{v}$ lies in is taken as the required codeword. If $\boldsymbol{v}$ lies below the horizontal line then the decoder assumes that $\boldsymbol{v}$ contains more than t errors but is nevertheless correctable (see Section 2.6). An incomplete decoder, however, gives a decoding failure if $\boldsymbol{v}$ lies below the horizontal line, and in doing so the probability of a decoding error is less than that of a complete decoder for the array.

Table 2.7
Standard array for the (5, 2) code

0 0 0 0 0	0 1 1 0 1	1 0 1 1 1	1 1 0 1 0
0 0 0 0 1	0 1 1 0 0	1 0 1 1 0	1 1 0 1 1
0 0 0 1 0	0 1 1 1 1	1 0 1 0 1	1 1 0 0 0
0 0 1 0 0	0 1 0 0 1	1 0 0 1 1	1 1 1 1 0
0 1 0 0 0	0 0 1 0 1	1 1 1 1 1	1 0 0 1 0
1 0 0 0 0	1 1 1 0 1	0 0 1 1 1	0 1 0 1 0
1 0 0 0 1	1 1 1 0 0	0 0 1 1 0	0 1 0 1 1
1 0 1 0 0	1 1 0 0 1	0 0 0 1 1	0 1 1 1 0

Example 2.7
Using the standard array of the (6, 3) code (Table 2.6), and assuming an incomplete decoder, determine the codewords for $\boldsymbol{v}_1 = (1\ 0\ 0\ 0\ 1\ 1)$, $\boldsymbol{v}_2 = (1\ 1\ 1\ 1\ 1\ 0)$, and $\boldsymbol{v}_3 = (0\ 1\ 1\ 1\ 0\ 0)$.

(a) $\boldsymbol{v}_1 = (1\ 0\ 0\ 0\ 1\ 1)$ is in the first row of the standard array and is therefore already a codeword.

(b) $\boldsymbol{v}_2 = (1\ 1\ 1\ 1\ 1\ 0)$ lies above the horizontal line and in the column headed by (1 1 0 1 1 0), so this is taken as the required codeword. The row that $\boldsymbol{v}_2$ lies in has coset leader (0 0 1 0 0 0) and this is taken as the error pattern that (1 1 0 1 1 0) incurred to give $\boldsymbol{v}_2$.

(c) $\boldsymbol{v}_3 = (0\ 1\ 1\ 1\ 0\ 0)$ falls in the column headed by (0 1 0 1 0 1) but lies below the horizontal line so giving a decoding failure. □

The standard array provides a framework for visualizing the decoding of linear codes but clearly it is not practically realizable for codes with large blocklengths and information lengths. However, the cosets possess a structure that allows the size of the array to be reduced significantly (see Section 2.6).

2.4 Parity-check matrices

We have seen that the generator matrix of a linear code can be expressed as $\boldsymbol{G} = [\boldsymbol{I}_k|\boldsymbol{P}]$ where $\boldsymbol{I}_k$ is a k by k identity matrix and the parity matrix $\boldsymbol{P}$ is a k by $n-k$ matrix. From $\boldsymbol{G}$ a *parity-check matrix* $\boldsymbol{H}$ can be constructed as follows

$$\boldsymbol{H} = [\boldsymbol{P}^{\mathrm{T}}|\boldsymbol{I}_{n-k}] \tag{2.10}$$

where $\boldsymbol{P}^{\mathrm{T}}$ is the transpose of $\boldsymbol{P}$, and is therefore an $n-k$ by k matrix, and $\boldsymbol{I}_{n-k}$ is an $n-k$ by $n-k$ identity matrix. Note that whilst $\boldsymbol{G}$ is a k by n matrix, $\boldsymbol{H}$ is an $n-k$ by n matrix. The parity-check matrix is used at the decoding stage to determine the error syndrome of a word. We have already defined the error syndrome of a word in context of the (7, 4) code and as we shall see in Section 2.5 an error syndrome can be defined for any linear code. Here we take a look at the relationships between the parity-check matrix, codewords, generator matrices, and the error-control properties of a code.

Consider the parity-check matrix $\boldsymbol{H}$ of the (7, 4) code. Recall that the generator matrix for the (7, 4) code is

$$\boldsymbol{G} = \left[\begin{array}{cccc|ccc} 1 & 0 & 0 & 0 & 1 & 0 & 1 \\ 0 & 1 & 0 & 0 & 1 & 1 & 1 \\ 0 & 0 & 1 & 0 & 1 & 1 & 0 \\ 0 & 0 & 0 & 1 & 0 & 1 & 1 \end{array}\right]$$

and so the parity matrix is

$$\boldsymbol{P} = \begin{bmatrix} 1 & 0 & 1 \\ 1 & 1 & 1 \\ 1 & 1 & 0 \\ 0 & 1 & 1 \end{bmatrix}.$$

Taking the transpose (i.e. swapping the rows and columns) of $\boldsymbol{P}$ gives

$$\boldsymbol{P}^{\mathrm{T}} = \begin{bmatrix} 1 & 1 & 1 & 0 \\ 0 & 1 & 1 & 1 \\ 1 & 1 & 0 & 1 \end{bmatrix}.$$

To complete the parity-check matrix a 3 by 3 identity matrix

$$\boldsymbol{I}_3 = \begin{bmatrix} 1 & 0 & 0 \\ 0 & 1 & 0 \\ 0 & 0 & 1 \end{bmatrix}$$

is required, and so the parity-check matrix for the (7, 4) code is

$$\boldsymbol{H} = \left[\begin{array}{cccc|ccc} 1 & 1 & 1 & 0 & 1 & 0 & 0 \\ 0 & 1 & 1 & 1 & 0 & 1 & 0 \\ 1 & 1 & 0 & 1 & 0 & 0 & 1 \end{array}\right]. \tag{2.11}$$

Example 2.8

Determine the parity-check matrix of the (6, 3) linear code.

The generator matrix of the (6, 3) code is given by eqn 2.4

$$\boldsymbol{G} = \left[\begin{array}{ccc|ccc} 1 & 0 & 0 & 0 & 1 & 1 \\ 0 & 1 & 0 & 1 & 0 & 1 \\ 0 & 0 & 1 & 1 & 1 & 0 \end{array}\right]$$

which can be expressed as $[\boldsymbol{I}_3|\boldsymbol{P}]$ where $\boldsymbol{I}_3$ is a 3 by 3 identity matrix and

$$\boldsymbol{P} = \begin{bmatrix} 0 & 1 & 1 \\ 1 & 0 & 1 \\ 1 & 1 & 0 \end{bmatrix}.$$

The transpose of $\boldsymbol{P}$ is

$$\boldsymbol{P}^{\mathrm{T}} = \begin{bmatrix} 0 & 1 & 1 \\ 1 & 0 & 1 \\ 1 & 1 & 0 \end{bmatrix}$$

note that here we get $\boldsymbol{P}^{\mathrm{T}} = \boldsymbol{P}$. From eqn 2.10 the parity-check matrix is $\boldsymbol{H} = [\boldsymbol{P}^{\mathrm{T}}|\boldsymbol{I}_3]$ which gives

$$\boldsymbol{H} = \left[\begin{array}{ccc|ccc} 0 & 1 & 1 & 1 & 0 & 0 \\ 1 & 0 & 1 & 0 & 1 & 0 \\ 1 & 1 & 0 & 0 & 0 & 1 \end{array}\right]. \tag{2.12}$$

□

Consider now the product of the generator matrix $\boldsymbol{G}$ with the transpose of the parity-check matrix $\boldsymbol{H}$, of an (n, k) linear code, from eqns 2.5 and 2.10 we get

$$\boldsymbol{G}\boldsymbol{H}^{\mathrm{T}} = [\boldsymbol{I}_k|\boldsymbol{P}][\boldsymbol{P}^{\mathrm{T}}|\boldsymbol{I}_{n-k}]^{\mathrm{T}}.$$

Note that for two matrices $\boldsymbol{A}$ and $\boldsymbol{B}$ the product $\boldsymbol{C}=\boldsymbol{AB}$ is defined only if the number of columns in $\boldsymbol{A}$ equals the number of rows in $\boldsymbol{B}$. If $\boldsymbol{A}$ is a k by m matrix and $\boldsymbol{B}$ is an m by n matrix then $\boldsymbol{C}$ is a k by n matrix (we can think of this as $(k \times m) \times (m \times n) = k \times n$). The transpose of $[\boldsymbol{P}^{\mathrm{T}}|\boldsymbol{I}_{n-k}]$ is the matrix $\boldsymbol{P}$ augmented with a $n-k$ by $n-k$ identity matrix positioned along the last $n-k$ rows

$$[\boldsymbol{P}^{\mathrm{T}}|\boldsymbol{I}_{n-k}]^{\mathrm{T}} = \left[\frac{\boldsymbol{P}}{\boldsymbol{I}_{n-k}}\right]$$

and so

$$\begin{aligned}\boldsymbol{GH}^{\mathrm{T}} &= [\boldsymbol{I}_k|\boldsymbol{P}]\left[\frac{\boldsymbol{P}}{\boldsymbol{I}_{n-k}}\right]\\ &= \boldsymbol{I}_k\boldsymbol{P} + \boldsymbol{P}\boldsymbol{I}_{n-k}\\ &= \boldsymbol{P} + \boldsymbol{P}\\ &= \boldsymbol{0}.\end{aligned}$$

Note that $\boldsymbol{I}_k\boldsymbol{P}=\boldsymbol{P}$ where $(k \times k) \times (k \times (n-k)) = k \times (n-k)$ is the order of $\boldsymbol{P}$, as required. Likewise $\boldsymbol{P}\boldsymbol{I}_{n-k}=\boldsymbol{P}$ and again with order $(k \times (n-k)) \times ((n-k) \times (n-k)) = k \times (n-k)$. Furthermore because modulo-2 addition is used the resulting matrix $\boldsymbol{0}$ is a k by $n-k$ matrix whose elements are all 0. Therefore, for any linear code with generator matrix $\boldsymbol{G}$ and parity-check matrix $\boldsymbol{H}$

$$\boldsymbol{GH}^{\mathrm{T}} = \boldsymbol{0}. \tag{2.13}$$

Next consider a codeword $\boldsymbol{c}$ for the information word $\boldsymbol{i}$. From eqn 2.2, $\boldsymbol{c}=\boldsymbol{iG}$ and multiplying this by $\boldsymbol{H}^{\mathrm{T}}$ gives

$$\boldsymbol{cH}^{\mathrm{T}} = (\boldsymbol{iG})\boldsymbol{H}^{\mathrm{T}} = \boldsymbol{i}(\boldsymbol{GH}^{\mathrm{T}}).$$

However, from eqn 2.13, $\boldsymbol{GH}^{\mathrm{T}}=\boldsymbol{0}$ and so $\boldsymbol{cH}^{\mathrm{T}}=\boldsymbol{0}$ where $\boldsymbol{0}$ is now the all-zero word with $n-k$ components. Therefore, for any linear code with parity-check matrix $\boldsymbol{H}$

$$\boldsymbol{cH}^{\mathrm{T}} = \boldsymbol{0} \tag{2.14}$$

for all codewords $\boldsymbol{c}$ belonging to the code.

Example 2.9
Determine the parity-check matrix $\boldsymbol{H}$ for the (5, 3) code. Show that $\boldsymbol{GH}^{\mathrm{T}}=\boldsymbol{0}$ and $\boldsymbol{cH}^{\mathrm{T}}=\boldsymbol{0}$ for $\boldsymbol{c}=(1\ 1\ 0\ 1\ 0)$.

The generator matrix for the (5, 3) code is

$$\boldsymbol{G} = \left[\begin{array}{ccc|cc} 1 & 0 & 0 & 1 & 1\\ 0 & 1 & 0 & 0 & 1\\ 0 & 0 & 1 & 1 & 1\end{array}\right]$$

and from $\boldsymbol{G}$ we get

$$\boldsymbol{P} = \begin{bmatrix} 1 & 1\\ 0 & 1\\ 1 & 1\end{bmatrix}.$$

The parity-check matrix is given by $[\boldsymbol{P}^{\mathrm{T}}|\boldsymbol{I}]$, and so

$$\boldsymbol{H}=\left[\begin{array}{ccc|cc}1&0&1&1&0\\1&1&1&0&1\end{array}\right]. \tag{2.15}$$

Taking the product $\boldsymbol{GH}^{\mathrm{T}}$ gives

$$\boldsymbol{GH}^{\mathrm{T}}=\begin{bmatrix}1&0&0&1&1\\0&1&0&0&1\\0&0&1&1&1\end{bmatrix}\begin{bmatrix}1&1\\0&1\\1&1\\1&0\\0&1\end{bmatrix}$$

$$=\begin{bmatrix}1+1&1+1\\0&1+1\\1+1&1+1\end{bmatrix}$$

$$=\begin{bmatrix}0&0\\0&0\\0&0\end{bmatrix}$$

and therefore $\boldsymbol{GH}^{\mathrm{T}}=\boldsymbol{0}$ as required. Given $\boldsymbol{c}=(1\ 1\ 0\ 1\ 0)$ then

$$\begin{aligned}\boldsymbol{cH}^{\mathrm{T}}&=(1\cdot1+1\cdot0+0\cdot1+1\cdot1+0\cdot0, 1\cdot1+1\cdot1+0\cdot1+1\cdot0+0\cdot1)\\&=(1+1,1+1)\\&=(0,0)\end{aligned}$$

as required. □

Whilst the generator matrix is responsible for constructing codewords, it is the parity-check matrix that determines the error-control properties of a code. It is useful to distinguish between the two matrices along this line, even though each matrix can be obtained from the other and equally represent the code. To construct a single-error-correcting code the columns in $\boldsymbol{H}$ have to be nonzero and unique. Each column in $\boldsymbol{H}$ is responsible for detecting and correcting an error in a specified position of the word being decoded. A zero column is unable to detect errors in the position that the column addresses and if two columns are the same then an error in either of the corresponding positions will be indistinguishable. Hence, by the columns being nonzero single errors are guaranteed to be detected and furthermore by being unique all single errors can be corrected. An (n, k) single-error-correcting code needs to correct n single-bit errors and therefore requires a parity-check matrix $\boldsymbol{H}$ with n columns. Each row of $\boldsymbol{H}$ generates a parity-check sum, of which $n-k$ are required, and therefore $\boldsymbol{H}$ needs $n-k$ rows. From $n-k$ rows we can obtain at the most $2^{n-k}-1$ nonzero unique columns (just as an m-bit word gives 2^m-1 nonzero different patterns) and so $n\leq 2^{n-k}-1$. The upper limit therefore gives

$$n=2^{n-k}-1 \tag{2.16}$$

as the maximum number of nonzero unique columns that can be obtained from $n-k$ rows. In Section 1.6 we saw that for any integer $r \geq 3$ there exists an (n, k) single-error-correcting Hamming code with $n=2^r-1$ and $k=2^r-1-r$ where $r=n-k$. Hence the codes constructed here are actually the Hamming codes when eqn 2.16 is satisfied. An (n, k) Hamming code is therefore a single-error-correcting code with the largest blocklength that can be constructed with $n-k$ parity bits.

Consider the parity-check matrix for the (7, 4) code (eqn 2.11). This has 7 nonzero and unique columns and can therefore detect and correct the 7 single-bit errors in a 7-bit word. Furthermore its blocklength and information length satisfy eqn 2.16 and so the (7, 4) code is a Hamming code (as we already know). The parity-check matrix of the (6, 3) code has 6 nonzero unique columns and can therefore detect and correct single-bit errors over 6-bit words. However its blocklength and information length do not satisfy eqn 2.16 and so the (6, 3) code is not a Hamming code. Likewise the (5, 2) code with parity-check matrix

$$\boldsymbol{H}=\begin{bmatrix}1 & 1 & 1 & 0 & 0\\ 1 & 0 & 0 & 1 & 0\\ 1 & 1 & 0 & 0 & 1\end{bmatrix} \tag{2.17}$$

is a single-error-correcting code as it has 5 nonzero and unique columns. The (5, 3) linear code with

$$\boldsymbol{H}=\begin{bmatrix}1 & 0 & 1 & 1 & 0\\ 1 & 1 & 1 & 0 & 1\end{bmatrix}$$

has only 2 rows which give a maximum of 4 columns of which 3 are nonzero. Hence there are not enough rows to give 5 unique columns, and the code is therefore incapable of error correction. Nevertheless its 5 columns are nonzero and so single-bit errors can be detected.

Example 2.10

Construct the parity-check matrix for the (15, 11) Hamming code.

The code has 4 parity-check bits and a blocklength of 15, so the parity-check matrix $\boldsymbol{H}$ will have 4 rows and 15 columns. The 4 rows provide 15 nonzero unique words, namely (0 0 0 1), (0 0 1 0), (0 0 1 1), ... (1 1 1 1), which will form the columns of $\boldsymbol{H}$. Although the order in which the columns appear is arbitrary, it is nevertheless convenient to place (0 0 0 1), (0 0 1 0), (0 1 0 0), and (1 0 0 0) on the right-hand side of $\boldsymbol{H}$ as so to form an identity matrix (this will enable the generator matrix $\boldsymbol{G}$ of the code to be easily obtained). The resulting matrix is

$$\boldsymbol{H}=\left[\begin{array}{ccccccccccc|cccc}1 & 1 & 1 & 1 & 0 & 1 & 0 & 1 & 1 & 0 & 0 & 1 & 0 & 0 & 0\\ 0 & 1 & 1 & 1 & 1 & 0 & 1 & 0 & 1 & 1 & 0 & 0 & 1 & 0 & 0\\ 0 & 0 & 1 & 1 & 1 & 1 & 0 & 1 & 0 & 1 & 1 & 0 & 0 & 1 & 0\\ 1 & 1 & 1 & 0 & 1 & 0 & 1 & 1 & 0 & 0 & 1 & 0 & 0 & 0 & 1\end{array}\right]. \tag{2.18}$$

Recall that from a given generator matrix $\boldsymbol{G}$ the parity-check matrix $\boldsymbol{H}$ can be obtained using eqns 2.5 and 2.10. We can just as easily obtain $\boldsymbol{G}$ if $\boldsymbol{H}$ is given, so that here we get

$$G = \left[\begin{array}{ccccccccccc|cccc}
1 & 0 & 0 & 0 & 0 & 0 & 0 & 0 & 0 & 0 & 0 & 1 & 0 & 0 & 1 \\
0 & 1 & 0 & 0 & 0 & 0 & 0 & 0 & 0 & 0 & 0 & 1 & 1 & 0 & 1 \\
0 & 0 & 1 & 0 & 0 & 0 & 0 & 0 & 0 & 0 & 0 & 1 & 1 & 1 & 1 \\
0 & 0 & 0 & 1 & 0 & 0 & 0 & 0 & 0 & 0 & 0 & 1 & 1 & 1 & 0 \\
0 & 0 & 0 & 0 & 1 & 0 & 0 & 0 & 0 & 0 & 0 & 0 & 1 & 1 & 1 \\
0 & 0 & 0 & 0 & 0 & 1 & 0 & 0 & 0 & 0 & 0 & 1 & 0 & 1 & 0 \\
0 & 0 & 0 & 0 & 0 & 0 & 1 & 0 & 0 & 0 & 0 & 0 & 1 & 0 & 1 \\
0 & 0 & 0 & 0 & 0 & 0 & 0 & 1 & 0 & 0 & 0 & 1 & 0 & 1 & 1 \\
0 & 0 & 0 & 0 & 0 & 0 & 0 & 0 & 1 & 0 & 0 & 1 & 1 & 0 & 0 \\
0 & 0 & 0 & 0 & 0 & 0 & 0 & 0 & 0 & 1 & 0 & 0 & 1 & 1 & 0 \\
0 & 0 & 0 & 0 & 0 & 0 & 0 & 0 & 0 & 0 & 1 & 0 & 0 & 1 & 1
\end{array}\right]$$

as the generator matrix of the (15, 11) Hamming code. □

2.5 Error syndromes

The equation $\boldsymbol{c}\boldsymbol{H}^{\mathrm{T}} = \mathbf{0}$ provides a way of testing whether an arbitrary n-bit word $\boldsymbol{v}$ is a codeword, without having to refer to a table of codewords. Taking the product $\boldsymbol{v}\boldsymbol{H}^{\mathrm{T}}$ then

$$\begin{aligned} \boldsymbol{v}\boldsymbol{H}^{\mathrm{T}} &= \mathbf{0} \quad \text{if } \boldsymbol{v} \text{ is a codeword} \\ &\neq \mathbf{0} \quad \text{if } \boldsymbol{v} \text{ is not a codeword.} \end{aligned}$$

Assuming now that $\boldsymbol{v}$ represents a codeword $\boldsymbol{c}$ that has incurred a nonzero error pattern $\boldsymbol{e}$, then the received word is $\boldsymbol{v} = \boldsymbol{c} + \boldsymbol{e}$. Taking the product of $\boldsymbol{v}$ and $\boldsymbol{H}^{\mathrm{T}}$ gives

$$\begin{aligned} \boldsymbol{v}\boldsymbol{H}^{\mathrm{T}} &= (\boldsymbol{c} + \boldsymbol{e})\boldsymbol{H}^{\mathrm{T}} \\ &= \boldsymbol{c}\boldsymbol{H}^{\mathrm{T}} + \boldsymbol{e}\boldsymbol{H}^{\mathrm{T}} \\ &= \boldsymbol{e}\boldsymbol{H}^{\mathrm{T}} \end{aligned}$$

because $\boldsymbol{c}\boldsymbol{H}^{\mathrm{T}} = \mathbf{0}$. Therefore the resulting term depends solely on the error pattern $\boldsymbol{e}$ and not on the original codeword $\boldsymbol{c}$ that incurs the error. The product $\boldsymbol{v}\boldsymbol{H}^{\mathrm{T}}$ contains information relating to the errors incurred by a codeword and we define

$$\boldsymbol{s} = \boldsymbol{v}\boldsymbol{H}^{\mathrm{T}} \tag{2.20}$$

as the error syndrome of $\boldsymbol{v}$. If $\boldsymbol{v}$ contains no errors, then $\boldsymbol{e} = \mathbf{0}$ and the error syndrome is zero, again irrespective of the codeword $\boldsymbol{c}$.

Example 2.11
Find the error syndromes of $\boldsymbol{v}_1 = (1\,1\,0\,1\,1\,0\,1)$ and $\boldsymbol{v}_2 = (0\,1\,0\,0\,1\,0\,1)$ where $\boldsymbol{v}_1$ and $\boldsymbol{v}_2$ represent codewords of the (7, 4) code that have incurred errors.

Using eqn 2.20 and the parity-check matrix of the (7, 4) code, the error syndrome of $\boldsymbol{v}_1$ is

$$\begin{aligned} \boldsymbol{s}_1 &= \boldsymbol{v}_1\boldsymbol{H}^{\mathrm{T}} \\ &= (1\ 1\ 0\ 1\ 1\ 0\ 1)\begin{bmatrix} 1 & 0 & 1 \\ 1 & 1 & 1 \\ 1 & 1 & 0 \\ 0 & 1 & 1 \\ 1 & 0 & 0 \\ 0 & 1 & 0 \\ 0 & 0 & 1 \end{bmatrix} \\ &= (1\ 0\ 0). \end{aligned}$$

Likewise $\boldsymbol{v}_2 = (0\ 1\ 0\ 0\ 1\ 0\ 1)$ gives $\boldsymbol{s}_2 = \boldsymbol{v}_2\boldsymbol{H}^{\mathrm{T}} = (0\ 1\ 0)$. □

Not all error patterns give a nonzero error syndrome. If a codeword $\boldsymbol{c}$ incurs an error pattern $\boldsymbol{e}$ that resembles a codeword $\boldsymbol{c}'$, then

$$\begin{aligned} \boldsymbol{v} &= \boldsymbol{c} + \boldsymbol{e} \\ &= \boldsymbol{c} + \boldsymbol{c}' \\ &= \boldsymbol{c}'' \end{aligned}$$

where $\boldsymbol{c}'' = \boldsymbol{c} + \boldsymbol{c}'$ is some other codeword (recall that, for a linear code, the sum of two codewords gives another codeword). The error syndrome is

$$\begin{aligned} \boldsymbol{s} &= \boldsymbol{v}\boldsymbol{H}^{\mathrm{T}} \\ &= \boldsymbol{c}''\boldsymbol{H}^{\mathrm{T}} \\ &= \boldsymbol{0} \end{aligned}$$

and therefore a zero error syndrome can occur for a nonzero error pattern and does so whenever the error pattern resembles a codeword.

In Section 1.6 we defined the error syndrome, for the (7, 4) code, using the parity-check sums of the code (see eqns 1.19 and 1.20). The error syndrome given by $\boldsymbol{s} = \boldsymbol{v}\boldsymbol{H}^{\mathrm{T}}$ is the same as that given by the parity-check sums, except that we now have a matrix representation of the parity-check sums (i.e. the parity-check matrix $\boldsymbol{H}$). We have already shown that the parity-check equations for the (7, 4) code can be obtained from the code's generator matrix, we can likewise show that the parity-check sums can be obtained from the parity-check matrix. Given $\boldsymbol{v} = (v_1, v_2, \ldots, v_7)$ its error syndrome $\boldsymbol{s} = (s_1, s_2, s_3)$ is

$$\boldsymbol{s} = \boldsymbol{v}\boldsymbol{H}^{\mathrm{T}} = (v_1 + v_2 + v_3 + v_5, v_2 + v_3 + v_4 + v_6, v_1 + v_2 + v_4 + v_7)$$

and so $s_1 = v_1 + v_2 + v_3 + v_5$, $s_2 = v_2 + v_3 + v_4 + v_6$ and $s_3 = v_1 + v_2 + v_4 + v_7$ as given by the parity-check sums (eqns 1.19).

To construct the syndrome table using the matrix representation of a linear code we determine the error syndrome $\boldsymbol{s} = \boldsymbol{e}\boldsymbol{H}^{\mathrm{T}}$ for all the correctable error patterns $\boldsymbol{e}$.

For the single-error-correcting (7, 4) code the correctable errors are the 7 single-bit-error patterns

$$\begin{matrix}(0\ 0\ 0\ 0\ 0\ 0\ 1)\\(0\ 0\ 0\ 0\ 0\ 1\ 0)\\ \vdots \qquad \vdots\\(1\ 0\ 0\ 0\ 0\ 0\ 0)\end{matrix}$$

and the transpose of $\boldsymbol{H}$ is

$$\boldsymbol{H}^{\mathrm{T}}=\begin{bmatrix}1&0&1\\1&1&1\\1&1&0\\0&1&1\\1&0&0\\0&1&0\\0&0&1\end{bmatrix}.$$

For a single-error-correcting code the error syndromes, in the syndrome table, are given by the rows of $\boldsymbol{H}^{\mathrm{T}}$. When $\boldsymbol{e}=(0\,0\,0\,0\,0\,0\,1)$ the product $\boldsymbol{eH}^{\mathrm{T}}$ gives the bottom row of $\boldsymbol{H}^{\mathrm{T}}$, and so the required error syndrome is $\boldsymbol{s}=(0\,0\,1)$. For $\boldsymbol{e}=(0\,0\,0\,0\,0\,1\,0)$ the error syndrome is given by the second row from the bottom and so $\boldsymbol{s}=(0\ 1\ 0)$. Likewise we obtain (1 0 0), (0 1 1), (1 1 0), and (1 1 1) as the error syndromes corresponding to (0 0 0 0 1 0 0), (0 0 0 1 0 0 0), (0 0 1 0 0 0 0), and (0 1 0 0 0 0 0) respectively. The first row of $\boldsymbol{H}^{\mathrm{T}}$ gives (1 0 1) as the error syndrome of (1 0 0 0 0 0 0). All 7 error syndromes are in agreement with those obtained using the parity-check sums (see Table 1.7).

Example 2.12
Determine the syndrome table for the (6, 3) single-error correcting code.

The parity-check matrix of the (6, 3) code is

$$\boldsymbol{H}=\left[\begin{array}{ccc|ccc}0&1&1&1&0&0\\1&0&1&0&1&0\\1&1&0&0&0&1\end{array}\right].$$

The code has blocklength 6 and so we need to determine the error syndromes $\boldsymbol{s}=\boldsymbol{eH}^{\mathrm{T}}$ corresponding to the 6 single-bit error patterns $\boldsymbol{e}=(0\ 0\ 0\ 0\ 0\ 1)$, (0 0 0 0 1 0)... (1 0 0 0 0 0). The transpose of $\boldsymbol{H}$ is

$$\boldsymbol{H}^{\mathrm{T}}=\begin{bmatrix}0&1&1\\1&0&1\\1&1&0\\1&0&0\\0&1&0\\0&0&1\end{bmatrix}$$

and as we have already seen the product $\boldsymbol{eH}^{\mathrm{T}}$ gives a single row of $\boldsymbol{H}$ when $\boldsymbol{e}$ is a single-bit error pattern. The error syndrome of $\boldsymbol{e}=(0\,0\,0\,0\,0\,1)$ is given by the bottom

Table 2.8
Syndrome table for the (6, 3) linear code

e	s
(0 0 0 0 0 0)	(0 0 0)
(0 0 0 0 0 1)	(0 0 1)
(0 0 0 0 1 0)	(0 1 0)
(0 0 0 1 0 0)	(1 0 0)
(0 0 1 0 0 0)	(1 1 0)
(0 1 0 0 0 0)	(1 0 1)
(1 0 0 0 0 0)	(0 1 1)

row of $\boldsymbol{H}^{\mathrm{T}}$ and so $\boldsymbol{s}=(0\,0\,1)$. The error syndrome of (0 0 0 0 1 0) is given by the second row from the bottom, i.e. (0 1 0), and so forth. Table 2.8 shows the resulting syndrome table. □

2.6 Error detection and correction

At the decoding stage of a linear code the parity-check matrix $\boldsymbol{H}$ is used to calculate the error syndrome $\boldsymbol{s}$ of the decoder input $\boldsymbol{v}$ using $\boldsymbol{s}=\boldsymbol{v}\boldsymbol{H}^{\mathrm{T}}$. Codewords $\boldsymbol{c}$ have the property that $\boldsymbol{c}\boldsymbol{H}^{\mathrm{T}}=\boldsymbol{0}$ and the error syndrome of a codeword is therefore zero. In the event of $\boldsymbol{v}$ being error free then $\boldsymbol{v}=\boldsymbol{c}$ and the error syndrome $\boldsymbol{s}=\boldsymbol{0}$. However, a zero error syndrome need not necessarily imply a zero error pattern because a zero error syndrome also occurs whenever an error pattern resembles a codeword. The probability of a codeword incurring no errors is greater than the probability of a codeword incurring errors that resemble a codeword, and therefore a zero error syndrome is more likely to be due to a zero error pattern.

Whenever a nonzero error syndrome is obtained the decoder knows that at least one error has occurred. The error pattern that is most likely to be responsible for the error syndrome is the error pattern corresponding to the error syndrome given in the syndrome table. Recall that the decoder's estimate or guess of the error pattern is denoted by $\hat{\boldsymbol{e}}$ to distinguish it from the actual error pattern $\boldsymbol{e}$ that has occurred. The error pattern $\hat{\boldsymbol{e}}$ is then used to correct $\boldsymbol{v}$ so giving a codeword $\hat{\boldsymbol{c}}$ that is the best estimate of the original codeword $\boldsymbol{c}$. We can summarize decoding in the three steps:

1. Calculate $\boldsymbol{s}=\boldsymbol{v}\boldsymbol{H}^{\mathrm{T}}$.
2. Obtain $\hat{\boldsymbol{e}}$ corresponding to $\boldsymbol{s}$ as given in the syndrome table.
3. The decoder's estimate of the codeword is $\hat{\boldsymbol{c}}=\boldsymbol{v}+\hat{\boldsymbol{e}}$.

Example 2.13
Consider the codeword $\boldsymbol{c}=(1\,1\,0\,1\,1\,0)$ belonging to the (6, 3) single-error-correcting code. Determine $\hat{\boldsymbol{c}}$, the decoder's estimate of $\boldsymbol{c}$, when $\boldsymbol{c}$ incurs the error patterns
(a) $\boldsymbol{e}_1=(0\ 0\ 1\ 0\ 0\ 0)$
(b) $\boldsymbol{e}_2=(0\ 1\ 1\ 0\ 1\ 1)$
(c) $\boldsymbol{e}_3=(0\ 0\ 0\ 1\ 1\ 1)$.

(a) The error pattern $\boldsymbol{e}_1 = (0\ 0\ 1\ 0\ 0\ 0)$ gives

$$\begin{aligned}\boldsymbol{v} &= \boldsymbol{c} + \boldsymbol{e}_1 \\ &= (1\ 1\ 0\ 1\ 1\ 0) + (0\ 0\ 1\ 0\ 0\ 0) \\ &= (1\ 1\ 1\ 1\ 1\ 0).\end{aligned}$$

Using the parity-check matrix for the (6, 3) code, the error syndrome is

$$\boldsymbol{s} = \boldsymbol{v}\boldsymbol{H}^{\mathrm{T}} = (1\ 1\ 1\ 1\ 1\ 0)\begin{bmatrix} 0 & 1 & 1 \\ 1 & 0 & 1 \\ 1 & 1 & 0 \\ 1 & 0 & 0 \\ 0 & 1 & 0 \\ 0 & 0 & 1 \end{bmatrix} = (1\ 1\ 0).$$

From Table 2.8 $\boldsymbol{s} = (1\ 1\ 0)$ gives $\hat{\boldsymbol{e}} = (0\ 0\ 1\ 0\ 0\ 0)$ as the decoder's estimate of the error pattern (which we know is the correct error pattern), and the decoder estimates the codeword to be

$$\begin{aligned}\hat{\boldsymbol{c}} &= \boldsymbol{v} + \hat{\boldsymbol{e}} \\ &= (1\ 1\ 1\ 1\ 1\ 0) + (0\ 0\ 1\ 0\ 0\ 0) \\ &= (1\ 1\ 0\ 1\ 1\ 0)\end{aligned}$$

which is the correct codeword.

(b) Here $\boldsymbol{e}_2 = (0\ 1\ 1\ 0\ 1\ 1)$ and so $\boldsymbol{v} = (1\ 0\ 1\ 1\ 0\ 1)$ which gives the error syndrome $\boldsymbol{s} = \boldsymbol{v}\boldsymbol{H}^{\mathrm{T}} = (0\ 0\ 0)$. The decoder therefore can only assume that no errors have occurred and that $\hat{\boldsymbol{c}} = \boldsymbol{v} = (1\ 0\ 1\ 1\ 0\ 1)$, which is incorrect. Note that the error pattern $\boldsymbol{e}_2$ resembles the codeword for the information word $\boldsymbol{i} = (0\ 1\ 1)$ (see Table 2.2) which is why the error syndrome is zero.

(c) For $\boldsymbol{e}_3 = (0\ 0\ 0\ 1\ 1\ 1)$ we get $\boldsymbol{v} = (1\ 1\ 0\ 0\ 0\ 1)$ and $\boldsymbol{s} = (1\ 1\ 1)$. Referring to Table 2.8 we can see that $\boldsymbol{s} = (1\ 1\ 1)$ is absent and therefore a decoding failure occurs. □

We have now seen two ways of decoding linear codes, one through the use of a syndrome table and the other using a standard array. As we saw in Section 2.3 the standard array is useful in helping to visualize minimum-distance decoding but is only practical for codes where n and k are small. Recall that in the standard array, words in the same coset consist of the same error pattern plus some codeword (where the codeword depends on the column that the word lies in) and therefore words in the same coset have the same error syndrome. Each coset could be replaced by its error syndrome resulting in a table of error patterns (i.e. the coset leaders) and corresponding error syndromes, which is nothing other than a syndrome table. Hence the syndrome table gives a concise representation of the standard array.

In Section 1.7 we considered how the minimum distance of a code affects its error-control capability, we now take another look at minimum distance and the error-control capabilities that go beyond that guaranteed by a code's minimum distance.

The minimum distance of an (n, k) linear block code satisfies the condition

$$d_{\min} \leq 1 + n - k \tag{2.21}$$

which is known as the *Singleton bound*. Therefore the minimum distance can never be greater than $1 + n - k$ and codes that have

$$d_{\min} = 1 + n - k \tag{2.22}$$

are known as *maximum-distance codes*. Rearranging eqn 2.22 to $(d_{\min} - 1)/2 = (n - k)/2$, and recalling that a code with minimum distance $d_{\min}$ can correct up to $t = (d_{\min} - 1)/2$ errors, we see therefore that

$$n - k = 2t. \tag{2.23}$$

Now $(n - k)$ is the number of parity bits and therefore eqn 2.23 tells us that a maximum-distance code capable of correcting t errors requires $2t$ parity bits, i.e. 2 parity bits are required to correct each error. Few codes achieve the Singleton Bound, and for such codes more than 2 parity bits are required for each correctable error. We might expect maximum-distance codes to always have good error-control capabilities, but this is not necessarily the case. The $(3, 1)$ repetition code has $d_{\min} = 3$, it therefore achieves the upper limit of the Singleton Bound and is a maximum-distance code, as are all the $(n, 1)$ repetition codes, yet the code is by no means a good code. One of the most important class of maximum-distance codes are the Reed Solomon codes, these are considered in Chapter 7.

Whilst the error detection limit ℓ of a linear code gives the maximum number of errors that the code is *guaranteed* to detect, there are however many more error patterns with more than ℓ errors that are also detectable. For example the single-parity-check codes have $d_{\min} = 2$ which guarantees single-error detection, but the codes can detect all error patterns with an odd number of errors. The $(7, 4)$ code has $d_{\min} = 3$ and yet we have already seen that the code can detect some 3-bit and 4-bit errors. In a linear code, error patterns that resemble a redundant word are detectable because they give a nonzero error syndrome (only codewords and error patterns that resemble codewords give a zero error syndrome). There are 2^k codewords and $2^n - 2^k$ redundant words in an (n, k) code, and therefore an (n, k) linear code is capable of detecting $2^n - 2^k$ error patterns. The $(7, 4)$ code has 16 codewords and 112 words that are redundant to the code. Error patterns that resemble any one of the 112 redundant words are detectable and therefore the $(7, 4)$ code can detect 112 error patterns and not just the 7 single-bit errors and 21 double-bit errors guaranteed by the code's minimum distance of 3. Likewise, the $(8, 7)$ even parity code has 128 codewords and 128 redundant words, it can therefore detect 128 error patterns and not just the 8 single-bit errors guaranteed by the code's minimum distance of 2.

The number of correctable error patterns for a linear code is likewise greater than that determined by the code's error-correction limit. Let's reconsider the standard array for a t-error-correcting code. As we have already seen, in Section 2.3, the coset leaders below the horizontal line in an array represent error patterns with more than t errors. Nevertheless a word $\boldsymbol{v}$ will be correctly decoded if the error pattern that $\boldsymbol{v}$ has incurred is the coset leader of the coset that $\boldsymbol{v}$ lies in (irrespective of whether $\boldsymbol{v}$ lies above or below the horizontal line). The coset leaders, of which there are 2^{n-k}, are all correctable error patterns and therefore an (n, k) linear code is capable of correcting

2^{n-k} error patterns. When constructing a standard array, error patterns with t or less errors always end up as coset leaders, whereas error patterns with more than t errors may or may not end up as coset leaders (recall that the construction of cosets below the horizontal line is arbitrary). Therefore the correctable error patterns below the horizontal line vary according to how the array is constructed.

2.7 Shortened and extended linear codes

One problem encountered in the application of error-control codes is that it may not be possible to find a code with suitable blocklength, information length and minimum distance. In which case, it may be necessary to select a code that is closest to that required and modify it to meet the requirements. Various modifications can be made to a linear code, here we consider modifications based on shortening or extending codes.

An (n, k) linear code can be shortened to an $(n - i, k - i)$ *shortened linear code* by setting the first (left-hand side) i information bits to zero. Encoding and decoding can then be carried out in the same way as for the (n, k) code, that is using the generator and parity-check matrices of the (n, k) code $\boldsymbol{G}$ and $\boldsymbol{H}$ respectively. An alternative is to construct a generator matrix and parity-check matrix, for the shortened code, from $\boldsymbol{G}$ and $\boldsymbol{H}$. The generator matrix of the shortened code is obtained by omitting the first i rows and columns of $\boldsymbol{G}$, and the parity-check matrix is obtained by omitting the first i columns of $\boldsymbol{H}$. Codewords then have $n - i$ bits but note that the number of parity bits remains the same beause $(n - i) - (k - i) = n - k$. Consider the (15, 11) Hamming code with generator matrix

$$\boldsymbol{G} = \left[\begin{array}{ccccccccccc|cccc}
1&0&0&0&0&0&0&0&0&0&0&1&0&0&1\\
0&1&0&0&0&0&0&0&0&0&0&1&1&0&1\\
0&0&1&0&0&0&0&0&0&0&0&1&1&1&1\\
0&0&0&1&0&0&0&0&0&0&0&1&1&1&0\\
0&0&0&0&1&0&0&0&0&0&0&0&1&1&1\\
0&0&0&0&0&1&0&0&0&0&0&1&0&1&0\\
0&0&0&0&0&0&1&0&0&0&0&0&1&0&1\\
0&0&0&0&0&0&0&1&0&0&0&1&0&1&1\\
0&0&0&0&0&0&0&0&1&0&0&1&1&0&0\\
0&0&0&0&0&0&0&0&0&1&0&0&1&1&0\\
0&0&0&0&0&0&0&0&0&0&1&0&0&1&1
\end{array}\right].$$

Removing the first 3 rows and columns gives

$$\boldsymbol{G} = \left[\begin{array}{cccccccc|cccc}
1&0&0&0&0&0&0&0&1&1&1&0\\
0&1&0&0&0&0&0&0&0&1&1&1\\
0&0&1&0&0&0&0&0&1&0&1&0\\
0&0&0&1&0&0&0&0&0&1&0&1\\
0&0&0&0&1&0&0&0&1&0&1&1\\
0&0&0&0&0&1&0&0&1&1&0&0\\
0&0&0&0&0&0&1&0&0&1&1&0\\
0&0&0&0&0&0&0&1&0&0&1&1
\end{array}\right]. \tag{2.24}$$

as the generator matrix of the resulting (12, 8) shortened code. The parity-check matrix of the (15, 11) code is

$$\boldsymbol{H}=\left[\begin{array}{ccccccccccc|cccc} 1&1&1&1&0&1&0&1&1&0&0&1&0&0&0\\ 0&1&1&1&1&0&1&0&1&1&0&0&1&0&0\\ 0&0&1&1&1&1&0&1&0&1&1&0&0&1&0\\ 1&1&1&0&1&0&1&1&0&0&1&0&0&0&1 \end{array}\right]$$

and removing the first 3 columns gives

$$\boldsymbol{H}=\left[\begin{array}{cccccccc|cccc} 1&0&1&0&1&1&0&0&1&0&0&0\\ 1&1&0&1&0&1&1&0&0&1&0&0\\ 1&1&1&0&1&0&1&1&0&0&1&0\\ 0&1&0&1&1&0&0&1&0&0&0&1 \end{array}\right] \tag{2.25}$$

as the parity-check matrix of the (12, 8) code. Note that we could have used $\boldsymbol{H}=[\boldsymbol{P}^{\mathrm{T}}|\boldsymbol{I}]$ to determine the parity-check matrix of the shortened code. The minimum distance of a shortened $(n-i, k-i)$ code is greater than or equal to the minimum distance of the (n, k) code from which it is derived. Hence by shortening a code we know that we are not reducing its error control capability. Here the parity-check matrix $\boldsymbol{H}$ of the shortened code has 12 nonzero unique columns and so the code is capable of correcting single-bit errors in 12-bit words. Therefore the (12, 8) code constructed here is single-error correcting, just like the (15, 11) code from which it was constructed.

A common way of modifying an (n, k) linear code is by adding an overall parity-check bit to produce an *extended* $(n+1, k)$ *code*. The parity-check bit is added as so to give even parity codewords, as this will preserve the linearity of the code. The Hamming codes are often extended in this way, recall that for all integers $r \geq 3$ there exists a $(2^r-1,\ 2^r-1-r)$ single-error-correcting Hamming code with minimum distance 3. The inclusion of an overall parity-check bit gives the $(2^r,\ 2^r-1-r)$ *extended Hamming codes* with minimum distance 4, which are single-error-correcting and double-error-detecting codes. Consider the (8, 4) extended Hamming code, to construct the codewords we take the codewords of the (7, 4) code and add an overall parity-check bit. If $(i_1, i_2, i_3, i_4, p_1, p_2, p_3)$ is a codeword of the (7, 4) code then

$$\boldsymbol{c}=(i_1, i_2, i_3, i_4, p_1, p_2, p_3, p) \tag{2.26}$$

is a codeword of the (8, 4) code, where $p_1=i_1+i_2+i_3$, $p_2=i_2+i_3+i_4$ and $p_3=i_1+i_2+i_4$ are the parity bits given by the (7, 4) code and

$$p=i_1+i_2+i_3+i_4+p_1+p_2+p_3 \tag{2.27}$$

is the overall parity-check bit. Substituting p_1, p_2, and p_3 into p gives

$$p=i_1+i_2+i_3+i_4+(i_1+i_2+i_3)+(i_2+i_3+i_4)+(i_1+i_2+i_4)$$

which reduces to

$$p=i_1+i_3+i_4. \tag{2.28}$$

Recall that the generator matrix G of the (7, 4) code in its systematic form is

$$G = \left[\begin{array}{cccc|ccc} 1 & 0 & 0 & 0 & 1 & 0 & 1 \\ 0 & 1 & 0 & 0 & 1 & 1 & 1 \\ 0 & 0 & 1 & 0 & 1 & 1 & 0 \\ 0 & 0 & 0 & 1 & 0 & 1 & 1 \end{array}\right]$$

and that every column in the parity matrix P produces a parity-check bit. To generate an overall parity-check bit, in accordance with eqn 2.28, we add a fourth column to P to give

$$G = \left[\begin{array}{cccc|cccc} 1 & 0 & 0 & 0 & 1 & 0 & 1 & 1 \\ 0 & 1 & 0 & 0 & 1 & 1 & 1 & 0 \\ 0 & 0 & 1 & 0 & 1 & 1 & 0 & 1 \\ 0 & 0 & 0 & 1 & 0 & 1 & 1 & 1 \end{array}\right] \tag{2.29}$$

as the generator matrix of the (8, 4) extended Hamming code. Note that the column that we have added to G is simply the sum of the other 7 columns (taken a row at a time). If we let $\boldsymbol{i} = (i_1, i_2, i_3, i_4)$ then the codewords of the (8, 4) code are given by

$$\begin{aligned} \boldsymbol{c} &= \boldsymbol{i}G \\ &= (i_1, i_2, i_3, i_4)\left[\begin{array}{cccc|cccc} 1 & 0 & 0 & 0 & 1 & 0 & 1 & 1 \\ 0 & 1 & 0 & 0 & 1 & 1 & 1 & 0 \\ 0 & 0 & 1 & 0 & 1 & 1 & 0 & 1 \\ 0 & 0 & 0 & 1 & 0 & 1 & 1 & 1 \end{array}\right] \\ &= (i_1, i_2, i_3, i_4, i_1 + i_2 + i_3, i_2 + i_3 + i_4, i_1 + i_2 + i_4, i_1 + i_3 + i_4) \\ &= (i_1, i_2, i_3, i_4, p_1, p_2, p_3, p) \end{aligned}$$

as required.

The parity-check matrix H of an extended $(n + 1, k)$ linear code is given by

$$H = \left[\begin{array}{cccc|c} & & & & 0 \\ & & & & 0 \\ & & H' & & \vdots \\ & & & & 0 \\ \hline 1 & 1 & \dots & 1 & 1 \end{array}\right] \tag{2.30}$$

where H' is the parity-check matrix of the (n, k) code. This assumes that the overall parity-check bit is on the right-hand side of codewords, as given in eqn 2.26. The resulting error syndrome is

$$\boldsymbol{s} = (s_1, s_2, s_3, \dots, s_{n-k}, s) \tag{2.31}$$

where $s_1, s_2, \dots, s_{n-k}$ are the parity-check sums given by H and s is the overall parity-check sum. Using eqn. 2.30 and the parity-check matrix for the (7, 4) code we find that the parity-check matrix for the extended (8, 4) code is

$$H = \begin{bmatrix} 1 & 1 & 1 & 0 & 1 & 0 & 0 & 0 \\ 0 & 1 & 1 & 1 & 0 & 1 & 0 & 0 \\ 1 & 1 & 0 & 1 & 0 & 0 & 1 & 0 \\ 1 & 1 & 1 & 1 & 1 & 1 & 1 & 1 \end{bmatrix}. \tag{2.32}$$

If needed, we can use elementary row operations to reduce $\boldsymbol{H}$ to a systematic form. Given $\boldsymbol{v}=(v_1,v_2,v_3,v_4,v_5,v_6,v_7,v_8)$ its error syndrome is $\boldsymbol{s}=(s_1,s_2,s_3,s)$ where

$$s = v_1 + v_2 + v_3 + v_4 + v_5 + v_6 + v_7 + v_8$$

is the overall parity-check sum and s_1, s_2, s_3 are the same as the parity-check sums given by the (7, 4) code. Recall that the rows of the transpose of $\boldsymbol{H}$ give the error syndromes in the syndrome table. The (8, 4) code has 8 correctable single-bit errors (000...01), (000...10), ..., (100...0) and corresponding error syndromes in its syndrome table (see Table 2.9). Note that in the syndrome table the overall parity bit s is always 1 because the occurrence of a single error gives incorrect parity.

Decoding a word $\boldsymbol{v}$ is achieved by determining the error syndrome using $\boldsymbol{s}=\boldsymbol{v}\boldsymbol{H}^{\mathrm{T}}$. If the parity of $\boldsymbol{v}$ is correct then $s=0$, otherwise $s=1$. The occurrence of a single error gives $s=1$, and the syndrome table can be used for single-error correction. If $s=0$ but s_1, s_2, or s_3 are nonzero then more than 1 error has occurred, and we assume that $\boldsymbol{v}$ contains an uncorrectable error pattern.

Example 2.14

Consider the (8, 4) extended Hamming code. Determine the outcome of a decoder when the codeword $\boldsymbol{c}=(0\,0\,1\,0\,1\,1\,0\,1)$ incurs the errors

(a) $\boldsymbol{e}_1=(0\,0\,1\,0\,0\,0\,0\,0)$

(b) $\boldsymbol{e}_2=(1\,0\,0\,0\,0\,1\,0\,0)$

(c) $\boldsymbol{e}_3=(0\,0\,0\,0\,0\,1\,1\,1)$.

(a) The received word is $\boldsymbol{v}=\boldsymbol{c}+\boldsymbol{e}_1=(0\,0\,0\,0\,1\,1\,0\,1)$ and using $\boldsymbol{s}=\boldsymbol{v}\boldsymbol{H}^{\mathrm{T}}$ (with $\boldsymbol{H}$ given by eqn (2.32) gives $\boldsymbol{s}=(1\,1\,0\,1)$. Hence $s=1$ and so we assume that a single error has occurred. From Table 2.9, $\boldsymbol{s}=(1\,1\,0\,1)$ gives $\hat{\boldsymbol{e}}=(0\,0\,1\,0\,0\,0\,0\,0)$, and carrying out error correction gives $\hat{\boldsymbol{c}}=\boldsymbol{v}+\hat{\boldsymbol{e}}=(0\,0\,1\,0\,1\,1\,0\,1)$, which is the correct codeword.

(b) Here $\boldsymbol{v}=\boldsymbol{c}+\boldsymbol{e}_2=(1\,0\,1\,0\,1\,0\,0\,1)$ which gives the error syndrome $\boldsymbol{s}=(1\,1\,1\,0)$. The parity-check sum $s=0$ is correct, however the other 3 check sums are 1. Therefore the decoder knows that more than 1 error has occurred and so this gives a decoding failure.

(c) Here $\boldsymbol{v}=(0\,0\,1\,0\,1\,0\,1\,0)$ gives $\boldsymbol{s}=(0\,1\,1\,1)$ and the decoder incorrectly assumes that a single error has occurred. From the syndrome table $\boldsymbol{s}=(0\,1\,1\,1)$

Table 2.9
Syndrome table for the (8, 4) extended code

$\boldsymbol{e}$	$\boldsymbol{s}$
(0 0 0 0 0 0 0 1)	(0 0 0 1)
(0 0 0 0 0 0 1 0)	(0 0 1 1)
(0 0 0 0 0 1 0 0)	(0 1 0 1)
(0 0 0 0 1 0 0 0)	(1 0 0 1)
(0 0 0 1 0 0 0 0)	(0 1 1 1)
(0 0 1 0 0 0 0 0)	(1 1 0 1)
(0 1 0 0 0 0 0 0)	(1 1 1 1)
(1 0 0 0 0 0 0 0)	(1 0 1 1)

gives $\hat{e} = (0\,0\,0\,1\,0\,0\,0\,0)$ and the decoder concludes that $\hat{c} = v + \hat{e} = (0\,0\,1\,1\,1\,0\,1\,0)$ is the required codeword, which is incorrect. □

Problems

2.1 Show that the sum of any two codewords belonging to the (6, 3) linear code (see Table 2.2) is also a codeword. Construct a table showing addition of codewords belonging to the (6, 3) code.

2.2 Let $\boldsymbol{c}_i$ and $\boldsymbol{c}'_i$ respectively be the systematic and nonsystematic codewords of the (5, 3) code given in Table 2.3. Determine the correspondence between $\boldsymbol{c}_i$ and $\boldsymbol{c}'_i$.

2.3 Show that the (7, 4) linear codes generated by

$$\boldsymbol{G}_1 = \begin{bmatrix} 1 & 0 & 0 & 0 & 1 & 0 & 1 \\ 0 & 1 & 0 & 0 & 1 & 1 & 1 \\ 0 & 0 & 1 & 0 & 1 & 1 & 0 \\ 0 & 0 & 0 & 1 & 0 & 1 & 1 \end{bmatrix}$$

and

$$\boldsymbol{G}_2 = \begin{bmatrix} 1 & 1 & 1 & 1 & 1 & 1 & 1 \\ 0 & 1 & 1 & 1 & 0 & 1 & 0 \\ 0 & 0 & 1 & 1 & 1 & 0 & 1 \\ 0 & 0 & 0 & 1 & 0 & 1 & 1 \end{bmatrix}$$

are equivalent.

2.4 Given the (7, 3) linear code with generator matrix

$$\boldsymbol{G} = \begin{bmatrix} 1 & 1 & 1 & 0 & 1 & 0 & 0 \\ 0 & 1 & 1 & 1 & 0 & 1 & 0 \\ 1 & 1 & 0 & 1 & 0 & 0 & 1 \end{bmatrix}$$

determine a systematic form of $\boldsymbol{G}$. Hence find a parity-check matrix for the code.

2.5 Given that $\boldsymbol{c} = (i_1, i_2, i_3, p_1, p_2, p_3)$ is a codeword of the (6, 3) linear code determine the parity-check equations for p_1, p_2, and p_3 in terms of i_1, i_2, and i_3. Hence show that

$$p_1 = i_1 + p$$
$$p_2 = i_2 + p$$
$$p_3 = i_3 + p$$

where p is the parity of $\boldsymbol{i} = (i_1, i_2, i_3)$ and therefore codewords can be expressed as $\boldsymbol{c} = (i_1, i_2, i_3, i_1 + p, i_2 + p, i_3 + p)$. Note that this gives a simple encoding rule for the (6, 3) code. If $\boldsymbol{i}$ has even parity ($p = 0$) then $p_1 = i_1$, $p_2 = i_2$, $p_3 = i_3$ and so $\boldsymbol{c} = (i_1, i_2, i_3, i_1, i_2, i_3)$. Otherwise for odd parity ($p = 1$) each parity bit is the inverse of the corresponding information bit, so giving $\boldsymbol{c} = (i_1, i_2, i_3, i_1 + 1, i_2 + 1, i_3 + 1)$.

2.6 Determine the minimum distance and error-control limits of the (5, 3) code given in Table 2.3.

2.7 Show that the generator matrix $\boldsymbol{G}$ (eqn 2.4) and parity-check matrix $\boldsymbol{H}$ (eqn 2.12) of the (6, 3) code satisfy $\boldsymbol{G}\boldsymbol{H}^{\mathrm{T}} = \mathbf{0}$ and that all the codewords $\boldsymbol{c}$ (Table 2.2) satisfy $\boldsymbol{c}\boldsymbol{H}^{\mathrm{T}} = 0$.

2.8 Construct the syndrome table for the (7, 3) linear code given in Problem 2.4. Given that $\boldsymbol{c} = (1\,0\,1\,0\,0\,1\,1)$ is a codeword determine the decoding decisions when $\boldsymbol{c}$ incurs the errors

(a) $\boldsymbol{e}_1 = (0\,0\,1\,0\,0\,0\,0)$
(b) $\boldsymbol{e}_2 = (1\,0\,0\,1\,0\,0\,0)$
(c) $\boldsymbol{e}_3 = (0\,0\,1\,1\,1\,0\,1)$.

2.9 Show that the $(n, 1)$ repetition codes are maximum-distance codes. Write down generator and parity-check matrices for the (7, 1) repetition code.

2.10 Given the parity-check matrix

$$\boldsymbol{H} = \begin{bmatrix} 1 & 1 & 1 & 0 & 1 & 0 & 0 \\ 0 & 1 & 1 & 1 & 0 & 1 & 0 \\ 1 & 1 & 0 & 1 & 0 & 0 & 1 \end{bmatrix}$$

of the (7, 4) Hamming code, determine systematic forms for the parity-check and generator matrices of the (8, 4) extended Hamming code.

2.11 Given that $\boldsymbol{c} = (0\,1\,1\,0\,0\,0\,1\,1)$ is a codeword of the (8, 4) extended Hamming code with even parity, determine the decoding decisions when $\boldsymbol{c}$ incurs the errors

(a) $\boldsymbol{e}_1 = (0\,0\,0\,0\,0\,1\,0\,0)$
(b) $\boldsymbol{e}_2 = (0\,1\,0\,1\,0\,0\,0\,0)$
(c) $\boldsymbol{e}_3 = (0\,0\,0\,1\,1\,0\,1\,0)$.

Cyclic codes 3

In going from arbitrary block codes to linear codes the characteristic of linearity enables the minimum distance and therefore error-control capabilities of codes to be easily determined, and encoding and decoding can be concisely described in terms of matrix algebra. A second characteristic is now considered, namely that codewords are cyclic, and in doing so further gains are made, in particular with regard to the implementation of encoding and decoding.

3.1 Definition of cyclic codes

With cyclic codes it is conventional to think of codewords as starting with the component c_{n-1} and ending with c_0, and as such codewords are represented as

$$\boldsymbol{c} = (c_{n-1}, c_{n-2}, \ldots, c_1, c_0).$$

This representation proves more convenient than that previously used, namely $\boldsymbol{c} = (c_1, c_2, \ldots, c_{n-1}, c_n)$, and should not cause confusion. The coefficients c_0 and c_{n-1} are now referred to as the least-order and highest-order coefficients respectively, and define the low-order and high-order sides of $\boldsymbol{c}$. A *cyclic shift* leftwards on a codeword $\boldsymbol{c}$ is one in which each bit is shifted one place towards the left with the bit c_{n-1} occupying the position vacated by c_0. The resulting word is

$$\boldsymbol{c}' = (c_{n-2}, c_{n-3}, \ldots, c_0, c_{n-1}).$$

Bits can be shifted towards the right in a similar manner, and shifts may be by more than one bit. So, for example, a cyclic shift on $\boldsymbol{c}'$ of 4 bits towards the right gives

$$\boldsymbol{c}'' = (c_2, c_1, c_0, c_{n-1}, \ldots, c_4, c_3).$$

A *cyclic code* is a linear code that has the property that a cyclic shift on any of its codewords produces another codeword. The cyclic shift may be leftwards or rightwards by any number of places. Hence if $\boldsymbol{c}$ given above is a codeword of an (n, k) cyclic code then $\boldsymbol{c}'$ and $\boldsymbol{c}''$ are also codewords.

Example 3.1
Returning to the (7, 4) Hamming code we find that all the codewords satisfy the cyclic requirement. For example, let's consider the codeword $\boldsymbol{c}_3 = (0\,0\,1\,1\,1\,0\,1)$ given in Table 1.6. A shift of 2 places to the left produces (1 1 1 0 1 0 0), which is the codeword $\boldsymbol{c}_{14}$. A further shift of 3 places to the left gives (0 1 0 0 1 1 1), which is $\boldsymbol{c}_4$ and a shift of 1 place to the right gives $\boldsymbol{c}_{10} = (1\,0\,1\,0\,0\,1\,1)$. □

The above example does not prove that the (7, 4) Hamming code is cyclic, it only shows that some of its codewords have the required cyclic property. However, the code is indeed cyclic, the reader may wish to consider cyclic shifts on other codewords. Not all linear codes are cyclic. A linear code may have some codewords that satisfy the cyclic requirement, but other codewords that fail to do so, such codes are therefore not cyclic.

Example 3.2
Consider the (5, 2) linear code with nonzero codewords

$$\begin{aligned} c_1 &= (0\ 1\ 1\ 0\ 1) \\ c_2 &= (1\ 0\ 1\ 1\ 1) \\ c_3 &= (1\ 1\ 0\ 1\ 0). \end{aligned}$$

A leftwards shift of 1 bit on c_1 produces c_3, but a further shift of 1 bit leftwards gives (1 0 1 0 1) which is not a codeword. The (5, 2) linear code is therefore not cyclic.

□

3.2 Polynomials

As cyclic codes are linear, encoding and decoding can be achieved in the same way as for linear codes, namely using matrices. However, with cyclic codes, a polynomial representation of codewords is often more suitable and encoding and decoding rules can be formulated in terms of polynomial algebra. Before proceeding to consider cyclic codes in terms of polynomials, it is useful to look at polynomial addition, subtraction, multiplication, and division, for these are required for encoding and decoding cyclic codes.

An n-bit word $(a_{n-1}, a_{n-2}, \ldots, a_2, a_1, a_0)$ can be represented by the polynomial

$$p(x) = a_{n-1}x^{n-1} + a_{n-2}x^{n-2} + \cdots + a_2x^2 + a_1x + a_0 \tag{3.1}$$

where the n components of the word $a_{n-1}, a_{n-2}, \ldots, a_2, a_1, a_0$ form the polynomial coefficients. Note that the highest possible power of x in the polynomial representation of an n-bit word is $n - 1$. For a binary word, the components a_i are binary and hence the polynomial coefficients are 0 or 1. For example, the words (0 1 1 0 0 1 0) and (0 0 1 0 1 1) give the polynomials

$$p_1(x) = x^5 + x^4 + x$$

and

$$p_2(x) = x^3 + x + 1$$

respectively. The *degree* of a polynomial is the largest power of x that has a nonzero coefficient. The above polynomials $p_1(x)$ and $p_2(x)$ have degrees 5 and 3 respectively. Two polynomials can be added together by adding the polynomial coefficients pairwise, so the sum of the two polynomials

$$p_a(x) = a_{n-1}x^{n-1} + a_{n-2}x^{n-2} + \cdots + a_2x^2 + a_1x + a_0$$

and

$$p_b(x) = b_{n-1}x^{n-1} + b_{n-2}x^{n-2} + \cdots + b_2x^2 + b_1x + b_0$$

is

$$p_a(x) + p_b(x) = (a_{n-1} + b_{n-1})x^{n-1} + (a_{n-2} + b_{n-2})x^{n-2}$$
$$+ \cdots + (a_2 + b_2)x^2 + (a_1 + b_1)x + (a_0 + b_0).$$

If the coefficients are binary then polynomial addition is simplified because modulo-2 arithmetic is used (see Table 1.4). Hence if $a_i = b_i$ then $a_i + b_i = 0$ modulo-2 and the resulting polynomial does not contain an x^i term. Consider again the polynomials $p_1(x) = x^5 + x^4 + x$ and $p_2(x) = x^3 + x + 1$, their sum is

$$\begin{aligned} p_1(x) + p_2(x) &= (x^5 + x^4 + x) + (x^3 + x + 1) \\ &= (x^5 + x^4 + x^3 + x(1 + 1) + 1 \\ &= x^5 + x^4 + x^3 + 1 \end{aligned}$$

as $1 + 1 = 0$ modulo-2. Polynomial subtraction is the same as polynomial addition because $-1 = +1$ modulo-2. Note that polynomial coefficients are assumed to be binary, and subject to modulo-2 addition, unless stated otherwise.

The product of two arbitrary polynomials is achieved by determining the products of all the terms taken pairwise and then cancelling like terms. The product of $p_1(x)$ and $p_2(x)$ is

$$\begin{aligned} p_1(x)p_2(x) &= (x^5 + x^4 + x)(x^3 + x + 1) \\ &= (x^8 + x^6 + x^5) + (x^7 + x^5 + x^4) + (x^4 + x^2 + x) \\ &= x^8 + x^7 + x^6 + x^2 + x. \end{aligned}$$

The degree of the resulting polynomial is clearly the sum of the degrees of $p_1(x)$ and $p_2(x)$.

Polynomial division can be carried out using the long-division method that is well known in ordinary arithmetic. If $p_1(x)$ is divided by $p_2(x)$ then $p_1(x)$ is referred to as the *dividend* and $p_2(x)$ as the *divisor*. As with ordinary division, polynomial division yields a quotient and remainder denoted by the polynomials $q(x)$ and $r(x)$ respectively. Polynomial division is achieved by successively removing factors of the divisor from the dividend until the resulting remainder has a degree less than that of the divisor. Note that in the ordinary long-division method, factors of the divisor are removed by subtracting them from the dividend, but as we are using modulo-2 arithmetic we add the factors to the dividend. Consider again $p_1(x) = x^5 + x^4 + x$ and $p_2(x) = x^3 + x + 1$ then $p_1(x)$ divided by $p_2(x)$ gives

$$\begin{array}{rll} & x^2 + x + 1 & \\ x^3 + x + 1 \,\big) & x^5 + x^4 + x & \\ & x^5 + x^3 + x^2 & \left.\right] \text{step 1} \\ - & x^4 + x^3 + x^2 + x & \\ & x^4 + x^2 + x & \left.\right] \text{step 2} \\ - & x^3 & \\ & x^3 + x + 1 & \left.\right] \text{step 3} \\ - & x + 1. & \end{array}$$

In Step 1 the divisor is multiplied by x^2 and then added to the dividend (underneath the square root sign) so removing the term with the highest power, x^5, from the dividend. This leaves a remainder with x^4 as the highest term. A note of x^2 is made above the dividend. In Step 2, a factor of x times the divisor is added to the remainder of Step 1, so removing the x^4 term. The factor x is added to the x^2 term above the dividend. Finally, Step 3 is required to remove the x^3 term from the remainder of Step 2, this is achieved by adding 1 times the divisor to the remainder of Step 2. Again the factor 1 is added to the x^2+x term above the dividend. After Step 3, the remainder has a degree less than that of the divisor, division is therefore complete and the remainder at the end of Step 3 is the required remainder $r(x)=x+1$. The quotient is given by the sum of the terms noted above the dividend, and so $q(x)=x^2+x+1$.

Example 3.3

Divide $x^{12}+x^7+x^4+x^3+1$ by x^3+x^2+1.

$$
\begin{array}{r l}
 & \underline{x^9+x^8+x^7+x^5+x^4+x^3} \\
x^3+x^2+1\,) & x^{12}+x^7+x^4+x^3+1 \\
 & \underline{x^{12}+x^{11}+x^9} \\
 & -\quad x^{11}+x^9+x^7+x^4+x^3+1 \\
 & \qquad \underline{x^{11}+x^{10}+x^8} \\
 & \qquad -\quad x^{10}+x^9+x^8+x^7+x^4+x^3+1 \\
 & \qquad\qquad \underline{x^{10}+x^9+x^7} \\
 & \qquad\qquad -\quad x^8+x^4+x^3+1 \\
 & \qquad\qquad\qquad \underline{x^8+x^7+x^5} \\
 & \qquad\qquad\qquad -\quad x^7+x^5+x^4+x^3+1 \\
 & \qquad\qquad\qquad\qquad \underline{x^7+x^6+x^4} \\
 & \qquad\qquad\qquad\qquad -\quad x^6+x^5+x^3+1 \\
 & \qquad\qquad\qquad\qquad\qquad \underline{x^6+x^5+x^3} \\
 & \qquad\qquad\qquad\qquad\qquad -\quad 1
\end{array}
$$

The quotient is therefore $q(x)=x^9+x^8+x^7+x^5+x^4+x^3$ and the remainder $r(x)=1$. □

Polynomial division is quite similar to ordinary division in many aspects. If we take an integer I_1 and divide it by another integer I_2 to give a quotient q and remainder r, then

$$I_1 = qI_2 + r$$

where $0 \le r < |I_2|$ and where the quotient q and remainder r are unique for every pair of integers I_1 and I_2. This is known as the *Euclidean division algorithm.* A simple example of this is 17 divided by 5, which gives the quotient 3 and remainder 2, and so according to the Euclidean division algorithm we can express 17 as $17=3\times 5+2$. The Euclidean division algorithm applies to any pair of integers, and whilst possibly appearing to be rather trivial and obvious, it is an important property of integers that carries over into polynomials. For polynomials, the Euclidean division algorithm tells us that given a polynomial $p_1(x)$ divided by some other polynomial $p_2(x)$, then

$$p_1(x) = q(x)p_2(x) + r(x) \tag{3.2}$$

where $q(x)$ and $r(x)$ are the quotient and remainder respectively. We have already seen that dividing $p_1(x) = x^5 + x^4 + x$ by $p_2(x) = x^3 + x + 1$ gives $q(x) = x^2 + x + 1$ and $r(x) = x + 1$. According to the Euclidean division algorithm $p_1(x) = q(x)p_2(x) + r(x)$, constructing the right-hand side of this expression gives

$$\begin{aligned} q(x)p_2(x) + r(x) &= (x^2 + x + 1)(x^3 + x + 1) + (x + 1) \\ &= (x^5 + x^3 + x) + (x^4 + x^2 + x) + (x^3 + x + 1) + (x + 1) \\ &= x^5 + x^4 + x^2 \\ &= p_1(x) \end{aligned}$$

as required.

Example 3.4

In Example 3.3 we saw that $x^{12} + x^7 + x^4 + x^3 + 1$ divided by $x^3 + x^2 + 1$ give the quotient and remainder

$$\begin{aligned} q(x) &= x^9 + x^8 + x^7 + x^5 + x^4 + x^3 \\ r(x) &= 1 \end{aligned}$$

respectively. Using the Euclidean division algorithm gives

$$\begin{aligned} &(x^9 + x^8 + x^7 + x^5 + x^4 + x^3)(x^3 + x^2 + 1) + 1 \\ &= (x^{12} + x^{11} + x^{10} + x^8 + x^7 + x^6) + (x^{11} + x^{10} + x^9 + x^7 + x^6 + x^5) \\ &\quad + (x^9 + x^8 + x^7 + x^5 + x^4 + x^3) + 1 \\ &= x^{12} + x^7 + x^4 + x^3 + 1 \end{aligned}$$

which is the required dividend. □

Later we shall see that encoding and decoding cyclic codes requires polynomial division, and in particular we will be interested in the remainder after division. The remainder can be expressed as

$$r(x) = R_{p(x)}[p_1(x)] \tag{3.3}$$

where the right-hand side is taken to mean the remainder of $p_1(x)$ divided $p(x)$. Alternatively we can use the modulo notation $p_1(x) = r(x)$ modulo-$p(x)$. If the polynomial $p(x)$ is a factor of $p_1(x)$, so that $p_1(x) = f(x)p(x)$, then dividing $p_1(x)$ by $p(x)$ gives the quotient $q(x) = f(x)$ and remainder $r(x) = 0$. Hence

$$R_{p(x)}[f(x)p(x)] = 0 \tag{3.4}$$

for any polynomial $p(x)$, we can also express this as

$$f(x)p(x) = 0 \text{ modulo-}p(x). \tag{3.5}$$

Consider, for example, $p_1(x)=x^4+x^2$ and $p_2(x)=x^2+1$ here $p_2(x)$ is a factor of $p_1(x)$ because $x^4+x^2=x^2(x^2+1)$, and so $p_1(x)=x^2p_2(x)$. Dividing $p_1(x)$ by $p_2(x)$ gives $q(x)=x^2$ and $r(x)=0$, as can be easily verified by long division, and therefore

$$R_{x^2+1}[x^4+x^2]=0.$$

It is not always obvious that one polynomial is a factor of another. To establish whether $p_2(x)$ is a factor of $p_1(x)$ we need to divide $p_1(x)$ by $p_2(x)$ and check if the remainder is zero.

Example 3.5

Determine whether $p(x)=x^3+1$ is a factor of $p_1(x)=x^5+x^3+x^2+1$ and of $p_2(x)=x^7+x^6+x^3+x^2+1$.

Dividing $p_1(x)$ by $p(x)$ gives

$$\begin{array}{r|l} & x^2+1 \\ \hline x^3+1 & x^5+x^3+x^2+1 \\ & x^5+x^2 \\ \hline & -\quad x^3+1 \\ & x^3+1 \\ \hline & -\quad 0. \end{array}$$

Hence the remainder is 0 and so $p(x)$ is a factor of $p_1(x)$. Furthermore because $q(x)=x^2+1$ we therefore see that $p_1(x)=(x^2+1)p(x)$.

Dividing $p_2(x)$ by $p(x)$ gives the quotient $q(x)=x^4+x^3+x$ and the remainder $r(x)=x^2+x+1$, and so $p(x)$ is not a factor of $p_2(x)$. □

A cyclic shift on an n-bit word can be expressed in terms of polynomials. Consider $(a_{n-1}, a_{n-2}, \ldots, a_2, a_1, a_0)$ shifted by 1 place to the left, to give $(a_{n-2}, a_{n-3}, \ldots, a_1, a_0, a_{n-1})$ and let

$$p(x)=a_{n-1}x^{n-1}+a_{n-2}x^{n-2}+\cdots+a_2x^2+a_1x+a_0$$

and

$$p'(x)=a_{n-2}x^{n-1}+a_{n-3}x^{n-2}+\cdots+a_1x^2+a_0x+a_{n-1}.$$

It can be shown that $p'(x)$ is given by the remainder of $xp(x)$ divided by x^n+1, so that

$$p'(x)=R_{x^n+1}[xp(x)]. \tag{3.6}$$

Assuming that $a_{n-1}=1$ then $xp(x)$ will contain an x^n-term, and taking $xp(x)$ modulo-(x^n+1) has the effect of replacing the x^n term by 1, so that overall the x^{n-1} term has 'cycled' around from the high-order side of the polynomial to the low-order side. For example consider a cyclic shift on (1 0 1 1 0 0) of 1 place to the left. Here $p(x)=x^5+x^3+x^2$ and $xp(x)=x^6+x^4+x^3$. The remainder of $x^6+x^4+x^3$ divided by x^6+1 is x^4+x^3+1, which gives the word (0 1 1 0 0 1) as required. If $a_{n-1}=0$ then eqn 3.6 gives $p'(x)=xp(x)$.

Example 3.6
Given that $p_1(x) = x^6 + x^2 + 1$ and $p_2(x) = x^3 + x$ represent two 7-bit words $\boldsymbol{w}_1$ and $\boldsymbol{w}_2$. Using eqn 3.6 find the resulting words when $\boldsymbol{w}_1$ and $\boldsymbol{w}_2$ are shifted leftwards by 1 place.

For $p_1(x) = x^6 + x^2 + 1$, the product $xp(x)$ gives $x^7 + x^3 + x$. Dividing $x^7 + x^3 + x$ by $x^7 + 1$ gives the remainder $x^3 + x + 1$ which corresponds to the 7-bit word (0 0 0 1 0 1 1). As a check, note that $p_1(x) = x^6 + x^2 + 1$ gives the 7-bit word (1 0 0 0 1 0 1), which when shifted 1 place towards the left gives (0 0 0 1 0 1 1) as already obtained.

Given $p_2(x) = x^3 + x$, then $xp(x) = x^4 + x^2$ and dividing by $x^7 + 1$ gives the remainder $x^4 + x^2$. Hence the required word is (0 0 1 0 1 0 0). We can again easily check that this is correct. Note that here the degree of $xp(x)$ is less than 7 and so dividing $xp(x)$ by $x^7 + 1$ gives $xp(x)$ as the remainder (with a zero quotient). □

We now reconsider polynomial multiplication. We have already considered the product of two arbitrary polynomials $p_1(x)$ and $p_2(x)$, and we have seen that the degree of $p(x) = p_1(x)p_2(x)$ is the sum of the degrees of $p_1(x)$ and $p_2(x)$. If $p_1(x)$ and $p_2(x)$ represent n-bit words then the degree of $p(x) = p_1(x)p_2(x)$ could be greater than $n - 1$, in which case $p(x)$ could not represent an n-bit word. To address this, we define a second kind of multiplication that ensures that the degree of the product of two polynomials does not exceed $n - 1$. Given the polynomials $p_1(x)$ and $p_2(x)$ then

$$p(x) = R_{x^n+1}[p_1(x)p_2(x)] \tag{3.7}$$

has degree less than or equal to $n - 1$ and can therefore represent an n-bit word. Consider, for example, $p_1 = x^4 + x$ and $p_2 = x^3 + x + 1$ representing two 5-bit words (1 0 0 1 0) and (0 1 0 1 1) respectively. The product $p_1(x)p_2(x) = x^7 + x^5 + x^2 + x$ has degree 7 and therefore cannot represent words with less than 8 bits. However the remainder of $p_1(x)p_2(x)$ divided by $x^5 + 1$ gives $x + 1$ which can be expressed as the 5-bit word (0 0 0 1 1).

3.3 Generator polynomials

Consider an (n, k) cyclic code with codeword

$$\boldsymbol{c} = (c_{n-1}, c_{n-2}, \ldots, c_2, c_1, c_0).$$

The coefficients of the codeword can be used to construct a *codeword polynomial*

$$c(x) = c_{n-1}x^{n-1} + c_{n-2}x^{n-2} + \cdots + c_2x^2 + c_1x + c_0$$

so giving a polynomial representation of $\boldsymbol{c}$. We now return to the (7, 4) Hamming code which, as we have already seen, is a cyclic code. The first nonzero codeword, in the list of codewords given in Table 1.6, is

$$\boldsymbol{c}_1 = (0\ 0\ 0\ 1\ 0\ 1\ 1)$$

which can be expressed as

$$c_1(x) = 0x^6 + 0x^5 + 0x^4 + 1x^3 + 0x^2 + 1x + 1$$

that is

$$c_1(x) = x^3 + x + 1.$$

Table 3.1 shows the codeword polynomials, and corresponding codewords, for all the codewords belonging to the (7,4) Hamming code.

Consider now the codewords $\boldsymbol{c}_1 = (0\,0\,0\,1\,0\,1\,1)$ and $\boldsymbol{c}_2 = (0\,0\,1\,0\,1\,1\,0)$ with codeword polynomials $c_1(x) = x^3 + x + 1$ and $c_2(x) = x^4 + x^2 + x$ respectively. The codeword $\boldsymbol{c}_2$ can be obtained from a single leftwards cyclic shift of $\boldsymbol{c}_1$, and the cyclic relationship between the 2 codewords can be represented as

$$\begin{aligned} c_2(x) &= x^4 + x^2 + x \\ &= x(x^3 + x + 1) \\ &= xc_1(x). \end{aligned}$$

Further cyclic shifts of $\boldsymbol{c}_1$ produce the codeword polynomials

$$c_5(x) = x^2 c_1(x)$$

and

$$c_{11}(x) = x^3 c_1(x).$$

Table 3.1
The (7,4) codewords and codeword polynomials

Codewords $\boldsymbol{c}$	*Codeword polynomials* $c(x)$
$\boldsymbol{c}_0 = (0\ 0\ 0\ 0\ 0\ 0\ 0)$	$c_0(x) = 0$
$\boldsymbol{c}_1 = (0\ 0\ 0\ 1\ 0\ 1\ 1)$	$c_1(x) = x^3 + x + 1$
$\boldsymbol{c}_2 = (0\ 0\ 1\ 0\ 1\ 1\ 0)$	$c_2(x) = x^4 + x^2 + x$
$\boldsymbol{c}_3 = (0\ 0\ 1\ 1\ 1\ 0\ 1)$	$c_3(x) = x^4 + x^3 + x^2 + 1$
$\boldsymbol{c}_4 = (0\ 1\ 0\ 0\ 1\ 1\ 1)$	$c_4(x) = x^5 + x^2 + x + 1$
$\boldsymbol{c}_5 = (0\ 1\ 0\ 1\ 1\ 0\ 0)$	$c_5(x) = x^5 + x^3 + x^2$
$\boldsymbol{c}_6 = (0\ 1\ 1\ 0\ 0\ 0\ 1)$	$c_6(x) = x^5 + x^4 + 1$
$\boldsymbol{c}_7 = (0\ 1\ 1\ 1\ 0\ 1\ 0)$	$c_7(x) = x^5 + x^4 + x^3 + x$
$\boldsymbol{c}_8 = (1\ 0\ 0\ 0\ 1\ 0\ 1)$	$c_8(x) = x^6 + x^2 + 1$
$\boldsymbol{c}_9 = (1\ 0\ 0\ 1\ 1\ 1\ 0)$	$c_9(x) = x^6 + x^3 + x^2 + x$
$\boldsymbol{c}_{10} = (1\ 0\ 1\ 0\ 0\ 1\ 1)$	$c_{10}(x) = x^6 + x^4 + x + 1$
$\boldsymbol{c}_{11} = (1\ 0\ 1\ 1\ 0\ 0\ 0)$	$c_{11}(x) = x^6 + x^4 + x^3$
$\boldsymbol{c}_{12} = (1\ 1\ 0\ 0\ 0\ 1\ 0)$	$c_{12}(x) = x^6 + x^5 + x$
$\boldsymbol{c}_{13} = (1\ 1\ 0\ 1\ 0\ 0\ 1)$	$c_{13}(x) = x^6 + x^5 + x^3 + 1$
$\boldsymbol{c}_{14} = (1\ 1\ 1\ 0\ 1\ 0\ 0)$	$c_{14}(x) = x^6 + x^5 + x^4 + x^2$
$\boldsymbol{c}_{15} = (1\ 1\ 1\ 1\ 1\ 1\ 1)$	$c_{15}(x) = x^6 + x^5 + x^4 + x^3 + x^2 + x + 1$

Other codeword polynomials can be generated from $c_1(x)$ by multiplying $c_1(x)$ with polynomials of degree less than or equal to 3. For example

$$(x+1)c_1(x) = x^4 + x^3 + x^2 + 1 = c_3(x)$$

and

$$(x^3 + x^2)c_1(x) = x^6 + x^5 + x^4 + x^2 = c_{14}(x).$$

Furthermore, the product of $c_1(x) = x^3 + x + 1$ with every polynomial of degree less than or equal to 3 produces the complete set of codeword polynomials of the $(7, 4)$ code. The polynomial $x^3 + x + 1$ is therefore able to generate all the codeword polynomials of the $(7, 4)$ code, it is called a *generator polynomial* and is denoted by $g(x)$. The existence of a generator polynomial is not restricted to the $(7, 4)$ cyclic code, but is an important property of all cyclic codes. In an (n, k) cyclic code there exists a unique generator polynomial from which all the codeword polynomials can be generated. The generator polynomial is unique within the code and so none of the other codeword polynomials are able to generate the complete set of codeword polynomials. The generator polynomial can be likened to the generator matrix of a linear code, in that they are capable of generating all the codewords of a code, and in Section 3.9 the relationship between generator polynomials and matrices will be considered. For an (n, k) binary cyclic code the generator polynomial has the form

$$g(x) = g_{n-k}x^{n-k} + g_{n-k-1}x^{n-k-1} + \cdots + g_2x^2 + g_1x + g_0 \tag{3.8}$$

where the coefficients $g_{n-k-1}, g_{n-k-2}, \ldots, g_2, g_1$ are 0 or 1. The coefficients of the first and last terms are always equal to 1, and so

$$g_{n-k} = 1$$
$$g_0 = 1.$$

The x^{n-k} term is always present in the generator polynomial and there are no higher order terms, therefore the degree of the generator polynomial $g(x)$ of an (n, k) cyclic code is $n-k$. Furthermore, the product of $g(x)$ with every polynomial of degree $k-1$ or less generates all the codeword polynomials of an (n, k) cyclic code. The generator polynomial $g(x)$ is the smallest degree codeword polynomial of a cyclic code and no other codeword polynomial has degree less than or equal to $n-k$. Note that the degree of $g(x)$ is equal to the number of parity-check bits of the code.

3.4 Encoding cyclic codes

We have seen that the codeword polynomials of a cyclic code have the code's generator polynomial $g(x)$ as a factor. Hence to encode an information word we must first express the information word as a polynomial and then construct some suitable polynomial that has $g(x)$ as a factor. As with linear codes, there are two ways in which encoding can be achieved, namely nonsystematic and systematic encoding, we consider nonsystematic encoding first.

Nonsystematic encoding is the easiest way of encoding, but not necessarily the preferred method because the resulting codeword polynomials represent nonsystematic codewords. We have seen that the product of a generator polynomial $g(x)$ with any polynomial of degree $k-1$ or less produces a codeword polynomial for an (n, k) cyclic code and it is this that forms the basis for nonsystematic encoding. The information word $\boldsymbol{i}$ to be encoded is first expressed as an *information polynomial* $i(x)$, so if

$$\boldsymbol{i} = (i_{k-1}, i_{k-2}, \ldots, i_2, i_1, i_0)$$

then

$$i(x) = i_{k-1}x^{k-1} + i_{k-2}x^{k-2} + \cdots + i_2x^2 + i_1x + i_0.$$

We then simply take the product of $i(x)$ and $g(x)$ to give the codeword polynomial

$$c(x) = i(x)g(x). \tag{3.9}$$

Taking the product of $g(x)$ with all the information polynomials $i(x)$ generates the complete set of codewords. Note that there are 2^k polynomials of degree $k-1$ or less (i.e. the same number as different patterns in a k-bit word). To illustrate why encoding using eqn 3.9 is nonsystematic we reconsider the (7, 4) code. The generator polynomial of the (7, 4) code is

$$g(x) = x^3 + x + 1 \tag{3.10}$$

as already seen in Section 3.3 and taking say $\boldsymbol{i} = (0\,1\,0\,1)$ gives $i(x) = x^2 + 1$. From eqn 3.9 the corresponding codeword polynomial is

$$\begin{aligned} c(x) &= i(x)g(x) \\ &= (x^2 + 1)(x^3 + x + 1) \\ &= x^5 + x^3 + x^2 + x^3 + x + 1 \\ &= x^5 + x^2 + x + 1 \end{aligned}$$

giving the codeword $\boldsymbol{c} = (0\,1\,0\,0\,1\,1\,1)$, which is not the codeword for (0 1 0 1) but rather for (0 1 0 0) (see Table 1.6). The encoding rule $c(x) = i(x)g(x)$ therefore produces nonsystematic codewords.

Example 3.7
Determine nonsystematic codeword polynomials, for the (7, 4) code, given $\boldsymbol{i}_1 = (1\ 1\ 0\ 0)$ and $\boldsymbol{i}_2 = (1\ 0\ 0\ 1)$.

The information polynomial corresponding to $\boldsymbol{i}_1$ is $i(x) = x^3 + x^2$, and taking $g(x) = x^3 + x + 1$ gives

$$\begin{aligned} c(x) &= i(x)g(x) \\ &= (x^3 + x^2)(x^3 + x + 1) \\ &= x^6 + x^4 + x^3 + x^5 + x^3 + x^2 \\ &= x^6 + x^5 + x^4 + x^2 \end{aligned}$$

as the codeword polynomial. Note that the corresponding codeword is $\boldsymbol{c} = (1\,1\,1\,0\,1\,0\,0)$ and encoding is therefore nonsystematic.

Given $\boldsymbol{i}_2 = (1\ 0\ 0\ 1)$, then $i(x) = x^3 + 1$ and this gives $c(x) = x^6 + x^4 + x + 1$. Again this gives a nonsystematic codeword for $\boldsymbol{i}_2$. □

Table 3.2 shows the set of codeword polynomials generated by $c(x) = i(x)g(x)$. The table contains the same set of codeword polynomials shown in Table 3.1, except that the order in which the polynomials occur in the tables is different.

The encoding procedure for producing systematic codewords for cyclic codes is slightly more complicated than that for nonsystematic codewords. Given an (n, k) cyclic code with generator polynomial $g(x)$, an information polynomial $i(x)$ is encoded into a systematic codeword using the following three steps

1. Multiply $i(x)$ by x^{n-k} to give

$$i(x)x^{n-k}.$$

2. Divide $i(x)x^{n-k}$ by $g(x)$ to give the remainder

$$r(x) = R_{g(x)}[i(x)x^{n-k}].$$

3. Add $r(x)$ to $i(x)x^{n-k}$ to give the codeword

$$c(x) = i(x)x^{n-k} + r(x). \tag{3.11}$$

As an example, let's again consider the (7, 4) code with $\boldsymbol{i} = (0\ 1\ 0\ 1)$ and information polynomial $i(x) = x^2 + 1$. In Step 1, $i(x)$ is multiplied by $x^{n-k} = x^3$ to give

$$i(x)x^{n-k} = (x^2 + 1)x^3 = x^5 + x^3.$$

In Step 2, $x^5 + x^3$ is divided by $g(x) = x^3 + x + 1$ which gives the quotient x^2 and remainder x^2, and so

$$r(x) = R_{g(x)}[x^5 + x^3] = x^2.$$

Table 3.2
Codeword polynomials generated by $c(x) = i(x)g(x)$ for the (7, 4) code

$i(x)$	$i(x)g(x)$	$c_i(x)$
1	x^3+x+1	$c_1(x)$
x	x^4+x^2+x	$c_2(x)$
$x+1$	$x^4+x^3+x^2+1$	$c_3(x)$
x^2	$x^5+x^3+x^2$	$c_5(x)$
x^2+1	x^5+x^2+x+1	$c_4(x)$
x^2+x	$x^5+x^4+x^3+x$	$c_7(x)$
x^2+x+1	x^5+x^4+1	$c_6(x)$
x^3	$x^6+x^4+x^3$	$c_{11}(x)$
x^3+1	x^6+x^4+x+1	$c_{10}(x)$
x^3+x	$x^6+x^3+x^2+x$	$c_9(x)$
x^3+x+1	x^6+x^2+1	$c_8(x)$
x^3+x^2	$x^6+x^5+x^4+x^2$	$c_{14}(x)$
x^3+x^2+1	$x^6+x^5+x^4+x^3+x^2+x+1$	$c_{15}(x)$
x^3+x^2+x	x^6+x^5+x	$c_{12}(x)$
x^3+x^2+x+1	$x^6+x^5+x^3+1$	$c_{13}(x)$

Finally in Step 3, $r(x)$ is added to $i(x)x^{n-k}$ to give the codeword polynomial

$$c(x) = x^5 + x^3 + x^2$$

which corresponds to the codeword $\boldsymbol{c} = (0\,1\,0\,1\,1\,0\,0)$ and is the systematic codeword for $\boldsymbol{i} = (0\,1\,0\,1)$.

Example 3.8

Assuming the (7, 4) cyclic code determine the systematic codeword polynomials for $\boldsymbol{i}_1 = (1\,0\,0\,1)$ and $\boldsymbol{i}_2 = (1\,1\,1\,0)$.

The information polynomial for $\boldsymbol{i}_1$ is $i(x) = x^3 + 1$ and carrying out systematic encoding gives

Step 1: Multiplying $(x^3 + 1)$ by $x^{n-k} = x^3$ gives $x^6 + x^3$.

Step 2: Dividing $x^6 + x^3$ by $g(x) = x^3 + x + 1$ gives the remainder $r(x) = x^2 + x$.

Step 3: Adding $r(x)$ to $x^6 + x^3$ gives $c(x) = x^6 + x^3 + x^2 + x$.

Hence the codeword is $\boldsymbol{c} = (1\,0\,0\,1\,1\,1\,0)$ which is the systematic codeword for $(1\,0\,0\,1)$.

For $\boldsymbol{i}_2$ we get $i(x) = x^3 + x^2 + x$ and $x^3 i(x) = x^6 + x^5 + x^4$, which when divided by $g(x)$ gives $r(x) = x^2$. Hence $c(x) = x^6 + x^5 + x^4 + x^2$ and $\boldsymbol{c} = (1\,1\,1\,0\,1\,0\,0)$ (again encoding is systematic). □

Let's now consider why the three steps given produce systematic codewords. The first step, multiplying $i(x)$ by x^{n-k}, shifts the information bits $n - k$ places towards the left-hand side of the word. This leaves the k information bits in the correct position, required for a systematic codeword, and furthermore leaves space for the $n - k$ parity bits. In the second and third steps the remainder $r(x)$ of $i(x)x^{n-k}$ divided by $g(x)$ is determined and added to $i(x)x^{n-k}$. Using the Euclidean division algorithm, we can write

$$i(x)x^{n-k} = q(x)g(x) + r(x)$$

where $q(x)$ is the quotient obtained when $i(x)x^{n-k}$ is divided by $g(x)$. Subtracting $r(x)$ from both sides of the above equation, and remembering that polynomial subtraction is the same as polynomial addition, gives

$$i(x)x^{n-k} + r(x) = q(x)g(x). \tag{3.12}$$

Now the term on the right-hand side of eqn 3.12 has $g(x)$ as a factor, and is therefore a codeword polynomial, and so $i(x)x^{n-k} + r(x)$ will also be a codeword polynomial. Hence Steps 1, 2, and 3 give systematic codeword polynomials. Equation 3.12 shows that the codeword polynomial $c(x)$ can be obtained by multiplying the quotient $q(x)$ by $g(x)$ instead of adding the remainder $r(x)$ to $i(x)x^{n-k}$, and so codewords can be expressed as

$$c(x) = q(x)g(x) \tag{3.13}$$

for systematic encoding. In Step 3 eqn 3.11 could be replaced by eqn 3.13, however because polynomial addition is easier than multiplication it is normal to use eqn 3.11

rather than eqn 3.13. Nevertheless it is sometimes useful to think of systematic encoding in terms of eqn 3.13. Comparing eqns 3.9 and 3.13, we see that for an (n, k) cyclic code with generator polynomial $g(x)$, the codeword polynomial $c(x)$ corresponding to the information polynomial $i(x)$ is

$$c(x) = f(x)g(x) \tag{3.14}$$

where $f(x) = q(x)$ and $f(x) = i(x)$ for systematic and nonsystematic encoding respectively, and where $q(x)$ is the quotient obtained when dividing $i(x)x^{n-k}$ by $g(x)$.

Example 3.9
Determine systematic and nonsystematic codewords for $\boldsymbol{i} = (0\ 1\ 1\ 1)$ given the (7, 4) code with $g(x) = x^3 + x + 1$.

The information polynomial corresponding to $\boldsymbol{i}$ is $i(x) = x^2 + x + 1$, and so the nonsystematic codeword polynomial is

$$\begin{aligned} c(x) &= i(x)g(x) \\ &= (x^2 + x + 1)(x^3 + x + 1) \\ &= x^5 + x^4 + 1. \end{aligned}$$

For the systematic codeword polynomial we get
Step 1: $i(x)x^{n-k} = (x^2 + x + 1)\,x^3 = x^5 + x^4 + x^3$.
Step 2: $r(x) = R_{x^3+x+1}\,[x^5 + x^4 + x^3] = x$.
Step 3: $c(x) = i(x)x^{n-k} + r(x) = x^5 + x^4 + x^3 + x$
The corresponding systematic and nonsystematic codewords are therefore (0 1 1 1 0 1 0) and (0 1 1 0 0 0 1) respectively. Note also that in Step 2 the quotient is $q(x) = x^2 + x$, and replacing Step 3 with $c(x) = q(x)g(x)$ gives

$$\begin{aligned} c(x) &= q(x)g(x) \\ &= (x^2 + x)(x^3 + x + 1) \\ &= x^5 + x^4 + x^3 + x \end{aligned}$$

as given in Step 3. □

3.5 Decoding cyclic codes

We have already seen how with linear codes a syndrome table can be constructed and used for error correction. Here the same approach is adopted, but viewed in terms of polynomials. The implementation of decoding algorithms for cyclic codes can be achieved with linear-feedback shift registers, this is considered in Chapter 4.

Whether systematic or nonsystematic encoding is used, a codeword polynomial of a cyclic code always has the generator polynomial $g(x)$ as a factor. To test whether some arbitrary polynomial $v(x)$, of degree less than n, is a codeword polynomial of an (n, k) cyclic code the remainder of $v(x)$ divided by $g(x)$ is determined. If the remainder is zero then $v(x)$ is a codeword polynomial, otherwise it is not. A codeword polynomial will only give a nonzero remainder if it incurs errors, and as with linear codes an error syndrome can be defined as that depends solely on the presence of errors. The *syndrome polynomial* of $v(x)$ is defined as

$$s(x) = R_{g(x)}[v(x)] \tag{3.15}$$

recalling that the notation on the right-hand side denotes the remainder obtained when $v(x)$ is divided by $g(x)$. For a codeword polynomial $c(x)$

$$R_{g(x)}[c(x)] = 0 \tag{3.16}$$

because $c(x)$ has $g(x)$ as a factor and therefore the syndrome polynomial of a codeword polynomial is $s(x)=0$. If $v(x)$ represents a codeword polynomial $c(x)$ containing errors then

$$v(x) = c(x) + e(x)$$

where the *error polynomial* $e(x)$ is the polynomial representation of the error pattern. Using eqn 3.15 the syndrome polynomial is

$$\begin{aligned} s(x) &= R_{g(x)}[v(x)] \\ &= R_{g(x)}[c(x) + e(x)] \\ &= R_{g(x)}[c(x)] + R_{g(x)}[e(x)] \end{aligned}$$

and using eqn 3.16 gives

$$s(x) = R_{g(x)}[e(x)]. \tag{3.17}$$

Therefore, the syndrome polynomial depends solely on the error polynomial and not on the codeword polynomial incurring the errors. From eqn 3.17 a syndrome table, giving the syndrome polynomials of all the correctable error polynomials, can easily be constructed.

As an example, let's construct the syndrome table for the single-error-correcting (7, 4) code. We need to determine the syndrome polynomials for the 7 single-error patterns (0 0 0 0 0 0 1), (0 0 0 0 0 1 0), ..., (1 0 0 0 0 0 0). Starting with (0 0 0 0 0 0 1) the corresponding error polynomial is $e(x)=1$ and using eqn 3.17, with $g(x)=x^3+x+1$, gives the remainder $r(x)=R_{x^3+x+1}[1]=1$ and therefore the syndrome polynomial is $s(x)=1$. The error pattern (0 0 0 0 0 1 0) gives $e(x)=x$ and syndrome polynomial $s(x)=x$, and continuing through to (1 0 0 0 0 0 0) gives $e(x)=x^6$ with $s(x)=x^2+1$. Table 3.3 shows the resulting syndrome table. The table is equivalent to the syndrome table previously given for the (7, 4) Hamming code (see Table 1.7), except that here we have a polynomial representation of error patterns and error syndromes. Either table can be converted into the other simply by changing to or from a polynomial representation, as required. It is instructive, however, to see the construction of Table 3.3 from the perspective of cyclic codes and not just by a change of notation.

Error detection and correction is carried out in a manner similar to that already described for linear codes. Consider a codeword polynomial $c(x)$ incurring an error polynomial $e(x)$ such that the input to the decoder is $v(x)$. The syndrome polynomial $s(x)$ is determined using $s(x)=R_{g(x)}[v(x)]$ and the error polynomial, denoted by $\hat{e}(x)$, corresponding to $s(x)$ is read from the syndrome table. Adding $\hat{e}(x)$ to $v(x)$ gives

$$\hat{c}(x) = v(x) + \hat{e}(x) \tag{3.18}$$

Table 3.3
Syndrome table for the (7, 4) cyclic code

e	$e(x)$	$s(x)$
(0 0 0 0 0 0 0)	0	0
(0 0 0 0 0 0 1)	1	1
(0 0 0 0 0 1 0)	x	x
(0 0 0 0 1 0 0)	x^2	x^2
(0 0 0 1 0 0 0)	x^3	$x+1$
(0 0 1 0 0 0 0)	x^4	x^2+x
(0 1 0 0 0 0 0)	x^5	x^2+x+1
(1 0 0 0 0 0 0)	x^6	x^2+1

as the estimated codeword polynomial for $c(x)$. If the number of errors in $e(x)$ is not greater than the error-correction limit of the code then $\hat{e}(x) = e(x)$, $\hat{c}(x) = c(x)$ and decoding is correct. As already mentioned for decoding linear codes, the use of a syndrome table is not always practical, especially for codes with large blocklengths and information lengths. Furthermore, with cyclic codes this approach does not utilize the cyclic nature of the codes. In Chapter 4 we consider the implementation of encoding and decoding, using linear-feedback shift registers, and we will see how the cyclic property of the codes is utilized.

Example 3.10
Consider decoding when the codeword $\boldsymbol{c} = (1\,0\,1\,0\,0\,1\,1)$, belonging to the (7, 4) code, incurs the error $\boldsymbol{e} = (1\,0\,0\,0\,0\,0\,0)$.

The codeword $(1\,0\,1\,0\,0\,1\,1)$ gives $\boldsymbol{c}(x) = x^6 + x^4 + x + 1$ and $\boldsymbol{e} = (1\,0\,0\,0\,0\,0\,0)$ gives $e(x) = x^6$. Hence the word to be decoded is

$$v(x) = c(x) + e(x) = x^4 + x + 1.$$

The generator polynomial for the (7, 4) code is $g(x) = x^3 + x + 1$ and the syndrome polynomial is

$$s(x) = R_{x^3+x+1}[x^4 + x + 1] = x^2 + 1.$$

Referring to Table 3.3, we see that $s(x) = x^2 + 1$ corresponds to the error polynomial $\hat{e}(x) = x^6$. Therefore the estimated codeword polynomial is

$$\begin{aligned}\hat{c}(x) &= v(x) + \hat{e}(x)\\ &= (x^4 + x + 1) + x^6\\ &= x^6 + x^4 + x + 1\end{aligned}$$

which gives $\hat{\boldsymbol{c}} = (1\,0\,1\,0\,0\,1\,1)$. The decoder therefore correctly concludes that $\boldsymbol{v}$ was the codeword $(1\,0\,1\,0\,0\,1\,1)$. □

3.6 Factors of $x^n + 1$

We now ask whether any polynomial can be used to generate a cyclic code, and if not, then what characteristics must a polynomial have such that it can generate a cyclic code. The answer to this question lies in the factorization of the polynomial

$$p(x) = x^n + 1. \tag{3.19}$$

A polynomial $g(x)$ with degree r generates an (n, k) cyclic code, where $k = n - r$, if $g(x)$ divides $x^n + 1$. Note that by 'divides' it is implied that the remainder is zero and so $g(x)$ has to be a factor of $x^n + 1$ if it is to generate a cyclic code. Consider the $(7, 4)$ code with $g(x) = x^3 + x + 1$. In Section 3.1 it was claimed that the $(7, 4)$ code is cyclic and examples of cyclic shifts were given. This though does not prove that the code is cyclic but only illustrates the cyclic relationship between some of its codewords. To show that $g(x)$ generates a cyclic code we need to divide $x^7 + 1$ by $g(x)$ and check that the remainder is zero. Dividing $x^7 + 1$ by $g(x)$ gives

$$\begin{array}{r l}
 & x^4 + x^2 + x + 1 \\
x^3 + x + 1 \,\big) & x^7 + 1 \\
 & x^7 + x^5 + x^4 \\
 & - \quad x^5 + x^4 + 1 \\
 & \quad\;\; x^5 + x^3 + x^2 \\
 & \quad - \quad x^4 + x^3 + x^2 + 1 \\
 & \quad\quad\quad\;\; x^4 + x^2 + x \\
 & \quad\quad - \quad x^3 + x + 1 \\
 & \quad\quad\quad\quad\; x^3 + x + 1 \\
 & \quad\quad\quad\quad\quad\quad\quad 0
\end{array}$$

and so the remainder is zero. Hence, the polynomial $g(x) = x^3 + x + 1$ divides $x^7 + 1$ and therefore generates a $(7, 4)$ cyclic code. Note that the degree of $g(x)$ is $r = 3$ and so the information length of the code is $k = n - r = 4$. Furthermore the quotient obtained is $x^4 + x^2 + x + 1$, and if we let $h(x) = x^4 + x^2 + x + 1$ then

$$g(x)h(x) = x^7 + 1. \tag{3.20}$$

From eqn 3.20 we see that $h(x)$ also divides $x^7 + 1$ and therefore $h(x) = x^4 + x^2 + x + 1$ likewise generates a cyclic code of blocklength 7 but with information length 3 (because the degree of $h(x)$ is 4). Hence the polynomial $x^4 + x^2 + x + 1$ generates a $(7, 3)$ cyclic code (note that $h(x)$ is referred to as the parity-check polynomial of the $(7, 4)$ code, this is considered further in Section 3.7).

We can go a step further in factorizing $x^7 + 1$ because $x = 1$ is a root of $h(x)$ (letting $x = 1$ gives $h(1) = 1 + 1 + 1 + 1 = 0$) and so $(x + 1)$ is a factor of $h(x)$. Dividing $h(x)$ by $(x + 1)$ gives the quotient $x^3 + x^2 + 1$ and a zero remainder. Therefore $h(x) = x^4 + x^2 + x + 1 = (x + 1)\ (x^3 + x^2 + 1)$ and substituting this into eqn 3.20, along with $g(x) = x^3 + x + 1$, gives

$$(x + 1)(x^3 + x^2 + 1)(x^3 + x + 1) = x^7 + 1. \tag{3.21}$$

There are no other factors of $x^7 + 1$ as neither $x^3 + x^2 + 1$ or $x^3 + x + 1$ can be factorized. Any of the 3 factors of $x^7 + 1$ or products of the factors will generate a cyclic code with blocklength 7. Equation 3.21 therefore shows that there are 2 polynomials that generate $(7, 4)$ codes, the polynomial $x^3 + x + 1$ that we have already considered and $x^3 + x^2 + 1$ (the codes are however equivalent). Taking the

Table 3.4
Cyclic codes constructed from $x^7 + 1$

Factors	$g(x)$	(n, k) *code*
$f_1(x)$	$x+1$	$(7, 6)$
$f_2(x)$	x^3+x^2+1	$(7, 4)$
$f_3(x)$	x^3+x+1	$(7, 4)$
$f_1(x)\,f_2(x)$	$x^4+x^3+x^2+1$	$(7, 3)$
$f_1(x)\,f_3(x)$	x^4+x^2+x+1	$(7, 3)$
$f_2(x)\,f_3(x)$	$x^6+x^5+x^4+x^3+x^2+x+1$	$(7, 1)$

Note: $x^7+1 = f_1(x)\,f_2(x)\,f_3(x)$

product of the factors $(x + 1)$ and $(x^3 + x + 1)$ gives

$$(x + 1)(x^3 + x + 1) = x^4 + x^3 + x^2 + 1$$

which generates a (7, 3) cyclic code and is equivalent to the (7, 3) code generated by $x^4 + x^2 + x + 1$. The product of $(x^3 + x + 1)$ and $(x^3 + x^2 + 1)$ gives

$$(x^3 + x + 1)(x^3 + x^2 + 1) = x^6 + x^5 + x^4 + x^3 + x^2 + x + 1.$$

This has degree $r = 6$ and so the cyclic code constructed has $k = 1$ (recall that $k = n - r$) and is therefore the (7, 1) repetition code. The code is clearly cyclic, as a cyclic shift on either of its 2 codewords, (0 0 0 0 0 0 0) or (1 1 1 1 1 1 1), gives the same codeword. The factor $(x + 1)$ taken on its own generates a code with $k = 6$ and therefore gives the (7, 6) single-parity-check code. Again this is clearly a cyclic code because a codeword that is subjected to a cyclic shift preserves its parity and is therefore still a codeword. Note that 3 factors give 8 different combinations, of which we have seen 6 (giving the two (7, 4) codes, the two (7, 3) codes, the (7, 1) code and the (7, 6) code). The remaining 2 combinations give 2 trivial codes with $g(x) = 1$ (and $n = k = 7$) and $g(x) = x^7 + 1$ (with $n = 7$ and $k = 0$). Table 3.4 shows the 6 nontrivial cyclic codes that can be obtained from $x^7 + 1$.

Example 3.11
Given that $x^9 + 1 = (x + 1)\ (x^2 + x + 1)\ (x^6 + x^3 + 1)$ determine the cyclic codes with blocklength 9.

Let the individual factors be denoted by

$$f_1(x) = x + 1$$
$$f_2(x) = x^2 + x + 1$$
$$f_3(x) = x^6 + x^3 + 1.$$

Here $f_1(x) = x + 1$ has degree 1 and therefore generates a (9, 8) single-parity-check code, while $f_2(x)$ and $f_3(x)$ generate (9, 7) and (9, 3) cyclic codes respectively. Taking the product of $f_1(x)$ with $f_2(x)$ and then with $f_3(x)$ gives

$$f_1(x)f_2(x) = x^3 + 1$$
$$f_1(x)f_3(x) = x^7 + x^6 + x^4 + x^3 + x + 1$$

which give (9, 6) and (9, 2) cyclic codes respectively.

The product of $f_2(x)$ and $f_3(x)$ gives

$$f_2(x)f_3(x) = x^8 + x^7 + x^6 + x^5 + x^4 + x^3 + x^2 + x + 1$$

which has degree 8 and therefore gives the (9, 1) repetition code.

This accounts for 6 codes, the remaining 2 codes are trivial codes with $g(x) = 1$ $(n = k = 9)$ and $g(x) = x^9 + 1$ $(n = 9, k = 0)$. □

3.7 Parity-check polynomials

If $g(x)$ is the generator polynomial of an (n, k) cyclic code, then the polynomial $h(x)$ that satisfies

$$g(x)h(x) = x^n + 1 \tag{3.22}$$

is known as the *parity-check polynomial* of the code, and rearranging this gives

$$h(x) = \frac{x^n + 1}{g(x)}. \tag{3.23}$$

The degree of $h(x)$ is n minus the degree of $g(x)$, and as $g(x)$ has degree $n - k$ the parity-check polynomial $h(x)$ therefore has degree $n - (n - k) = k$. The parity-check polynomial can be written as

$$h(x) = h_k x^k + h_{k-1}x^{k-1} + \cdots + h_2x^2 + h_1x + h_0 \tag{3.24}$$

where the coefficients $h_0 = h_k = 1$ and the remaining coefficients $h_1, h_2, \ldots, h_{k-2}, h_{k-1}$ are 0 or 1. The (7, 4) code with $g(x) = x^3 + x + 1$ has the parity-check polynomial $h(x) = (x^7 + 1)/(x^3 + x + 1)$ which gives

$$h(x) = x^4 + x^2 + x + 1. \tag{3.25}$$

The parity-check polynomial is analogous to the parity-check matrix $\boldsymbol{H}$ used in linear codes, just as the generator polynomial is analogous to the generator matrix $\boldsymbol{G}$. Note however that we have already considered the decoding of cyclic codes, and as we saw this can be achieved using the generator polynomial (the parity-check polynomial was not required for decoding the (7, 4) code). The relevance of the parity-check polynomial to decoding is considered later. In Section 3.9 we consider how the generator and parity-check matrices can be obtained directly from the generator and parity-check polynomials respectively.

It is interesting to compare the relationships between codeword, generator, and parity-check polynomials with that of codewords, generator, and parity-check matrices (used in cyclic codes and linear codes respectively). First consider the construction of codeword polynomials. We have seen that the codeword polynomial $c(x)$ for an information polynomial $i(x)$ is given by

$$c(x) = i(x)g(x)$$

which can be compared to $\boldsymbol{c} = \boldsymbol{iG}$ used to construct the codeword $\boldsymbol{c}$, given an information word $\boldsymbol{i}$, of a linear code with generator matrix $\boldsymbol{G}$ (note that whilst $c(x) = i(x)g(x)$ gives nonsystematic codeword polynomials, systematic polynomials nevertheless also contain $g(x)$ as a factor). To see the relationship of $g(x)$ to $h(x)$ compared to that of $\boldsymbol{G}$ to $\boldsymbol{H}$ (the parity-check matrix of a linear code), we consider eqn 3.22 divided by $x^n + 1$. This gives a zero remainder and so

$$R_{x^n+1}[g(x)h(x)] = 0 \tag{3.26}$$

which can be likened to $\boldsymbol{GH}^{\mathrm{T}} = 0$ given previously for linear codes. Multiplying $c(x) = i(x)g(x)$ by $h(x)$ gives $c(x)h(x) = i(x)g(x)h(x)$ and using eqn 3.22 we get $c(x)h(x) = i(x)(x^n + 1)$. Dividing this by $x^n + 1$ gives

$$R_{x^n+1}[c(x)h(x)] = 0 \tag{3.27}$$

which can be compared to $\boldsymbol{cH}^{\mathrm{T}} = 0$. Moving on to error syndromes, the syndrome polynomial $s(x)$ is given by

$$s(x) = R_{g(x)}[v(x)]$$

and this can be likened to the definition of an error syndrome, in a linear code, $\boldsymbol{s} = \boldsymbol{vH}^{\mathrm{T}}$ of a word $\boldsymbol{v}$. Note here that there is a difference between the expressions for $s(x)$ and $\boldsymbol{s}$ in the sense that $s(x)$ depends on the *generator* polynomial while $\boldsymbol{s}$ depends on the *parity-check* matrix. However, the syndrome polynomial $s(x)$ can be expressed in terms of the parity-check polynomial as follows. The syndrome polynomial is obtained by first dividing $v(x)$ by $g(x)$, and then taking the remainder. However, from eqn 3.22, we have $g(x) = (x^n + 1)/h(x)$ and therefore dividing by $g(x)$ is the same as multiplying by $h(x)$ and dividing by $(x^n + 1)$. Hence the syndrome polynomial can be obtained by multiplying $v(x)$ by $h(x)$, dividing by $(x^n + 1)$ and then taking the remainder, so that

$$s(x) = R_{x^n+1}[v(x)h(x)] \tag{3.28}$$

and we see then that $s(x)$ can be expressed in terms of $h(x)$. Furthermore, we can now see why $g(x)$ and not $h(x)$ is used for decoding cyclic codes, it is easier to divide $v(x)$ by $g(x)$ and then take the remainder, rather than first multiplying $v(x)$ by $h(x)$, dividing by $(x^n + 1)$ and then taking the remainder. Table 3.5 summarizes the comparisons made between $c(x)$, $g(x)$, $h(x)$, $\boldsymbol{c}$, $\boldsymbol{G}$, and $\boldsymbol{H}$.

Table 3.5
Comparison of $c(x)$, $g(x)$, $h(x)$, $\boldsymbol{c}$, $\boldsymbol{G}$, and $\boldsymbol{H}$

Linear codes	*Cyclic codes*
$\boldsymbol{c} = \boldsymbol{iG}$	$c(x) = i(x)g(x)$
$\boldsymbol{GH}^{\mathrm{T}} = 0$	$R_{x^n+1}[g(x)h(x)] = 0$
$\boldsymbol{cH}^{\mathrm{T}} = 0$	$R_{x^n+1}[c(x)h(x)] = 0$
$\boldsymbol{s} = \boldsymbol{vH}^{\mathrm{T}}$	$s(x) = R_{x^n+1}[v(x)h(x)]$
	$(s(x) = R_{g(x)}[v(x)])$

3.8 Dual cyclic codes

The parity-check polynomial $h(x)$ of a code C can be used to generate another code referred to as the *dual code* of C. This is achieved by first defining the *reciprocal polynomial* $h^*(x)$ of $h(x)$ as

$$h^*(x) = x^r h(1/x) \tag{3.29}$$

where r is the degree of $h(x)$. The reciprocal polynomial forms the generator polynomial $g(x)$ of the dual code and therefore

$$g(x) = h^*(x).$$

Replacing each x term in $h(x)$ by $1/x$ and then multiplying through by x^r has the effect of reversing the order of the coefficients of $h(x)$. So, for example if

$$h(x) = h_3x^3 + h_2x^2 + h_1x + h_0$$

then the corresponding reciprocal polynomial is obtained by replacing x with $1/x$, to give

$$h\left(\frac{1}{x}\right) = \frac{h_3}{x^3} + \frac{h_2}{x^2} + \frac{h_1}{x} + h_0$$

and multiplying by x^3 gives

$$h^*(x) = h_3 + h_2x + h_1x^2 + h_0x^3.$$

Rearranging the order of the terms on the right-hand side gives

$$h^*(x) = h_0x^3 + h_1x^2 + h_2x + h_3$$

and comparing this to $h(x)$ we can see that the coefficients of $h^*(x)$ are in the reverse order to those in $h(x)$.

Consider the $(7, 4)$ code with parity-check polynomial $h(x) = x^4 + x^2 + x + 1$. The reciprocal polynomial is given by

$$\begin{aligned} h^*(x) &= x^4 h(1/x) \\ &= x^4\left(\frac{1}{x^4} + \frac{1}{x^2} + \frac{1}{x} + 1\right) \\ &= 1 + x^2 + x^3 + x^4 \end{aligned}$$

which gives

$$g(x) = x^4 + x^3 + x^2 + 1 \tag{3.30}$$

as the generator polynomial of the code that is dual to the $(7, 4)$ code. The blocklength of a code and its dual code are the same, so the dual code has blocklength 7. The degree of $g(x)$ is 4 and therefore the information length of the dual code is 3 (recall that the degree of a generator polynomial for an (n, k) cyclic code is $n - k$). Hence the code that is dual to the $(7, 4)$ cyclic code is a $(7, 3)$ code. The $(7, 3)$

code constructed here is clearly the (7, 3) cyclic code with $g(x) = x^4 + x^3 + x^2 + 1$ considered in Section 3.6. Systematic codewords generted by $g(x) = x^4 + x^3 + x^2 + 1$ are given in Table 3.6(a).

It is interesting to consider the code generated by the parity-check polynomial $h(x)$. Taking the parity-check polynomial of the (7, 4) code $h(x) = x^4 + x^2 + x + 1$ as a generator polynomial gives the systematic codewords shown in Table 3.6(b). This is the same set of codewords generated by $h^*(x)$ except that the order of the bits are reversed. Each codeword in Table 3.6(a) corresponds to a codeword in Table 3.6(b), but with the bits reversed. For example the 2nd codeword (0 0 1 1 1 0 1) in Table 3.6(a) is the 6th codeword (1 0 1 1 1 0 0) in Table 3.6(b) with the bits reversed. Likewise the 3rd codeword (0 1 0 0 1 1 1) in Table 3.6(a) corresponds to the last codeword (1 1 1 0 0 1 0) in Table 3.6(b). Hence each codeword $(c_0, c_1, c_2, c_3, c_4, c_5, c_6)$ produced by $h^*(x)$ has an equivalent codeword $(c_6, c_5, c_4, c_3, c_2, c_1, c_0)$ produced by $h(x)$, and so the codes produced by $h^*(x)$ and $h(x)$ are equivalent. It is for this reason that the parity-check polynomial is often said to generate the dual code but, strictly speaking, it is the reciprocal polynomial $h^*(x)$ that generates the dual code, the code generated by $h(x)$ is equivalent to the dual code.

Example 3.12

Given the (9, 6) cyclic code with generator polynomial $g(x) = x^3 + 1$ construct its dual code.

The parity-check ploynomial of the (9, 6) code is given by $h(x) = (x^9 + 1)/(x^3 + 1) = x^6 + x^3 + 1$. From $h(x)$ the reciprocal polynomial is obtained by replacing x by $1/x$ and multiplying by x^6, so that

$$h^*(x) = x^6\left(\frac{1}{x^6} + \frac{1}{x^3} + 1\right) = 1 + x^3 + x^6$$

and the generator polynomial of the dual code is therefore

$$g(x) = x^6 + x^3 + 1.$$

The blocklength of the dual code is 9 and given that the degree of $g(x)$ is 6 then the information length is 3. Therefore the code that is dual to the (9, 6) cyclic code is the (9, 3) code with $g(x) = x^6 + x^3 + 1$. □

Table 3.6
The (7, 3) codes generated $h^*(x)$ and $h(x)$

(a) $g(x) = h^*(x) = x^4 + x^3 + x^2 + 1$	(b) $g(x) = h(x) = x^4 + x^2 + x + 1$
(0 0 0 0 0 0 0)	(0 0 0 0 0 0 0)
(0 0 1 1 1 0 1)	(0 0 1 0 1 1 1)
(0 1 0 0 1 1 1)	(0 1 0 1 1 1 0)
(0 1 1 1 0 1 0)	(0 1 1 1 0 0 1)
(1 0 0 1 1 1 0)	(1 0 0 1 0 1 1)
(1 0 1 0 0 1 1)	(1 0 1 1 1 0 0)
(1 1 0 1 0 0 1)	(1 1 0 0 1 0 1)
(1 1 1 0 1 0 0)	(1 1 1 0 0 1 0)

3.9 Generator and parity-check matrices of cyclic codes

The generator matrix and parity-check matrix of a cyclic code can be obtained directly from the code's generator polynomial and parity-check polynomial respectively. The generator polynomial of an (n, k) cyclic code is given by eqn 3.8, namely

$$g(x) = g_{n-k}x^{n-k} + g_{n-k-1}x^{n-k-1} + \cdots + g_2x^2 + g_1x + g_0.$$

The corresponding generator matrix $\boldsymbol{G}$ is a k by n matrix whose first row contains the $n-k+1$ coefficients of $g(x)$ followed by $k-1$ zeros. The second row of $\boldsymbol{G}$ is obtained by shifting the first row one place to the right and placing a 0 in the position vacated on the left-hand side. The third and subsequent rows are likewise constructed by shifting the previous row by 1 place to the right and placing a 0 in the vacated position. This is continued until all the k rows are filled, at which point the coefficient $g_0 = 1$ will be at the far right-hand side of the bottom row. The resulting matrix is

$$\boldsymbol{G} = \begin{bmatrix} g_{n-k} & g_{n-k-1} & \cdots & g_2 & g_1 & g_0 & 0\ldots & & 0 \\ 0 & g_{n-k} & g_{n-k-1} & \cdots & g_2 & g_1 & g_0 & 0\ldots & 0 \\ \cdot & & & \cdots & & & & & \cdot \\ \cdot & & & & \cdots & & & & \cdot \\ \cdot & & & & & & \cdots & & \cdot \\ 0\ldots & 0 & g_{n-k} & g_{n-k-1} & \cdots & g_2 & g_1 & g_0 & 0 \\ 0 & \cdots & 0 & g_{n-k} & g_{n-k-1} & \cdots & g_2 & g_1 & g_0 \end{bmatrix}. \tag{3.31}$$

The (7, 4) Hamming code has generator polynomial $g(x) = x^3 + x + 1$, and so $g_3 = 1$, $g_2 = 0$, $g_1 = 1$ and $g_0 = 1$. Hence its generator matrix is given by

$$\boldsymbol{G} = \begin{bmatrix} g_3 & g_2 & g_1 & g_0 & 0 & 0 & 0 \\ 0 & g_3 & g_2 & g_1 & g_0 & 0 & 0 \\ 0 & 0 & g_3 & g_2 & g_1 & g_0 & 0 \\ 0 & 0 & 0 & g_3 & g_2 & g_1 & g_0 \end{bmatrix}$$

and substituting the values of g_3, g_2, g_1, and g_0 gives

$$\boldsymbol{G} = \begin{bmatrix} 1 & 0 & 1 & 1 & 0 & 0 & 0 \\ 0 & 1 & 0 & 1 & 1 & 0 & 0 \\ 0 & 0 & 1 & 0 & 1 & 1 & 0 \\ 0 & 0 & 0 & 1 & 0 & 1 & 1 \end{bmatrix}.$$

The resulting generator matrix is in a nonsystematic form, but can be easily changed to a systematic form using elementary row operations (see Section 5.5). Adding rows

3 and 4 to row 1 gives

$$G = \left[\begin{array}{cccc|ccc} 1 & 0 & 0 & 0 & 1 & 0 & 1 \\ 0 & 1 & 0 & 1 & 1 & 0 & 0 \\ 0 & 0 & 1 & 0 & 1 & 1 & 0 \\ 0 & 0 & 0 & 1 & 0 & 1 & 1 \end{array}\right]$$

where the matrix is now shown to be augmented. Adding row 4 to row 2 gives

$$G = \left[\begin{array}{cccc|ccc} 1 & 0 & 0 & 0 & 1 & 0 & 1 \\ 0 & 1 & 0 & 0 & 1 & 1 & 1 \\ 0 & 0 & 1 & 0 & 1 & 1 & 0 \\ 0 & 0 & 0 & 1 & 0 & 1 & 1 \end{array}\right]$$

which is the systematic form of $\boldsymbol{G}$ given in Section 2.2.

We have already seen how to obtain the parity-check matrix $\boldsymbol{H}$ from $\boldsymbol{G}$, let's now consider how $\boldsymbol{H}$ is obtained from the parity-check polynomial $h(x)$. The parity-check polynomial of an (n, k) cyclic code is given by eqn 3.24, namely

$$h(x) = h_k x^k + h_{k-1} x^{k-1} + \cdots + h_2 x^2 + h_1 x + h_0.$$

The parity-check matrix $\boldsymbol{H}$ is an $n-k$ by n matrix and is constructed in the same way as the generator matrix, except that the polynomial coefficients are in the reverse order. Therefore the first row of $\boldsymbol{H}$ contains the $k+1$ coefficients of $h(x)$, starting with h_0 and ending with h_k, followed by $n-k-1$ zeros. The second and subsequent rows are constructed by shifting the previous row by 1 place to the right and placing a 0 in the vacated position. This is continued until there are $n-k$ rows, at which point the coefficient $h_k = 1$ is at the far right-hand side of the bottom row, and the resulting matrix is

$$H = \begin{bmatrix} h_0 & h_1 & h_2 & \ldots & h_{k-1} & h_k & 0 \ldots & & 0 \\ 0 & h_0 & h_1 & h_2 & \ldots & h_{k-1} & h_k & 0 \ldots & 0 \\ \cdot & & & & & \ldots & & & \cdot \\ \cdot & & & & & & \ldots & & \cdot \\ \cdot & & & & & & & \ldots & \cdot \\ 0 \ldots & 0 & h_0 & h_1 & h_2 & \ldots & h_{k-1} & h_k & 0 \\ 0 & \ldots & 0 & h_0 & h_1 & h_2 & \ldots & h_{k-1} & h_k \end{bmatrix}. \qquad (3.32)$$

The parity-check matrix for the (7, 4) code has $n-k=3$ rows and $n=7$ columns. We have already seen that the code's parity-check polynomial is $h(x) = x^4 + x^2 + x + 1$, and so $h_0 = 1$, $h_1 = 1$, $h_2 = 1$, $h_3 = 0$, and $h_4 = 1$. Therefore the parity-check matrix is given by

$$H = \begin{bmatrix} h_0 & h_1 & h_2 & h_3 & h_4 & 0 & 0 \\ 0 & h_0 & h_1 & h_2 & h_3 & h_4 & 0 \\ 0 & 0 & h_0 & h_1 & h_2 & h_3 & h_4 \end{bmatrix}$$

and substituting the coefficients gives

$$H = \begin{bmatrix} 1 & 1 & 1 & 0 & 1 & 0 & 0 \\ 0 & 1 & 1 & 1 & 0 & 1 & 0 \\ 0 & 0 & 1 & 1 & 1 & 0 & 1 \end{bmatrix}.$$

Recall that in its systematic form, $\boldsymbol{H}$ is an augmented matrix with the identity matrix on the right-hand side. To change $\boldsymbol{H}$ to its systematic form simply requires adding row 1 to row 3 to give

$$H = \left[\begin{array}{cccc|ccc} 1 & 1 & 1 & 0 & 1 & 0 & 0 \\ 0 & 1 & 1 & 1 & 0 & 1 & 0 \\ 1 & 1 & 0 & 1 & 0 & 0 & 1 \end{array}\right]$$

which is the parity-check matrix for the (7, 4) code as given previously.

Problems

3.1 Show that the following linear codes are not cyclic
(a) the (6, 3) code given in Table 2.2
(b) the (5, 2) code with generator matrix

$$G = \begin{bmatrix} 1 & 1 & 0 & 1 & 0 \\ 0 & 1 & 1 & 0 & 1 \end{bmatrix}.$$

3.2 Given the two polynomials $p_1(x) = x^4 + x + 1$ and $p_2(x) = x^7 + x^3 + 1$ find the quotient $q(x)$ and remainder $r(x)$ of $p_2(x) \div p_1(x)$. Check your answer using the Euclidean division algorithm.

3.3 Show that
(a) $x^4 + x^2 + x + 1 = 0$ modulo-$(x + 1)$
(b) $R_{(x^2+x+1)}\,[x^5 + x^3 + x^2 + x + 1] = x$.

3.4 Given the (7, 4) cyclic code with $g(x) = x^3 + x + 1$ determine systematic and nonsystematic codeword polynomials for the information polynomial $i(x) = x^3 + x^2 + x + 1$. Show that the systematic codeword can be obtained using the quotient of $x^3 i(x) \div g(x)$ instead of the remainder.

3.5 The codeword polynomials $x^6 + x^3 + x^2 + x$ and $x^5 + x^3 + x^2$, belonging to the (7, 4) code with generator polynomial $g(x) = x^3 + x + 1$, incur the error pattern $e(x) = x^3$. Show that the resulting syndrome polynomials are the same.

3.6 A codeword polynomial $c(x)$, belonging to the (7, 4) code with $g(x) = x^3 + x + 1$, incurs errors so giving the received polynomial $v(x)$. Find $c(x)$ when
(a) $v(x) = x^5 + x^2 + 1$
(b) $v(x) = x^6 + x^3 + x$.

3.7 Given that

$$x^{15} + 1 = (x + 1)(x^2 + x + 1)(x^4 + x + 1)(x^4 + x^3 + 1)(x^4 + x^3 + x^2 + x + 1)$$

determine

(a) the number of cyclic codes with blocklength 15

(b) the number of (15,11) cyclic codes

(c) the generator polynomials for the (15, 7) cyclic codes.

3.8 The polynomial $x^7 + 1$ can be factorised as

$$x^7 + 1 = (x + 1)(x^6 + x^5 + x^4 + x^3 + x^2 + x + 1).$$

Construct the generator matrices for the codes generated by $(x + 1)$ and $(x^6 + x^5 + x^4 + x^3 + x^2 + x + 1)$, and show that the polynomials generate the (7, 6) single-parity-check code and the (7, 1) repetition code respectively.

3.9 Determine the parity-check polynomial of the (15, 5) cyclic code with generator polynomial

$$g(x) = x^{10} + x^8 + x^5 + x^4 + x^2 + x + 1.$$

Hence find the generator polynomial, the blocklength and the information length of the cyclic code that is dual to the (15, 5) code.

Linear-feedback shift registers for encoding and decoding cyclic codes

4

Cyclic codes and linear-feedback shift registers lend themselves to similar mathematical treatment and as we shall see the encoding and decoding of cyclic codes can be realized in terms of linear-feedback shift registers. The task is simplified by the fact that part of the decoding circuit is the same as that of the encoding circuit, this being the polynomial-division register. As we shall see, once the polynomial-division register is established, it is relatively simple to construct circuits for encoding and for decoding.

4.1 Linear-feedback shift registers

Figure 4.1 shows an *r-stage shift register* capable of storing r bits in its stages $b_0, b_1, \ldots, b_{r-1}$. Each stage is a 1-bit memory element which, when *shifted*, forwards its contents to the next stage and is then updated by the contents of the previous stage. Each time the register is shifted a bit b_{in} enters the register and all the bits in the register move along by one stage, with the bit in the last stage b_{r-1} leaving the register and forming the output b_{out}. At each shift the contents of the ith stage is fed to the $(i+1)$th stage and then the ith stage is updated with the contents of the $(i-1)$th stage. This is expressed as

$$\begin{aligned} b_{i+1} &= b_i \\ b_i &= b_{i-1} \end{aligned} \tag{4.1}$$

where b_i is used to represent the ith stage or the contents of the ith stage (this should not cause confusion).

The order in which the two expressions in eqns 4.1 are carried out is significant. If the order is reversed, so that the stage b_i is first updated with the contents of b_{i-1} and then b_{i+1} is updated, the original contents of b_i will be lost and b_{i+1} will end up containing the contents of b_{i-1}. For an r-stage shift register eqns 4.1 can be extended to the whole register to describe the changes in the stages at each shift, this gives

$$\begin{aligned} b_{\text{out}} &= b_{r-1} \\ b_{r-1} &= b_{r-2} \\ \vdots\ \ & \quad \vdots \\ b_2 &= b_1 \\ b_1 &= b_0 \\ b_0 &= b_{\text{in}} \end{aligned} \tag{4.2}$$

where b_{in} and b_{out} are the bits entering and leaving the shift register respectively. Equations 4.2 describe the operation of the register and give the *state* of the register

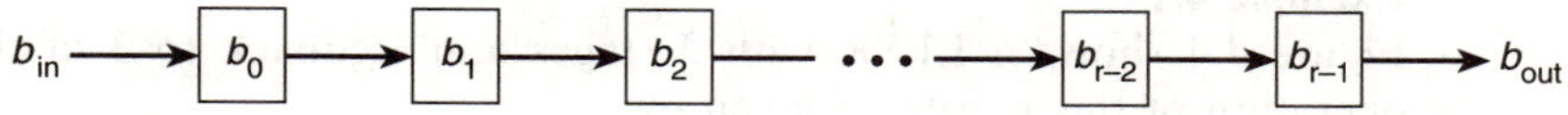

Fig. 4.1 An r-stage shift register.

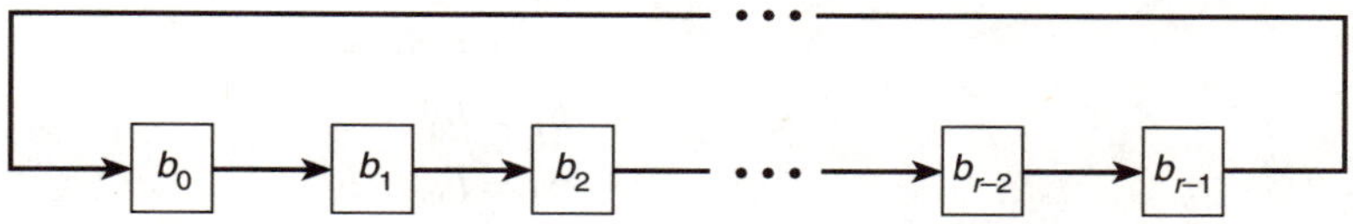

Fig. 4.2 An r-stage linear-feedback shift register.

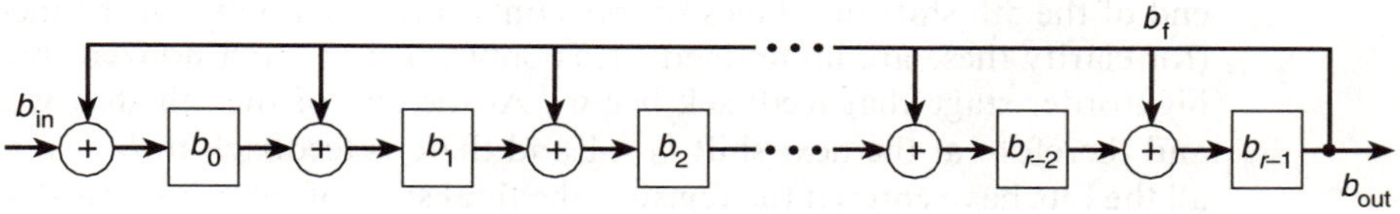

Fig. 4.3 An r-stage LFSR with feedback to all stages.

at each shift. The stages b_0 and b_{r-1} are referred to as the *low-order* and *high-order* sides of the register respectively.

In a *linear-feedback shift register* (*LFSR*) the output b_{out} is fed back into the register. The simplest example of such a register is shown in Fig. 4.2 which consists of r stages. Here the output bit is fed back into the first stage b_0 and so the stage contents are cyclically moved around the register as the register is shifted. If the stages are initialized with a codeword $\boldsymbol{c}$ from a cyclic code of blocklength r, then $r - 1$ shifts will produce the $r - 1$ codewords that are cyclic shifts of the codeword $\boldsymbol{c}$. At the rth shift the stages will again contain $\boldsymbol{c}$.

The feedback within a register can be used to alter any of the stages directly and not just b_0. Figure 4.3 shows a linear-feedback shift register with r stages and feedback, via modulo-2 adders, to all stages. Each modulo-2 adder has two, or more, inputs and one output that is the sum modulo-2 of the inputs. In practice some stages will have feedback and others will not, depending upon requirements. Denoting the feedback bit by b_f then stages with a feedback link change according to

$$b_i = b_{i-1} + b_f \tag{4.3}$$

where addition is modulo-2, and

$$b_i = b_{i-1} \tag{4.4}$$

describes changes to stages without a feedback link.

Example 4.1
Figure 4.4 shows a LFSR with 5 stages and feedback to 3 of the stages. The operation of this register is given by

$$
\begin{aligned}
b_{\text{out}} &= b_{\text{f}} = b_4 \\
b_4 &= b_3 \\
b_3 &= b_2 + b_{\text{f}} \\
b_2 &= b_1 + b_{\text{f}} \\
b_1 &= b_0 \\
b_0 &= b_{\text{in}} + b_{\text{f}}.
\end{aligned}
$$

Table 4.1 gives the operation of the register when the input is the 10-bit word (1 0 1 0 1 0 0 0 1 1) and where the left-hand bit enters the register first. The stages are initialized to zero before the first bit enters the register. For the first 5 shifts the feedback is zero as the input has yet to reach the high-order stage b_4, and so at the end of the 5th shift the stages simply contain the first 5 bits of the incoming word (for clarity these are underlined). It is only when the first nonzero bit reaches the high-order stage that feedback begins. At the end of the 5th shift we have $b_4=1$ and therefore at the next shift $b_{\text{f}}=1$ and there is feedback to b_3, b_2, and b_0. Once all the bits have entered the register, the final state of the register is (0 0 1 0 0) (also shown underlined). □

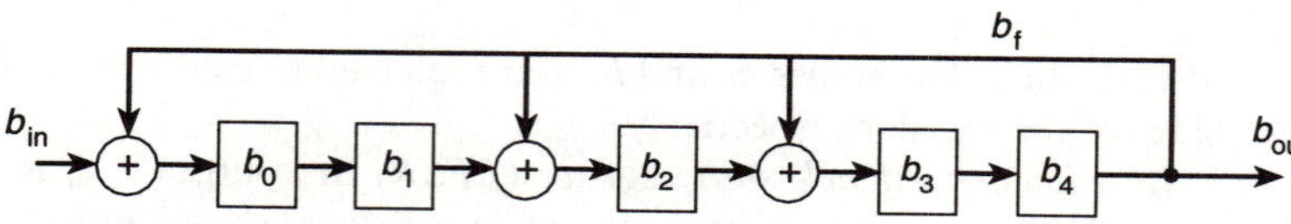

Fig. 4.4 A 5-stage LFSR.

Table 4.1
Step-by-step operation of the LFSR in Fig. 4.4

Input $v=(1\ 0\ 1\ 0\ 1\ 0\ 0\ 0\ 1\ 1)$

Shift	b_{in}	b_0	b_1	b_2	b_3	b_4	b_{f}	
		0	0	0	0	0	0	←Initial state
1	1	1	0	0	0	0	0	
2	0	0	1	0	0	0	0	
3	1	1	0	1	0	0	0	
4	0	0	1	0	1	0	0	
5	1	1	0	1	0	1	0	
6	0	1	1	1	0	0	1	
7	0	0	1	1	1	0	0	
8	0	0	0	1	1	1	0	
9	1	0	0	1	0	1	1	
10	1	0	0	1	0	0	1	←Final state

4.2 The polynomial-division register

The encoding and decoding of cyclic codes involves the division of a polynomial $p(x)$ by a generator polynomial $g(x)$ giving a remainder $r(x)$. When encoding, $p(x)$ is an information polynomial $i(x)$ and $r(x)$ gives the parity-check bits required to construct codewords. At the decoding stage $p(x)$ is the received polynomial $v(x)$ and $r(x)$ gives the syndrome polynomial. Encoding and decoding use the same *polynomial-division register* and so we consider this before addressing encoding and decoding specifically.

Figure 4.5 shows a linear-feedback shift register for dividing an arbitrary polynomial

$$p(x) = p_{n-1}x^{n-1} + p_{n-2}x^{n-2} + \cdots + p_2x^2 + p_1x + p_0$$

by a fixed polynomial

$$g(x) = g_rx^r + g_{r-1}x^{r-1} + \cdots + g_2x^2 + g_1x + g_0$$

where $p(x)$ and $g(x)$ are referred to as the dividend and divisor respectively. The switches in Fig. 4.5 represent the polynomial coefficients g_i and are open or closed depending on the coefficients. If $g_i = 1$ then the corresponding switch is closed so giving a feedback link to the ith stage, and changes in the ith stage are given by $b_i = b_{i-1} + b_f$ (eqn 4.3). If $g_i = 0$ then there is no feedback to the ith stage, the adder feeding into b_i can then be omitted, as it has only 1 input, and so $b_i = b_{i-1}$ (eqn 4.4). The feedback link g_r is always present in the register, as a polynomial $g(x)$ of degree r has $g_r = 1$ by definition.

The polynomial coefficients p_i are shifted into the register starting with the high-order coefficient p_{n-1}, followed by p_{n-2}, p_{n-3}, and so forth. For the first r shifts the feedback and output are zero as the dividend has yet to reach the high-order stage b_{r-1}. Once the first nonzero bit of the dividend reaches b_{r-1} feedback commences and it is at this point that division starts. The high-order stage b_{r-1} represents an x^{r-1} term within $p(x)$, and a 1 leaving this stage means that an x^r term needs to be subtracted from $p(x)$. This is achieved by the feedback circuit whose links to the stages represent the divisor $g(x)$. This is analogous to the way in which the divisor is subtracted from the dividend when carrying out long division. If the output of b_{r-1} is zero then no subtraction takes place. On shifting the last bit p_0 into the register the remainder of $p(x)$ divided by $g(x)$ is left in the stages. Note that because the stages $b_0, b_1, \ldots, b_{r-2}, b_{r-1}$

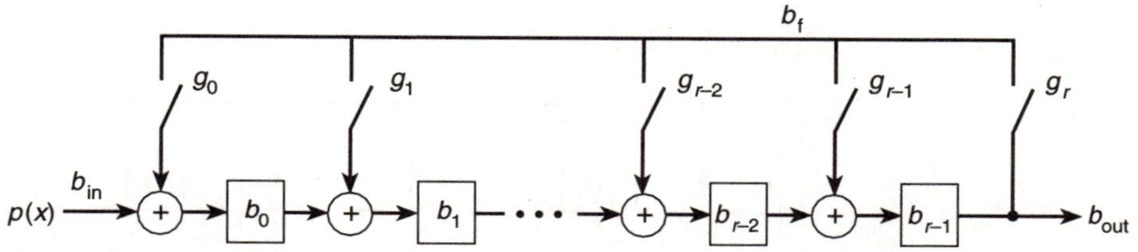

Fig. 4.5 A polynomial-division register.

contain the remainder of the input divided by $g(x)$ we refer to them as the *remainder stages*. If we let

$$r(x) = b_{r-1}x^{r-1} + b_{r-2}x^{r-2} + \cdots + b_2x^2 + b_1x + b_0$$

be a polynomial representing the stages then, using the notation introduced in Section 3.2, we have

$$r(x) = R_{g(x)}[p(x)].$$

Furthermore, the output from the register gives the quotient $q(x)$, where the first nonzero output determines the degree of $q(x)$. Let's consider dividing

$$p(x) = x^7 + x^6 + x^2 + x + 1$$

by

$$g(x) = x^4 + x^2 + 1.$$

The coefficients of the divisor $g(x)$ are $g_4 = 1$, $g_3 = 0$, $g_2 = 1$, $g_1 = 0$, and $g_0 = 1$. The linear-feedback shift register required is shown in Fig. 4.6. The dashed lines show the absence of feedback links due to $g_1 = 0$ and $g_3 = 0$. With each shift the stages change according to

$$\begin{aligned} b_{\text{out}} &= b_{\text{f}} = b_3 \\ b_3 &= b_2 \\ b_2 &= b_1 + b_{\text{f}} \\ b_1 &= b_0 \\ b_0 &= b_{\text{in}} + b_{\text{f}}. \end{aligned}$$

The coefficients of the dividend are $p_7 = 1$, $p_6 = 1$, $p_5 = 0$, $p_4 = 0$, $p_3 = 0$, $p_2 = 1$, $p_1 = 1$, and $p_0 = 1$, and these enter the register high-order bits first, p_7 followed by p_6 and so forth. Table 4.2 shows the division step-by-step for all the 8 shifts required to feed $p(x)$ into the register, the polynomial representation of the remainder stages, $r(x)$, is also shown in Table 4.2. The input is underlined for the first 4 shifts so that we can clearly see its progress into the register. It takes the first 4 shifts to move p_7 to the high-order stage b_3, during this time there is no feedback or output. On the 5th shift p_7 leaves b_3 therefore giving a nonzero feedback and output, so that $b_{\text{f}} = 1$ and $b_{\text{out}} = 1$, division has now started. By the end of the 8th shift all the polynomial coefficients have entered the register and the stages contain the remainder

$$r(x) = b_3x^3 + b_2x^2 + b_1x + b_0.$$

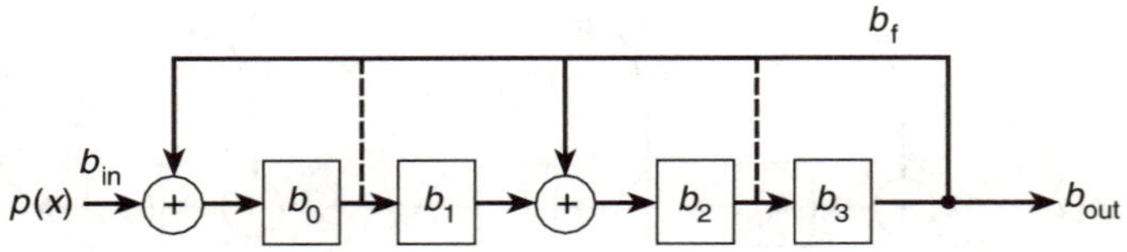

Fig. 4.6 An LFSR for dividing by $x^4 + x^2 + 1$.

Table 4.2
Operation of the LFSR in Fig. 4.6

Dividing $x^7+x^6+x^2+x+1$ by x^4+x^2+1

Shift	b_{in}	b_0	b_1	b_2	b_3	b_{f}	$r(x)$
		0	0	0	0	0	0
1	1	$\underline{\underline{1}}$	0	0	0	0	1
2	1	$\underline{\underline{1}}$	$\underline{\underline{1}}$	0	0	0	$x+1$
3	0	$\underline{\underline{0}}$	$\underline{\underline{1}}$	$\underline{\underline{1}}$	0	0	x^2+x
4	0	$\underline{\underline{0}}$	$\underline{\underline{0}}$	$\underline{\underline{1}}$	$\underline{\underline{1}}$	0	x^3+x^2
5	0	1	0	1	1	1	x^3+x^2+1
6	1	0	1	1	1	1	x^3+x^2+x
7	1	0	0	0	1	1	x^3
8	1	$\underline{\underline{0}}$	$\underline{\underline{0}}$	$\underline{\underline{1}}$	$\underline{\underline{0}}$	1	x^2

As we can see from Table 4.2 (last row, underlined) only b_2 is nonzero and therefore $r(x)=x^2$ is the remainder of $p(x)=x^7+x^6+x^2+x+1$ divided by $g(x)=x^4+x^2+1$. Furthermore the register output is (0 0 0 0 1 1 1 1) which gives the quotient $q(x)=x^3+x^2+x+1$. We can check the answer by constructing $p(x)$ from $r(x)$, $q(x)$, and $g(x)$. Using the Euclidean division algorithm

$$p(x) = q(x)g(x) + r(x)$$

and substituting $q(x)$, $g(x)$, and $r(x)$ gives

$$\begin{aligned} p(x) &= (x^3+x^2+x+1)(x^4+x^2+1)+x^2 \\ &= x^7+x^6+x^2+x+1 \end{aligned}$$

as required. We could have checked the answer by dividing $p(x)$ by $g(x)$ using the long division method, however the Euclidean division algorithm involves multiplication, and not division, and is therefore a bit easier.

Example 4.2
By considering the step-by-step operation of the register shown in Fig. 4.7 find the remainder and quotient when the input to the register is $p(x)=x^6+x^5+x^4+x+1$.

From Fig. 4.7 we can see that the register has 3 stages and therefore the divisor has degree 3 and is of the form $g(x)=x^3+g_2x^2+g_1x+g_0$. Stages b_0 and b_1 have feedback links whereas b_2 does not and so $g_0=g_1=1$, $g_2=0$ which gives $g(x)=x^3+x+1$ as the divisor. Also from Fig. 4.7 we get

$$\begin{aligned} b_{\text{out}} &= b_{\text{f}} = b_2 \\ b_2 &= b_1 \\ b_1 &= b_0 + b_{\text{f}} \\ b_0 &= b_{\text{in}} + b_{\text{f}}. \end{aligned}$$

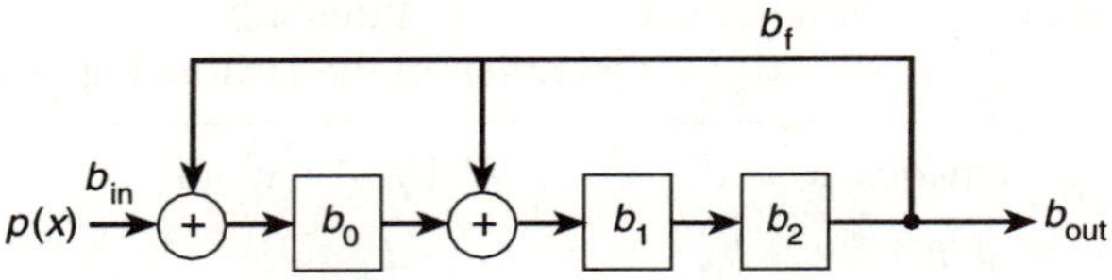

Fig. 4.7 An LFSR for dividing by $g(x)=x^3+x+1$.

Table 4.3
Operation of the LFSR in Fig. 4.7

Dividing $x^6+x^5+x^4+x+1$ by x^3+x+1

Shift	b_{in}	b_0	b_1	b_2	b_f	$r(x)$
	0	0	0	0	0	
1	1	1	0	0	0	1
2	1	1	1	0	0	$x+1$
3	1	1	1	1	0	x^2+x+1
4	0	1	0	1	1	x^2+1
5	0	1	0	0	1	1
6	1	1	1	0	0	$x+1$
7	1	1	1	1	0	x^2+x+1

Table 4.3 shows the step-by-step operation as $p(x)$ enters the register. After the 7th shift $p(x)$ has completely entered the register, and the remainder stages are (1 1 1) giving the remainder $r(x)=x^2+x+1$. The output, as given by the feedback bit, gives the quotient $q(x)=x^3+x^2$. The reader may wish to check the answers, using either the long division method or the Euclidean division algorithm. □

In the polynomial-division register the remainder at each shift can be determined from the previous remainder. If $r_i(x)$ is the remainder after the ith shift, then at the next shift the contents of the stages move 1 place towards the high-order side of the register, which is equivalent to multiplying $r_i(x)$ by x, and the feedback circuit divides by $g(x)$. The remainder after the $(i+1)$th shift is therefore

$$r_{i+1}(x) = R_{g(x)}[xr_i(x)] \tag{4.5}$$

where it is assumed that the input to the register is 0. If the input is 1, then 1 is added to the remainder. The degree of $r_i(x)$ is always less than or equal to $r-1$ and therefore the degree of $xr(x)$ cannot exceed the degree r of the divisor $g(x)$. Whenever $xr(x)$ results in an x^r term, so that it has the same degree as $g(x)$, the feedback circuit has the effect of adding (or subtracting, remember that we are using modulo-2 addition) $g(x)$ to $xr(x)$ and therefore removing the x^r term, this corresponds to division by $g(x)$. Equation 4.5 can be evaluated by multiplying $r_i(x)$ by x, and then adding $g(x)$ if $xr_i(x)$ contains an x^r term. If the register input is 1, then this is added to the remainder.

The polynomial-division register shown in Fig. 4.5 requires r shifts before the dividend reaches the high-order side of the register, which represents idle or wasted time as no division takes place during these shifts. This can be avoided by feeding the input directly into the high-order side of the register. By doing so feedback, and therefore division, commences immediately with the input of the first nonzero bit. However, this does not give the remainder of the input $p(x)$ divided by $g(x)$ but rather of $p'(x)=x^r p(x)$ divided by $g(x)$. This can be useful when carrying out encoding, as it corresponds to multiplication by x^{n-k} required when generating systematic codewords. At the decoding stage there is no requirement for multiplication by x^r and a register with input to the high-order side will give the wrong error-syndrome. Despite this, however, it is still possible to feed the input into the high-order side when decoding (this is discussed further in Section 4.5). Registers with inputs to the high-order side and low-order side are referred to as having *high-order input* and *low-order input* respectively.

A linear-feedback shift register with high-order input has the same feedback circuit as that with the input to the low-order side. It is only the input to b_0 and the feedback bit b_f that are affected. The feedback bit now depends on the output from the high-order stage b_{r-1} and the input bit b_{in}, while the input to b_0 is taken solely from the feedback. Hence $b_f=b_{r-1}+b_{in}$ and $b_0=b_f$ compared to $b_f=b_{r-1}$ and $b_0=b_f+b_{in}$ when the input is to the low-order side. The inputs to all the other stages remain the same and bits are still shifted from the low-order side of the register to the high-order side. Let's reconsider the problem considered in Example 4.2, namely dividing $p(x)=x^6+x^5+x^4+x+1$ by $g(x)=x^3+x+1$ but this time using a linear-feedback shift register with a high-order input. The linear-feedback shift register required to achieve this is shown in Fig. 4.8. The operation of the register is now described by

$$\begin{aligned} b_{out} &= b_f = b_2 + b_{in} \\ b_2 &= b_1 \\ b_1 &= b_0 + b_f \\ b_0 &= b_f \end{aligned}$$

and using these or eqn 4.5 gives the results shown in Table 4.4. Note that if eqn. 4.5 is used to determine the remainder at each shift, then each nonzero input contributes an x^3 term to the remainder. During the first shift $p_6=1$ enters the register giving a nonzero feedback $b_f=1$ and therefore division starts immediately. Note that because division starts with the first shift we cannot identify the information bits entry into the stages, so only the final remainder is underlined in Table 4.4. By the

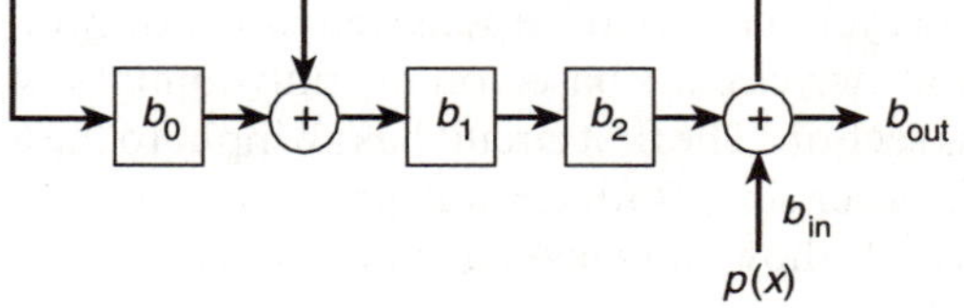

Fig. 4.8 An LFSR, with high-order input, for dividing by $g(x)=x^3+x+1$.

Table 4.4
Operation of the LFSR in Fig. 4.8

Dividing $x^6+x^5+x^4+x+1$ *by* $g(x)=x^3+x+1$

Shift	b_{in}	b_0	b_1	b_2	b_f	$r(x)$
	0	0	0	0	0	
1	1	1	1	0	1	$x+1$
2	1	1	0	1	1	x^2+1
3	1	0	1	0	0	x
4	0	0	0	1	0	x^2
5	0	1	1	0	1	$x+1$
6	1	1	0	1	1	x^2+1
7	1	0	1	0	0	x

end of the 7th shift the register contents are (0 1 0) giving the remainder $r(x)=x$ and the output gives the quotient $q(x)=x^6+x^5+x^2+x$. The remainder and quotient obtained are different from those obtained in Example 4.2 because $r(x)$ and $q(x)$ obtained here are not the remainder and quotient of $p(x)=x^6+x^5+x^4+x+1$ divided by $g(x)=x^3+x+1$, but rather of the polynomial $p'(x)=x^r p(x)=x^3 p(x)$ divided by $g(x)$. We can illustrate this by using the Euclidean division algorithm to construct $p'(x)$ from $r(x)$, $q(x)$, and $g(x)$:

$$\begin{aligned} p'(x) &= q(x)g(x)+r(x) \\ &= (x^6+x^5+x^2+x)(x^3+x+1)+x \\ &= x^9+x^8+x^7+x^4+x^3. \end{aligned}$$

Furthermore taking out a factor of x^3 gives

$$p'(x) = x^3(x^6+x^5+x^4+x+1)$$

and so

$$p'(x) = x^3 p(x)$$

as required. Therefore the results obtained with the register shown in Fig. 4.8 are the remainder and quotient of $x^3 p(x)$ divided by $g(x)$, and not $p(x)$ divided by $g(x)$.

4.3 Registers for encoding

Encoding cyclic codes can be achieved in a variety of ways using shift registers. Here we look at two circuits based on the polynomial-division register considered in the previous section. The first circuit has an input to the low-order side and the second to the high-order side. Both circuits produce systematic codewords.

Figure 4.9 shows an encoder for an (n, k) cyclic code with generator polynomial

$$g(x) = g_{n-k}x^{n-k} + g_{n-k-1}x^{n-k-1} + \cdots + g_2x^2 + g_1x + g_0$$

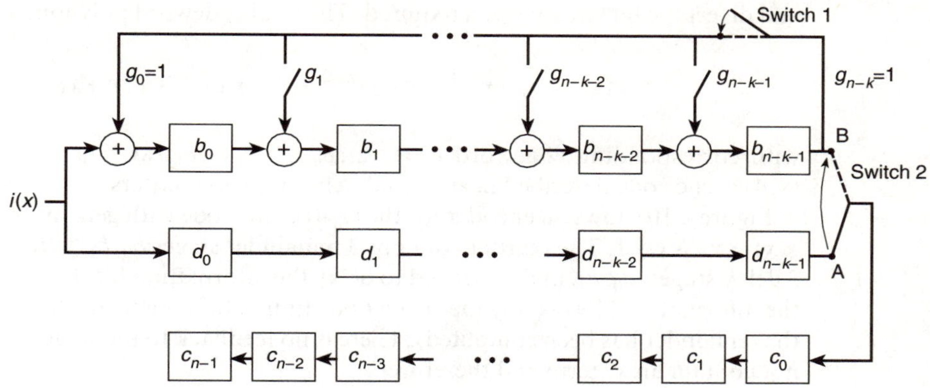

Fig. 4.9 An encoder for an (n, k) cyclic code.

where $g_{n-k} = 1$ and $g_0 = 1$. The encoder consists of a polynomial-division register, as discussed in Section 4.2, along with an additional $n - k$ *delay stages* $d_0, d_1, \ldots, d_{n-k-2}, d_{n-k-1}$ and n *codeword stages* $c_0, c_1, \ldots, c_{n-2}, c_{n-1}$. The codeword stages are for storing the final codeword and the delay stages are for delaying the information bits' entry into the codeword stages (this ensures that the remainder and the information bits enter the codeword stages in correct synchronization).

Given an information word $\boldsymbol{i} = (i_{k-1}, i_{k-2}, \ldots, i_2, i_1, i_0)$ to be encoded the corresponding information polynomial is

$$i(x) = i_{k-1}x^{k-1} + i_{k-2}x^{k-2} + \cdots + i_2x^2 + i_1x + i_0$$

where i_{k-1} is the first bit to enter the register followed by i_{k-2}, i_{k-3} and so forth. After k shifts the information bits have entered the register and the remainder stages contain

$$r(x) = R_{g(x)}[i(x)] \tag{4.6}$$

the remainder of $i(x)$ divided by $g(x)$. However, recall that for a systematic codeword we require the remainder of $x^{n-k}i(x)$ divided by $g(x)$. This is achieved by shifting the register an additional $n - k$ times, for this has the effect of moving $i(x)$ by $n - k$ places towards higher orders (which is equivalent to multiplying $i(x)$ by x^{n-k}) and dividing by $g(x)$. Therefore after $k + (n - k) = n$ shifts the remainder stages contain

$$r(x) = R_{g(x)}[x^{n-k}i(x)] \tag{4.7}$$

as required. A further $n - k$ shifts are then required to shift the remainder into the codeword stages, and so a total of $k + (n - k) + (n - k) = 2n - k$ are required to generate the codeword. Two switches are also needed to disable the feedback circuit and divert the bit stream when required. The final codeword polynomial is given by

$$c(x) = c_{n-1}x^{n-1} + c_{n-2}x^{n-2} + \cdots + c_2x^2 + c_1x + c_0$$

with corresponding codeword $\mathbf{c} = (c_{n-1}, c_{n-2}, \ldots, c_2, c_1, c_0)$. The encoder for the (7, 4) cyclic code described next should help to clarify matters.

Figure 4.10 shows an encoder for the (7, 4) cyclic code with generator polynomial $g(x) = x^3 + x + 1$. The register contains 3 remainder stages b_0, b_1, and b_2 along with 3 delay stages, d_0, d_1, and d_2 needed to delay the information bits by 3 shifts (so that the information bits occupy the correct position in the codeword stages by the time the remainder has been computed). There is no feedback to the stage b_2 as $g(x)$ does not contain an x^2 term and therefore

$$\begin{aligned} b_{\text{out}} &= b_{\text{f}} = b_2 \\ b_2 &= b_1 \\ b_1 &= b_0 + b_{\text{f}} \\ b_0 &= b_{\text{in}} + b_{\text{f}}. \end{aligned} \tag{4.8}$$

Seven codeword stages are needed to store the computed codeword polynomial which will be given by

$$c(x) = c_6x^6 + c_5x^5 + c_4x^4 + c_3x^3 + c_2x^2 + c_1x + c_0$$

with $\mathbf{c} = (c_6, c_5, c_4, c_3, c_2, c_1, c_0)$ as the corresponding codeword. Table 4.5 shows the step-by-step operation when the information word $\mathbf{i} = (1\ 0\ 1\ 0)$, with corresponding polynomial $i(x) = x^3 + x$, is encoded. Recall that for the (7, 4) code we need the remainder of $x^3 i(x)$ to construct systematic codewords. The operation of the register is best considered in three stages. Note that in Table 4.5 the information bits are underlined and the remainder (obtained after the 7th shift) is furthermore embolded.

1. Shifts 1 to 4: During the first 4 shifts the information bits enter the register with the high-order bits entering first, so the first bit in is $i_3 = 1$ followed by $i_2 = 0$, $i_1 = 1$,

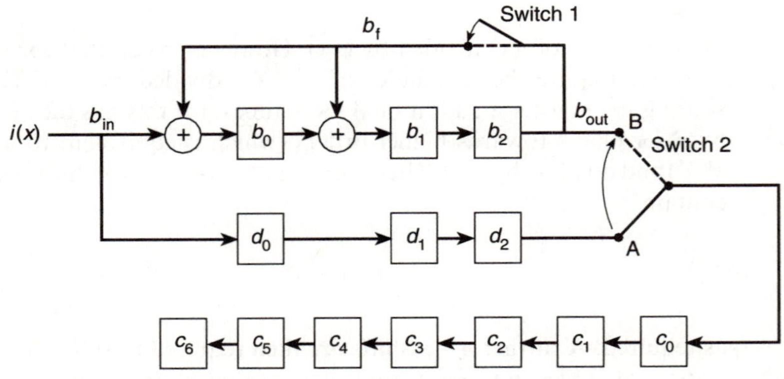

Fig. 4.10 An encoder for the (7, 4) cyclic code with $g(x) = x^3 + x + 1$.

Table 4.5
Operation of the encoder in Fig. 4.10

Input $i(x) = x^3 + x$

Shift	b_{in}	b_0	b_1	b_2	b_{f}	d_0	d_1	d_2	c_0	c_1	c_2	c_3	c_4	c_5	c_6	$r(x)$
	0	0	0	0	0	0	0	0	0	0	0	0	0	0	0	0
1	1	1	0	0	0	1	0	0	0	0	0	0	0	0	0	1
2	0	0	1	0	0	0	1	0	0	0	0	0	0	0	0	x
3	1	1	0	1	0	1	0	1	0	0	0	0	0	0	0	x^2+1
4	0	1	0	0	1	0	1	0	1	0	0	0	0	0	0	1
5	—	0	1	0	0	0	0	1	0	1	0	0	0	0	0	x
6	—	0	0	1	0	0	0	0	1	0	1	0	0	0	0	x^2
7	—	1	1	0	1	0	0	0	0	1	0	1	0	0	0	$x+1$
8	—	0	1	1	—	0	0	0	0	0	1	0	1	0	0	x^2+x
9	—	0	0	1	—	0	0	0	1	0	0	1	0	1	0	x
10	—	0	0	0	—	0	0	0	1	1	0	0	1	0	1	0

and $i_0 = 0$. The information bits are simultaneously fed into the delay stages. During this time switch 1 is closed to allow feedback, and switch 2 is in position A to allow the information bits that leave the delay stages to go into the codeword stages. By the end of the 4th shift, the remainder stages contain (1 0 0) and the remainder is therefore $r(x) = 1$. However, this is the remainder of $i(x)$ divided by $g(x)$ and not $x^3 i(x)$ divided by $g(x)$ as required to construct a systematic codeword.

2. Shifts 5, 6 and 7: Each additional shift has the effect of multiplying $i(x)$ by x and dividing by $g(x)$. So that

$$\text{shift 5 gives } R_{g(x)}[xi(x)] = x$$
$$\text{shift 6 gives } R_{g(x)}[x^2 i(x)] = x^2$$

and shift 7 gives the required remainder

$$R_{g(x)}[x^3 i(x)] = x + 1.$$

The information bits have now left the delay stages and occupy the first 4 low-order codeword stages. During these three shifts switches 1 and 2 remain closed and in position A respectively.

3. Shifts 8, 9 and 10: Finally the remainder has to be shifted into the codeword stages. Switch 1 now needs to be open to prevent further feedback, and switch 2 set to position B to divert the remainder into the codeword stages. By the end of the 10th shift the codeword stages contain (1 1 0 0 1 0 1) which gives the codeword polynomial

$$c(x) = x^6 + x^4 + x + 1$$

and codeword $\boldsymbol{c} = (1\ 0\ 1\ 0\ 0\ 1\ 1)$. Note that a total of 10 shifts are required to compute $\boldsymbol{c}$.

To summarize, an encoder for an (n, k) cyclic code as shown in Fig. 4.9 requires $2n-k$ shifts for encoding, where

(1) the initial k shifts are needed to enter $i(x)$ and determine $r(x) = R_{g(x)}[i(x)]$;
(2) the next $n-k$ to obtain $r(x) = R_{g(x)}[x^{n-k}i(x)]$;
(3) the last $n-k$ to shift $r(x)$ into the codeword stages.

Example 4.3
Here we consider an encoder for the $(7, 3)$ code with generator polynomial $g(x) = x^4 + x^3 + x^2 + 1$ and input $i(x) = x^2 + x$.

Figure 4.11 shows the shift register for generating systematic codewords. Four parity-check bits are needed and so there are 4 remainder stages and 4 delay stages. The only zero coefficient in $g(x)$ is g_1, therefore all the stages except b_1 have feedback added to their input. Hence

$$\begin{aligned} b_{\text{out}} &= b_{\text{f}} = b_3 \\ b_3 &= b_2 + b_{\text{f}} \\ b_2 &= b_1 + b_{\text{f}} \\ b_1 &= b_0 \\ b_0 &= b_{\text{in}} + b_{\text{f}}. \end{aligned}$$

Table 4.6 shows the step-by-step operation when $\boldsymbol{i} = (1\ 1\ 0)$ is encoded. Note here that as in Table 4.5, the information bits are shown underlined to show their progress through the stages, and the remainder (obtained after the 7th shift) is also embolded.

The input to the register is $i(x) = x^2 + x$. After the 3rd shift the information bits have entered the register and the remainder stages contain the remainder $R_{g(x)}[i(x)] = x^2 + x$. Four more shifts are then needed to give the required remainder $R_{g(x)}[x^4 i(x)] = x^3 + 1$.

At this point the stages c_0, c_1, and c_2 contain the information bits and 4 final shifts are needed to move these towards higher orders and to move the remainder into the

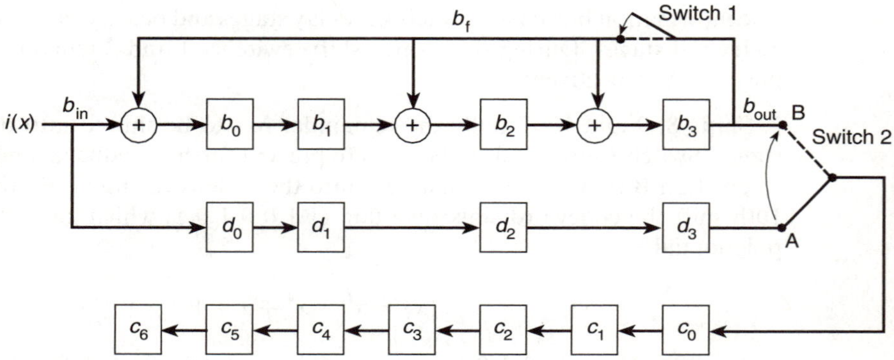

Fig. 4.11 An encoder for the $(7, 3)$ cyclic code with $g(x) = x^4 + x^3 + x^2 + 1$.

Table 4.6
Operation of the encoder in Fig. 4.11

Input $i(x)=x^2+x$

Shift	b_{in}	b_0	b_1	b_2	b_3	b_f	d_0	d_1	d_2	d_3	c_0	c_1	c_2	c_3	c_4	c_5	c_6	$r(x)$
	0	0	0	0	0	0	0	0	0	0	0	0	0	0	0	0	0	0
1	1	1	0	0	0	0	1	0	0	0	0	0	0	0	0	0	0	1
2	1	1	1	0	0	0	1	1	0	0	0	0	0	0	0	0	0	$x+1$
3	0	0	1	1	0	0	0	1	1	0	0	0	0	0	0	0	0	x^2+x
4	—	0	0	1	1	0	0	0	1	1	0	0	0	0	0	0	0	x^3+x^2
5	—	1	0	1	0	1	0	0	0	1	1	0	0	0	0	0	0	x^2+1
6	—	0	1	0	1	0	0	0	0	0	1	1	0	0	0	0	0	x^3+x
7	—	**1**	**0**	**0**	**1**	1	0	0	0	0	0	1	1	0	0	0	0	x^3+1
8	—	0	**1**	**0**	**0**	—	0	0	0	0	1	0	1	1	0	0	0	x
9	—	0	0	**1**	**0**	—	0	0	0	0	**0**	1	0	1	1	0	0	x^2
10	—	0	0	0	**1**	—	0	0	0	0	**0**	**0**	1	0	1	1	0	x^3
11	—	0	0	0	0	—	0	0	0	0	**1**	**0**	**0**	**1**	0	1	1	0

codeword stages. The final codeword polynomial is $c(x)=x^6+x^5+x^3+1$ which gives $\boldsymbol{c}=(1\ 1\ 0\ 1\ 0\ 0\ 1)$. A total of 11 shifts have been required. □

When encoding using the shift register shown in Fig. 4.9 the first $n-k$ shifts are needed to move the information bits towards the high-order side of the register. During this time the output from the high-order stage b_{n-k-1} is zero and therefore the feedback is zero. This can be clearly seen in the operation of the encoder for the (7, 4) code, the first 3 rows of Table 4.5 show the incoming bits progressing towards b_2 and there is no feedback ($b_f=0$). The first 4 rows of Table 4.6, which show the operation of the encoder for the (7, 3) code, show the same trend. Feedback, and therefore polynomial division, commences when the first nonzero information bit leaves the high-order stage, the earliest point at which this can occur is at the $(n-k+1)$th shift.

This completes our look at encoding when the input is to the low-order side of the register and we next consider encoding using registers with high-order input. With such an approach feedback commences immediately with the first nonzero input and therefore less shifts are required than when the input is to the low-order side of a register. We will consider the (7, 4) code with generator polynomial $g(x)=x^3+x+1$ and the information polynomial $i(x)=x^3+x$ as previously considered. Figure 4.12 shows an encoder for the (7, 4) code where the input is to the high-order side of the register. Comparing this with Fig. 4.10 we see that there are no delay stages in Fig. 4.12. This is because by the time all the information bits have entered the codeword stages, the remainder is ready for shifting into the codeword stages, therefore there are no synchronization problems and so no delay stages are needed. The feedback links in the register are the same as before, for we are using the same generator polynomial. However the adder which b_{in} feeds into has moved over to the high-order side of the register, resulting in the feedback changing to $b_f=b_2+b_{in}$ and $b_0=b_f$ as the input to the low-order side of the

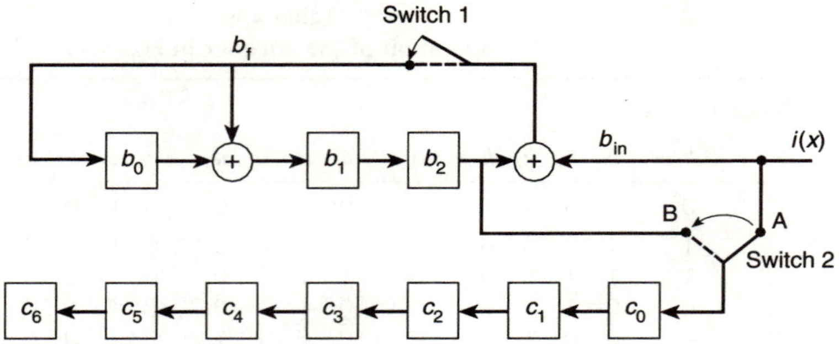

Fig. 4.12 An encoder with high-order input for the (7, 4) code with $g(x)=x^3+x+1$.

Table 4.7
Operation of the encoder in Fig. 4.12

High-order input $i(x)=x^3+x$

Shift	b_{in}	b_0	b_1	b_2	b_f	c_0	c_1	c_2	c_3	c_4	c_5	c_6	$r(x)$
		0	0	0	0	0	0	0	0	0	0	0	0
1	1	1	1	0	1	1	0	0	0	0	0	0	$x+1$
2	0	0	1	1	0	0	1	0	0	0	0	0	x^2+x
3	1	0	0	1	0	1	0	1	0	0	0	0	x^2
4	0	1	1	0	1	0	1	0	1	0	0	0	$x+1$
5	—	0	1	1	—	0	0	1	0	1	0	0	x^2+x
6	—	0	0	1	—	1	0	0	1	0	1	0	x^2
7	—	0	0	0	—	1	1	0	0	1	0	1	0

register. Hence, for the encoder in Fig. 4.12 we get

$$\begin{aligned} b_f &= b_2 + b_{in} \\ b_2 &= b_1 \\ b_1 &= b_0 + b_f \\ b_0 &= b_f. \end{aligned} \tag{4.9}$$

Table 4.7 gives the resulting step-by-step operation of the encoder when the input is $i(x)=x^3+x$. Note that because feedback and polynomial division start immediately with the first nonzero input, the input cannot be seen in the remainder stages. The first bit to enter is $b_{in}=i_3=1$, the feedback bit is $b_f=1$ and therefore polynomial division has commenced immediately at the first shift. Note also that because the register has an high-order input each nonzero input contributes an x^3 term to the remainder. By the end of the 4th shift all the information bits have entered the register and occupy the 4 codeword stages c_0, c_1, c_2 and c_3. Furthermore the remainder stages contain the remainder $R_{g(x)}[x^3 i(x)]$ as required. During these first 4 shifts switch 1 is kept closed, to enable feedback, and switch 2 is in position A so that the information bits can enter the codeword stages. After the 4th shift all that remains is to disable the feedback by opening switch 1, set switch 2 to position B and shift the register 3 times so moving the contents of c_0, c_1, c_2, c_3 into c_3, c_4, c_5, c_6 and b_0, b_1, b_2 into c_0, c_1, c_2. The resulting codeword polynomial is $c(x)=x^6+x^4+x+1$ as previously obtained. A total of 7 shifts have been used, instead of the 10 needed when the input is to the low-order side of the register. An encoder for an (n, k) code with high-order input requires n shifts, compared to the $2n-k$ shifts required by an encoder with input to the low-order side, and therefore encoding takes less time.

Feeding the input $i(x)$ into the high-order side of a register, with r stages, is equivalent to multiplying $i(x)$ by x^r. Hence for an encoder with $n-k$ remainder stages the effect of entering $i(x)$ at the high-order side is to multiply $i(x)$ by x^{n-k}, which is quite convenient because this is required for generating systematic codewords. However, when decoding a cyclic code there is no requirement for multiplication by x^{n-k} and so entering the input at the high-order side can be a disadvantage, this is considered further in Section 4.5.

Example 4.4
Consider an encoder with input to the high-order side, for the (7, 3) code with generator polynomial $g(x)=x^4+x^3+x^2+1$ and $i(x)=x^2+x$.

Figure 4.13 shows a register for generating systematic codewords. The (7, 3) code has 4 parity-check bits and therefore 4 remainder stages are required. The generator polynomial has $g_1=0$ and so there is no feedback to b_1. Delay stages are not required as the information bits are fed to the high-order side of the register. The input $i(x)=x^2+x$ is the same as that considered in Example 4.3. The results are shown in Table 4.8. During the first 3 shifts the information bits enter the codeword stages c_0, c_1, c_2 and are divided by $g(x)$ to give the required remainder. Note that each nonzero input contributes an x^4 term to the remainder. The next 4 shifts are needed to move the remainder into the codeword stages. By the end of the 7th shift the stages c_0, c_1, c_2, c_3, c_4, c_5, c_6 contain (1 0 0 1 0 1 1) and so the required codeword polynomial is $c(x)=x^6+x^5+x^3+1$. Switches 1 and 2 need to be respectively closed and in position *A* during the initial 3 shifts, and open and in position B for the final 4 shifts. A total of 7 shifts have been used, compared to 11 had the input been fed into the low-order side of the register (see Example 4.3). □

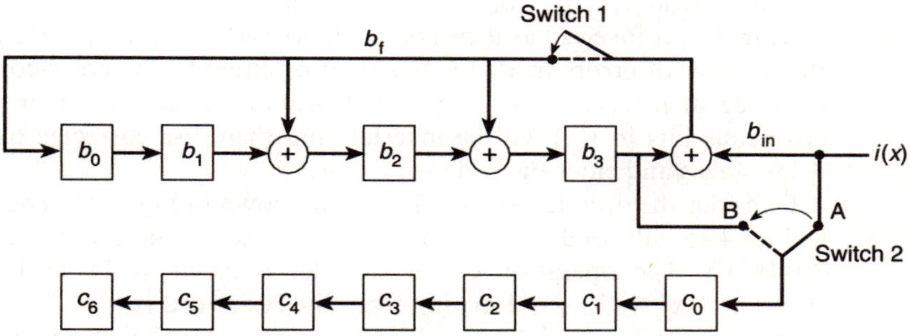

Fig. 4.13 An encoder with high-order input for the (7, 3) code with $g(x)=x^4+x^3+x^2+1$.

Table 4.8
Operation of the encoder in Fig. 4.13

High-order input $i(x) = x^2 + x$

Shift	b_{in}	b_0	b_1	b_2	b_3	b_{f}	c_0	c_1	c_2	c_3	c_4	c_5	c_6	$r(x)$
	0	0	0	0	0	0	0	0	0	0	0	0	0	0
1	1	1	0	1	1	1	1	0	0	0	0	0	0	x^3+x^2+1
2	1	0	1	0	1	0	1	1	0	0	0	0	0	x^3+x
3	0	1	0	0	1	1	0	1	1	0	0	0	0	x^3+1
4	—	0	1	0	0	—	1	0	1	1	0	0	0	x
5	—	0	0	1	0	—	0	1	0	1	1	0	0	x^2
6	—	0	0	0	1	—	0	0	1	0	1	1	0	x^3
7	—	0	0	0	0	—	1	0	0	1	0	1	1	0

4.4 Registers for error detection and correction

Decoders for cyclic codes require the same polynomial-division register as used for encoding, along with extra circuitry to carry out error detection and correction. Here we consider a decoder for error correction, based on a syndrome table, and a decoder for error detection. Both types of decoders can be implemented with the input to the low-order or high-order side of the polynomial-division register.

Figure 4.14 shows a decoder, for an (n, k) cyclic code, that employs a polynomial-division register along with a syndrome table. The remainder computed by the register is the syndrome polynomial $s(x)$ of the polynomial $v(x)$ to be decoded. The syndrome table needs to store every correctable error pattern along with the corresponding error syndromes. The decoder needs $n-k$ stages $b_0, b_1, \ldots, b_{n-k-2}, b_{n-k-1}$ to compute the error syndrome, n delay stages $d_0, d_1, \ldots, d_{n-2}, d_{n-1}$ to store $v(x)$ while the syndrome is being computed, n stages $e_0, e_1, \ldots, e_{n-2}, e_{n-1}$ to store the error pattern given by the syndrome table and n codeword stages $c_0, c_1, \ldots, c_{n-2}, c_{n-1}$ for the computed codeword. After $v(x)$ has entered the decoder the remainder stages will contain $s(x)$ and the delay stages contain $v(x)$. The error syndrome $s(x)$ forms the input to the syndrome table and the output is fed into $e_0, e_1, \ldots, e_{n-2}, e_{n-1}$. Error correction is carried out by shifting $e_0, e_1, \ldots, e_{n-2}, e_{n-1}$ and $d_0, d_1, \ldots, d_{n-2}, d_{n-1}$ through a modulo-2 adder. If $e_{n-1} = 1$ then the bit leaving d_{n-1} is inverted as it enters c_0, otherwise the bit leaving d_{n-1} is unaltered. If the number of errors in $v(x)$ is less than or equal to the error-correction limit of the code then $e_0, e_1, \ldots, e_{n-2}, e_{n-1}$ will contain the correct error pattern and the erroneous bits in $v(x)$ will be inverted and therefore corrected as they leave the delay stages and enter the codeword stages.

Consider the decoder for the $(7, 4)$ code shown in Fig. 4.15. The syndrome table in Fig. 4.15 will need to contain the error patterns and error syndromes given in Table 3.3. The operation of the decoder is given in Table 4.9 for the input $v(x) = x^6 + x^4 + x^3 + x + 1$ arising from the codeword $c(x) = x^6 + x^4 + x + 1$ incurring the error $e(x) = x^3$. The coefficients of $v(x)$ form the decoder input b_{in} (again with the high-order bits first) and 7 shifts are needed to input $v(x)$ (for clarity the

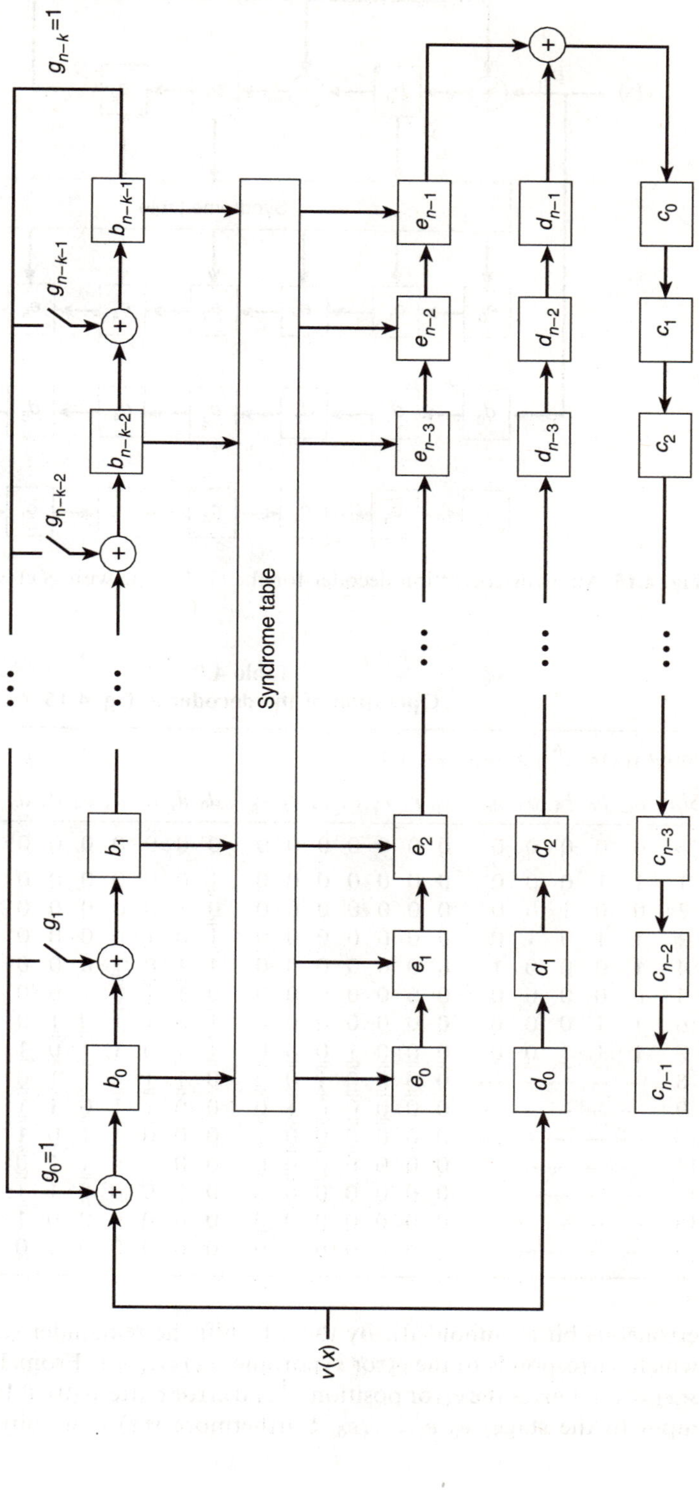

Fig. 4.14 A decoder using a syndrome table for error correction.

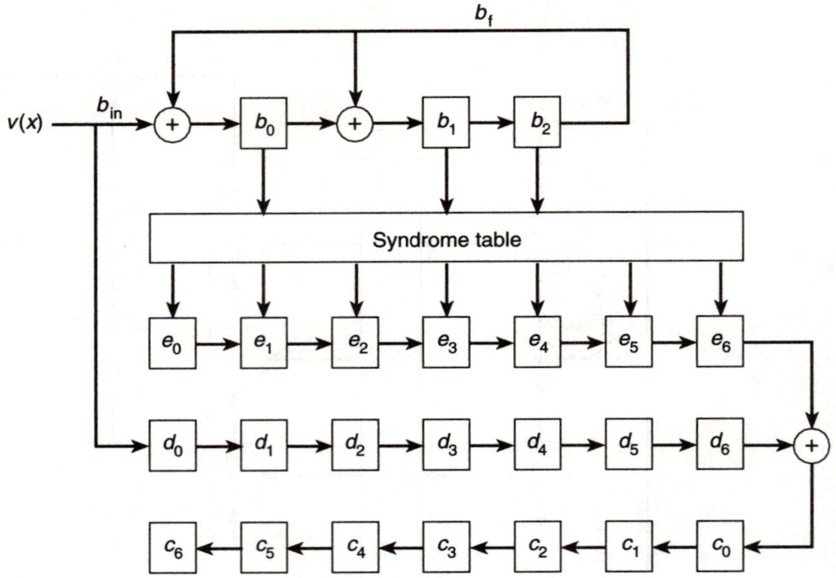

Fig. 4.15 An error-correction decoder for the (7, 4) code with $g(x) = x^3 + x + 1$.

Table 4.9
Operation of the decoder in Fig. 4.15

Input $v(x) = x^6 + x^4 + x^3 + x + 1$

Shift	b_{in}	b_0	b_1	b_2	b_f	e_0	e_1	e_2	e_3	e_4	e_5	e_6	d_0	d_1	d_2	d_3	d_4	d_5	d_6	c_0	c_1	c_2	c_3	c_4	c_5	c_6
	0	0	0	0	0	0	0	0	0	0	0	0	0	0	0	0	0	0	0	0	0	0	0	0	0	0
1	1	1	0	0	0	0	0	0	0	0	0	0	1	0	0	0	0	0	0	0	0	0	0	0	0	0
2	0	0	1	0	0	0	0	0	0	0	0	0	0	1	0	0	0	0	0	0	0	0	0	0	0	0
3	1	1	0	1	0	0	0	0	0	0	0	0	1	0	1	0	0	0	0	0	0	0	0	0	0	0
4	**1**	0	0	0	1	0	0	0	0	0	0	0	**1**	1	0	1	0	0	0	0	0	0	0	0	0	0
5	0	0	0	0	0	0	0	0	0	0	0	0	0	**1**	1	0	1	0	0	0	0	0	0	0	0	0
6	1	1	0	0	0	0	0	0	0	0	0	0	1	0	**1**	1	0	1	0	0	0	0	0	0	0	0
7	1	1	1	0	0	0	0	0	1	0	0	0	1	1	0	**1**	1	0	1	0	0	0	0	0	0	0
8	—	—	—	—	—	0	0	0	0	1	0	0	0	1	1	0	**1**	1	0	1	0	0	0	0	0	0
9	—	—	—	—	—	0	0	0	0	0	1	0	0	0	1	1	0	**1**	1	0	1	0	0	0	0	0
10	—	—	—	—	—	0	0	0	0	0	0	1	0	0	0	1	1	0	**1**	1	0	1	0	0	0	0
11	—	—	—	—	—	0	0	0	0	0	0	0	0	0	0	0	1	1	0	**0**	1	0	1	0	0	0
12	—	—	—	—	—	0	0	0	0	0	0	0	0	0	0	0	0	1	1	0	**0**	1	0	1	0	0
13	—	—	—	—	—	0	0	0	0	0	0	0	0	0	0	0	0	0	1	1	0	**0**	1	0	1	0
14	—	—	—	—	—	0	0	0	0	0	0	0	0	0	0	0	0	0	0	1	1	0	**0**	1	0	1

erroneous bit is embolded). By the 7th shift the remainder stages contain (1 1 0) which corresponds to the error syndrome $s(x) = x + 1$. From Table 3.3 we see that $s(x) = x + 1$ gives the error position x^3 and error pattern (0 0 0 1 0 0 0), this forms the input to the stages $e_0, e_1, \ldots, e_6$. Furthermore $v(x)$ now fully occupies the delay

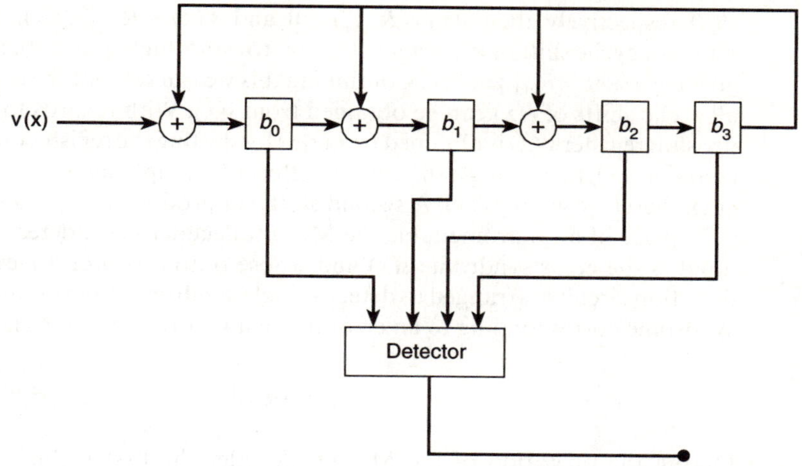

Fig. 4.16 An error-detection decoder for the (7, 3) code with $g(x)=x^4+x^2+x+1$.

stages. At each subsequent shift the output of the stages e_6 and d_6 are added together and fed into the codeword stages. By the 10th shift the erroneous bit and error position have reached the stages d_6 and e_6 respectively, and at the 11th shift the erroneous bit is corrected as it leaves d_6 and enters c_0. Shifts 12, 13, and 14 are required to move the remaining bits into the codeword stages. The final state of the codeword stages is (1 1 0 0 1 0 1) which gives $c(x)=x^6+x^4+x+1$ as required.

A decoder for error detection does not require a syndrome table but just a polynomial-division register, to determine the error syndrome $s(x)$, and a detector to indicate whether or not $s(x)$ is zero. If the detector indicates that $s(x)$ is nonzero then the input $v(x)$ contains errors, otherwise $v(x)$ is a codeword. Consider for example the (7, 3) code, with generator polynomial $g(x)=x^4+x^2+x+1$, used for error detection. Figure 4.16 shows a decoder consisting of a polynomial-division register with its 4 stages feeding into a detector. Here the detector is an OR gate with 4 inputs and whose output is 1 if any of the inputs are 1. If b_0, b_1, b_2, or b_3 are nonzero, after $v(x)$ has entered the register, the detector output will be 1, indicating that $s(x)$ is nonzero and that $v(x)$ therefore contains errors. The detector output is 0 only if b_0, b_1, b_2, and b_3 are 0, that is when $v(x)$ is a codeword polynomial.

4.5 The Meggitt decoder

The Meggitt decoder avoids the need for a syndrome table by computing the error syndromes of all correctable error patterns from a small number of error syndromes. In the case of a single-error-correcting code all the syndromes can be determined from any one error syndrome. Consider a cyclic code with generator polynomial $g(x)$ and let $e(x)$ be a correctable error pattern and $e'(x)$ a 1 bit cyclic shift of $e(x)$ towards high orders. If $s(x)$ and $s'(x)$ are the error syndromes of $e(x)$ and

$e'(x)$ respectively then $s(x) = R_{g(x)}[e(x)]$ and $s'(x) = R_{g(x)}[xs(x)]$. Furthermore, if $e''(x)$ is a cyclic shift of $e'(x)$, again by 1 bit towards high orders, then its syndrome is given by $s''(x) = R_{g(x)}[xs'(x)]$. Continuing this we can see that the error syndromes of all cyclic shifts of $e(x)$ can be obtained from $s(x)$. With regards to the polynomial-division register $s'(x)$ is obtained from $s(x)$ by shifting the register once. If the register contains $s(x)$, then a single shift has the effect of multiplying $s(x)$ by x and dividing by $g(x)$, therefore giving $s'(x)$. A second shift will produce $s''(x)$ and so forth.

In place of a syndrome table the Meggitt decoder has a detection circuit whose input is the error syndrome $s(x)$ and whose output is 0 or 1 (see Fig. 4.17). The detection circuit is arranged to detect a single syndrome $s_d(x)$ usually taken to be the syndrome corresponding to an error in the high-order bit of $v(x)$, so that

$$s_d(x) = R_{g(x)}[x^{n-1}].$$

During the operation of the Meggitt decoder the first n shifts are required for entering $v(x)$ into the register and into the stages $d_0, d_1, \ldots, d_{n-1}$. By the end of the nth shift the stages $b_0, b_1, \ldots, b_{n-k-1}$ contain the syndrome $s(x) = R_{g(x)}[v(x)]$ and $d_0, d_1, \ldots, d_{n-1}$ contain $v(x)$. If $s(x) = s_d(x)$ the output of the detector, denoted by s_d, is 1 otherwise it is 0. The output of stage d_{n-1} is added to the output of the detector s_d, so that the input to the stage c_0 is $d_{n-1} + s_d$. If the stage d_{n-1} contains an erroneous bit then $s(x) = s_d(x)$, $s_d = 1$ and the bit will be corrected as it leaves d_{n-1} to enter c_0. At the next shift the error syndrome will be $s'(x) = R_{g(x)}[xs(x)]$ and the process repeats itself. If $s'(x) = s_d(x)$, then $s_d = 1$ and the bit leaving d_{n-1} is inverted as it enters c_0, otherwise the bit is unaltered. This continues until $v(x)$ has left $d_0, d_1, \ldots, d_{n-1}$ and occupies the stages $c_0, c_1, \ldots, c_{n-1}$. A total of $2n$ shifts are required for decoding, where the first n shifts are for determining the error syndrome of $v(x)$ and the remaining n are for error correction.

Figure 4.18 shows a Meggitt decoder for the (7, 4) code with generator polynomial $g(x) = x^3 + x + 1$. Consider again the codeword $c(x) = x^6 + x^4 + x + 1$ incurring the

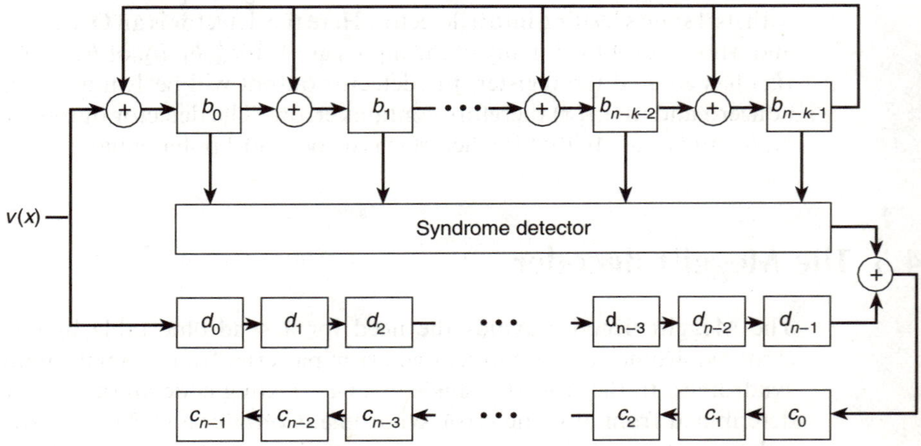

Fig. 4.17 A Meggitt decoder for an (n, k) cyclic code.

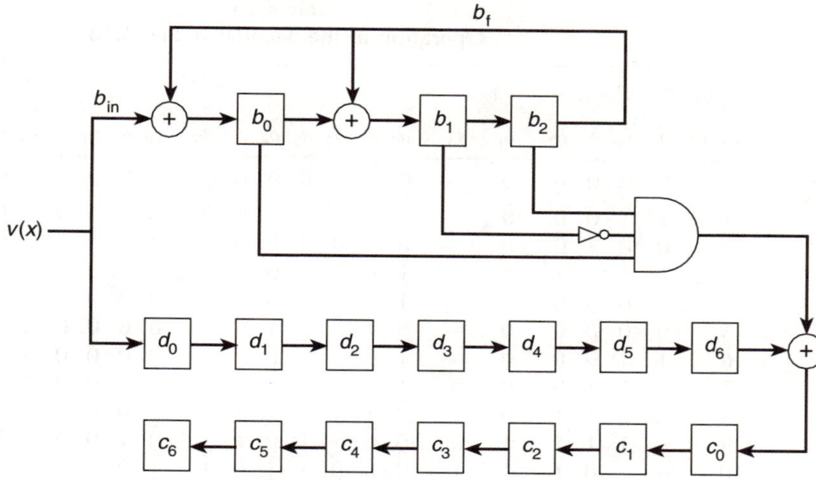

Fig. 4.18 A Meggitt decoder for the (7, 4) cyclic code with $g(x)=x^3+x+1$.

error x^3 so giving $v(x)=x^6+x^4+x^3+x+1$ as the word to be decoded (see Section 4.4 and Table 4.9). The input to the decoder is $(v_6, v_5, v_4, v_3, v_2, v_1, v_0)=(1\,0\,1\,1\,0\,1\,1)$ where v_3 is in error. The detector is arranged to detect the error syndrome of the highest-order bit, so that here we have

$$s_d(x) = R_{g(x)}[x^6]$$

and given the generator polynomial $g(x)=x^3+x+1$ this gives

$$s_d(x) = x^2 + 1.$$

Hence the detector needs to detect when $b_0=1, b_1=0$, and $b_2=1$. This is achieved by feeding b_0, b_1, and b_2 into a 3-input AND gate with the b_1 input passing through an inverter. When $b_0=1, b_1=0$, and $b_2=1$ all 3 inputs to the AND gate are 1 and so the gate's output $s_d=1$. At all other times the gate's output is 0.

The decoder's step-by-step operation is shown in Table 4.10. Fourteen shifts are required for decoding, of which the first 7 are needed to obtain the error syndrome of $v(x)$ and the last 7 for error correction. As before the coefficients of $v(x)$ are underlined to show their progress through the register, and furthermore the erroneous bit is embolded so that its progress can be easily seen. Bits enter the register high-order first, simultaneously feeding into the register and into the stages $d_0, d_1, \ldots, d_6$. By the end of the 7th shift the stages $d_0, d_1, \ldots, d_6$ contain $v_0, v_1, \ldots, v_6$ and $b_0=1, b_1=1, b_2=0$ which give the error syndrome $s(x)=x+1$. As $s(x)\neq s_d(x)$ the detector output $s_d=0$ and so v_6 will not be inverted as it leaves d_6 to enter c_0. After 3 more shifts the erroneous bit v_3 occupies d_6 and b_0, b_1, b_2 are 1, 0, 1 respectively, giving the error syndrome $s(x)=x^2+1=s_d(x)$ and $s_d=1$. Hence at the next shift v_3 is corrected as it is fed from d_6 and into c_0. Finally 3 more shifts are required to move the remaining bits into the codeword stages.

Table 4.10
Operation of the decoder in Fig. 4.18

Input $x^6+x^4+x^3+x+1$

Shift	b_{in}	b_0	b_1	b_2	b_f	s_d	d_0	d_1	d_2	d_3	d_4	d_5	d_6	c_0	c_1	c_2	c_3	c_4	c_5	c_6	$s(x)$
	0	0	0	0	0	—	0	0	0	0	0	0	0	0	0	0	0	0	0	0	0
1	1	1	0	0	0	—	1	0	0	0	0	0	0	0	0	0	0	0	0	0	1
2	0	0	1	0	0	—	0	1	0	0	0	0	0	0	0	0	0	0	0	0	x
3	1	1	0	1	0	—	1	0	1	0	0	0	0	0	0	0	0	0	0	0	x^2+1
4	**1**	0	0	0	1	—	**1**	1	0	1	0	0	0	0	0	0	0	0	0	0	0
5	0	0	0	0	0	—	0	**1**	1	0	1	0	0	0	0	0	0	0	0	0	0
6	1	1	0	0	0	—	1	0	**1**	1	0	1	0	0	0	0	0	0	0	0	1
7	1	1	1	0	0	0	1	1	0	**1**	1	0	1	0	0	0	0	0	0	0	$x+1$
8	—	0	1	1	0	0	0	1	1	0	**1**	1	0	1	0	0	0	0	0	0	x^2+x
9	—	1	1	1	1	0	0	0	1	1	0	**1**	1	0	1	0	0	0	0	0	x^2+x+1
10	—	1	0	1	1	1	0	0	0	1	1	0	**1**	1	0	1	0	0	0	0	x^2+1
11	—	1	0	0	1	0	0	0	0	0	1	1	0	**0**	1	0	1	0	0	0	1
12	—	0	1	0	0	0	0	0	0	0	0	1	1	0	**0**	1	0	1	0	0	x
13	—	0	0	1	1	0	0	0	0	0	0	0	1	1	0	**0**	1	0	1	0	x^2
14	—	1	1	0	1	0	0	0	0	0	0	0	0	1	1	0	**0**	1	0	1	$x+1$

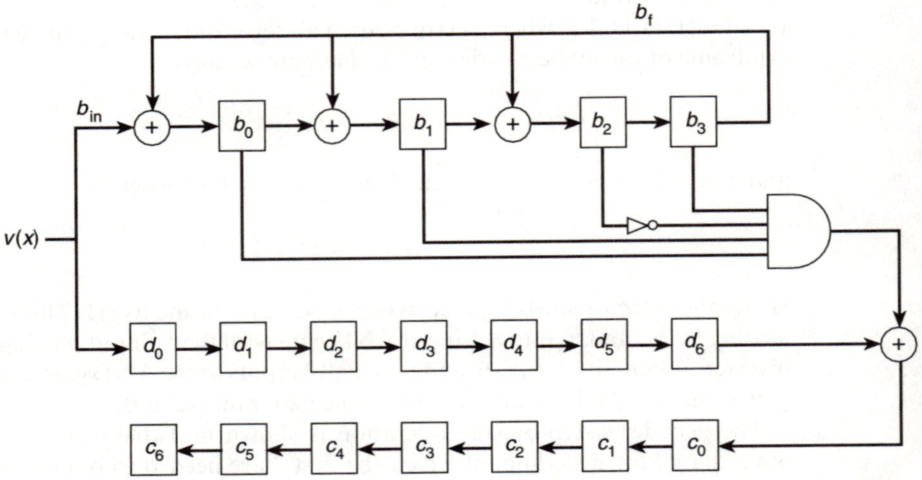

Fig. 4.19 A Meggitt decoder for the (7, 3) cyclic code with $g(x)=x^4+x^2+x+1$.

Example 4.5

Figure 4.19 shows a Meggitt decoder for the single-error correcting (7, 3) code with $g(x)=x^4+x^2+x+1$. Determine the step-by-step operation when $c(x)=x^6+x^5+x^4+x$ incurs the errors (a) $e(x)=1$ and (b) $e(x)=x^6$.

The detector needs to determine when the high-order bit of the input is in error and so it needs to detect when the remainder stages contain the error syndrome of x^6.

This gives

$$s_{\mathrm{d}}(x) = R_{g(x)}[x^6] = x^3 + x + 1$$

as the error syndrome that needs to be detected. Hence the detector needs to detect when $b_0 = 1$, $b_1 = 1$, $b_2 = 0$, and $b_3 = 1$, which can be achieved by feeding b_0, b_1, b_2, and b_3 into a 4-input AND gate with b_2 passing via an inverter (as shown in Fig. 4.19).

(a) When $c(x) = x^6 + x^5 + x^4 + x$ incurs the error $e(x) = 1$ then $v(x) = x^6 + x^5 + x^4 + x + 1$ and the decoder input is (1 1 1 0 0 1 1) as shown in Table 4.11(a).

Table 4.11
Operation of the Meggitt decoder in Fig. 4.19

(a) *Input* $v(x) = x^6 + x^5 + x^4 + x + 1$

Shift	b_{in}	b_0	b_1	b_2	b_3	b_{f}	s_{d}	d_0	d_1	d_2	d_3	d_4	d_5	d_6	c_0	c_1	c_2	c_3	c_4	c_5	c_6	$s(x)$
	0	0	0	0	0	0	—	0	0	0	0	0	0	0	0	0	0	0	0	0	0	0
1	1	1	0	0	0	0	—	1	0	0	0	0	0	0	0	0	0	0	0	0	0	1
2	1	1	1	0	0	0	—	1	1	0	0	0	0	0	0	0	0	0	0	0	0	$x+1$
3	1	1	1	1	0	0	—	1	1	1	0	0	0	0	0	0	0	0	0	0	0	x^2+x+1
4	0	0	1	1	1	0	—	0	1	1	1	0	0	0	0	0	0	0	0	0	0	x^3+x^2+x
5	0	1	1	0	1	1	—	0	0	1	1	1	0	0	0	0	0	0	0	0	0	x^3+x+1
6	1	0	0	0	0	1	—	1	0	0	1	1	1	0	0	0	0	0	0	0	0	0
7	**1**	1	0	0	0	0	0	**1**	1	0	0	1	1	1	0	0	0	0	0	0	0	1
8	—	0	1	0	0	0	0	0	**1**	1	0	0	1	1	1	0	0	0	0	0	0	x
9	—	0	0	1	0	0	0	0	0	**1**	1	0	0	1	1	1	0	0	0	0	0	x^2
10	—	0	0	0	1	0	0	0	0	0	**1**	1	0	0	1	1	1	0	0	0	0	x^3
11	—	1	1	1	0	1	0	0	0	0	0	**1**	1	0	0	1	1	1	0	0	0	x^2+x+1
12	—	0	1	1	1	0	0	0	0	0	0	0	**1**	1	0	0	1	1	1	0	0	x^3+x^2+x
13	—	1	1	0	1	1	1	0	0	0	0	0	0	**1**	1	0	0	1	1	1	0	x^2+x+1
14	—	1	0	0	0	1	0	0	0	0	0	0	0	0	**0**	1	0	0	1	1	1	1

(b) *Input* $v(x) = x^5 + x^4 + x$

Shift	b_{in}	b_0	b_1	b_2	b_3	b_{f}	s_{d}	d_0	d_1	d_2	d_3	d_4	d_5	d_6	c_0	c_1	c_2	c_3	c_4	c_5	c_6	$s(x)$
	0	0	0	0	0	0	—	0	0	0	0	0	0	0	0	0	0	0	0	0	0	0
1	**0**	0	0	0	0	0	—	**0**	0	0	0	0	0	0	0	0	0	0	0	0	0	0
2	1	1	0	0	0	0	—	1	**0**	0	0	0	0	0	0	0	0	0	0	0	0	1
3	1	1	1	0	0	0	—	1	1	**0**	0	0	0	0	0	0	0	0	0	0	0	$x+1$
4	0	0	1	1	0	0	—	0	1	1	**0**	0	0	0	0	0	0	0	0	0	0	x^2+x
5	0	0	0	1	1	0	—	0	0	1	1	**0**	0	0	0	0	0	0	0	0	0	x^3+x^2
6	1	0	1	1	1	1	—	1	0	0	1	1	**0**	0	0	0	0	0	0	0	0	x^3+x^2+x
7	0	1	1	0	1	1	1	0	1	0	0	1	1	**0**	0	0	0	0	0	0	0	x^3+x+1
8	—	1	0	0	0	1	0	0	0	1	0	0	1	1	**1**	0	0	0	0	0	0	1
9	—	0	1	0	0	0	0	0	0	0	1	0	0	1	1	**1**	0	0	0	0	0	x
10	—	0	0	1	0	0	0	0	0	0	0	1	0	0	1	1	**1**	0	0	0	0	x^2
11	—	0	0	0	1	0	0	0	0	0	0	0	1	0	0	1	1	**1**	0	0	0	x^3
12	—	1	1	1	0	1	0	0	0	0	0	0	0	1	0	0	1	1	**1**	0	0	x^2+x+1
13	—	0	1	1	1	0	0	0	0	0	0	0	0	0	1	0	0	1	1	**1**	0	x^3+x^2+x
14	—	1	1	0	1	1	0	0	0	0	0	0	0	0	0	1	0	0	1	1	**1**	x^3+x+1

The remainder at the 7th shift is $s(x)=1$ which is the error syndrome of $v(x)$. By the 13th shift the erroneous bit occupies the last stage of the delay stages, the remainder stages contain the syndrome $s_d(x)$ and $s_d=1$. Hence at the 14th shift the erroneous bit is corrected as it leaves d_6 and enters c_0.

(b) Here $v(x)=x^5+x^4+x$, the error is detected at the 7th shift and corrected at the 8th as shown in Table 4.11(b). □

We have seen that feeding the input to the high-order side of a polynomial-division register, for an (n,k) code, is equivalent to multiplying the input by x^{n-k}. This is useful when encoding, for multiplication by x^{n-k} is required for generating systematic codewords. At the decoding stage there is no such requirement and once $v(x)$ has entered the register the resulting remainder is $r(x)=R_{g(x)}[x^{n-k}v(x)]$ which is not the error syndrome required. However $r(x)$ still contains information that is solely dependent on errors within $v(x)$. Given a correctable error pattern $e(x)$ there is a unique correspondence between the error pattern and the remainder $r(x)$, and therefore

$$s'(x) = R_{g(x)}[x^{n-k}v(x)] \tag{4.10}$$

is an error syndrome of $v(x)$. The difference between $s'(x)$ and the 'normal' error syndrome $s(x)=R_{g(x)}[v(x)]$ is the correspondence between error syndromes and correctable error patterns. Consider for example the $(7,4)$ code with $g(x)=x^3+x+1$. The error syndrome $s(x)$ of, say, $e(x)=x^4$ is x^2+x, but using eqn 4.10 gives $s'(x)=1$. Table 4.12 gives the error syndromes $s(x)$ and $s'(x)$ for the $(7,4)$ code for all the correctable error patterns. From Table 4.12 we can see that the error syndromes $s'(x)$ are the same as $s(x)$ and differ only in their correspondence with the error patterns $e(x)$. Therefore there is no reason why $s'(x)$, instead of $s(x)$, cannot be used as the definition of the error syndrome of $v(x)$.

Consider now a Meggitt decoder for the $(7,4)$ code with input to the high-order side (see Fig. 4.20). Recall that the detector needs to detect an error in the high-order position x^6, and from Table 4.12 we can see that this means that we need to detect when the error syndrome $s'(x)=x^2$, that is when $b_0=0$, $b_1=0$, and $b_2=1$. A 3-input AND gate can be used for the detector with inverters on the b_0 and b_1 inputs to the gate. Decoding proceeds in the same way as that already described for a Meggitt

Table 4.12
Error syndromes for the (7, 4) code with $g(x)=x^3+x+1$

$e(x)$	$s(x)$	$s'(x)$
1	1	$x+1$
x	x	x^2+x
x^2	x^2	x^2+x+1
x^3	$x+1$	x^2+1
x^4	x^2+x	1
x^5	x^2+x+1	x
x^6	x^2+1	x^2

Note: $s(x)=R_{g(x)}[v(x)]$ and $s'(x)=R_{g(x)}[x^{n-k}v(x)]$

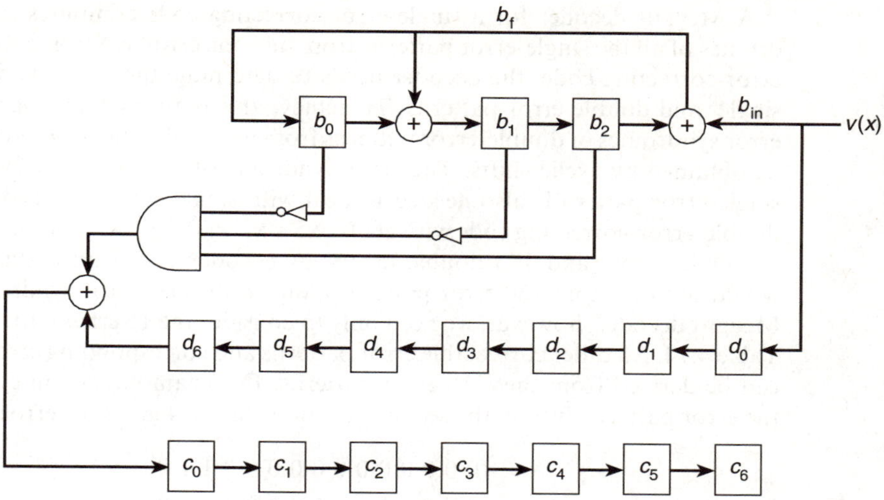

Fig. 4.20 A Meggitt decoder with high-order input for the (7, 4) code with $g(x)=x^3+x+1$.

Table 4.13
Operation of the Meggitt decoder in Fig. 4.20

High-order input $v(x)=x^6+x^4+x^3+x+1$

Shift	b_{in}	b_0	b_1	b_2	b_f	s_d	d_0	d_1	d_2	d_3	d_4	d_5	d_6	c_0	c_1	c_2	c_3	c_4	c_5	c_6	$r(x)$
	0	0	0	0	0	—	0	0	0	0	0	0	0	0	0	0	0	0	0	0	0
1	1	1	1	0	1	—	1	0	0	0	0	0	0	0	0	0	0	0	0	0	$x+1$
2	0	0	1	1	0	—	0	1	0	0	0	0	0	0	0	0	0	0	0	0	x^2+x
3	1	0	0	1	0	—	1	0	1	0	0	0	0	0	0	0	0	0	0	0	x^2
4	**1**	0	0	0	0	—	**1**	1	0	1	0	0	0	0	0	0	0	0	0	0	0
5	0	0	0	0	0	—	0	**1**	1	0	1	0	0	0	0	0	0	0	0	0	0
6	1	1	1	0	1	—	1	0	**1**	1	0	1	0	0	0	0	0	0	0	0	$x+1$
7	1	1	0	1	1	0	1	1	0	**1**	1	0	1	0	0	0	0	0	0	0	x^2+1
8	—	1	0	0	1	0	0	1	1	0	**1**	1	0	1	0	0	0	0	0	0	1
9	—	0	1	0	0	0	0	0	1	1	0	**1**	1	0	1	0	0	0	0	0	x
10	—	0	0	1	0	1	0	0	0	1	1	0	**1**	1	0	1	0	0	0	0	x^2
11	—	1	1	0	1	0	0	0	0	0	1	1	0	**0**	1	0	1	0	0	0	$x+1$
12	—	0	1	1	0	0	0	0	0	0	0	1	1	0	**0**	1	0	1	0	0	x^2+x
13	—	1	1	1	1	0	0	0	0	0	0	0	1	1	0	**0**	1	0	1	0	x^2+x+1
14	—	1	0	1	1	0	0	0	0	0	0	0	0	1	1	0	**0**	1	0	1	x^2+x

decoder with input to the low-order side. Table 4.13 shows the operation of the decoder for the input $v(x)=x^6+x^4+x^3+x+1$ (arising from the codeword polynomial $c(x)=x^6+x^4+x+1$ with the error $e(x)=x^3$). Note that 7 shifts are still required to obtain the remainder and 7 to correct the error.

A Meggitt decoder for a single-error-correcting code computes the error syndromes of all the single-error patterns from just one error syndrome. For a double-error-correcting code, the decoder needs to determine the error syndromes of all single- and double-error patterns. To achieve this requires a sufficient number of error syndromes of double-error patterns from which all other error syndromes can be obtained by cyclic shifts. The error syndrome of a single-error syndrome of a single-error pattern is also needed to deal with single errors. Consider the (15, 7) double-error-correcting code with $g(x) = x^8 + x^7 + x^6 + x^4 + 1$, this needs to correct 15 single errors and 105 double errors. A decoder based on a syndrome table would need to store 120 error patterns along with their error syndromes. With a Meggitt decoder, however, we need only to consider the 15 error patterns shown in Table 4.14. All other correctable error patterns and corresponding error syndromes can be derived from these 15 error patterns. For example, taking cyclic shifts of the error pattern given in the second row of Table 4.14 gives the error patterns

$$
\begin{array}{c}
(0\ 0\ 0\ 0\ 0\ 0\ 0\ 0\ 0\ 0\ 0\ 0\ 1\ 1\ 0)\\
(0\ 0\ 0\ 0\ 0\ 0\ 0\ 0\ 0\ 0\ 0\ 1\ 1\ 0\ 0)\\
(0\ 0\ 0\ 0\ 0\ 0\ 0\ 0\ 0\ 0\ 1\ 1\ 0\ 0\ 0)\\
\vdots \qquad\qquad\qquad\qquad \vdots\\
(0\ 1\ 1\ 0\ 0\ 0\ 0\ 0\ 0\ 0\ 0\ 0\ 0\ 0\ 0)\\
(1\ 1\ 0\ 0\ 0\ 0\ 0\ 0\ 0\ 0\ 0\ 0\ 0\ 0\ 0)\\
(1\ 0\ 0\ 0\ 0\ 0\ 0\ 0\ 0\ 0\ 0\ 0\ 0\ 0\ 1)
\end{array}
$$

which are all the different patterns with two successive bits in error. Each error pattern in Table 4.14 requires its own detection circuit and all error patterns are checked as decoding progresses. The decoder still operates in the manner described

Table 4.14
Syndrome table of a Meggitt decoder for the (15, 7) double-error-correcting code

e	*s*
(0 0 0 0 0 0 0 0 0 0 0 0 0 0 1)	(0 0 0 0 0 0 0 1)
(0 0 0 0 0 0 0 0 0 0 0 0 0 1 1)	(0 0 0 0 0 0 1 1)
(0 0 0 0 0 0 0 0 0 0 0 0 1 0 1)	(0 0 0 0 0 1 0 1)
(0 0 0 0 0 0 0 0 0 0 0 1 0 0 1)	(0 0 0 0 1 0 0 1)
(0 0 0 0 0 0 0 0 0 0 1 0 0 0 1)	(0 0 0 1 0 0 0 1)
(0 0 0 0 0 0 0 0 0 1 0 0 0 0 1)	(0 0 1 0 0 0 0 1)
(0 0 0 0 0 0 0 0 1 0 0 0 0 0 1)	(0 1 0 0 0 0 0 1)
(0 0 0 0 0 0 0 1 0 0 0 0 0 0 1)	(1 0 0 0 0 0 0 1)
(0 0 0 0 0 0 1 0 0 0 0 0 0 0 1)	(1 1 0 1 0 0 0 0)
(0 0 0 0 0 1 0 0 0 0 0 0 0 0 1)	(0 1 1 1 0 0 1 0)
(0 0 0 0 1 0 0 0 0 0 0 0 0 0 1)	(1 1 1 0 0 1 1 1)
(0 0 0 1 0 0 0 0 0 0 0 0 0 0 1)	(0 0 0 1 1 1 0 0)
(0 0 1 0 0 0 0 0 0 0 0 0 0 0 1)	(0 0 1 1 1 0 1 1)
(0 1 0 0 0 0 0 0 0 0 0 0 0 0 1)	(0 1 1 1 0 1 0 1)
(1 0 0 0 0 0 0 0 0 0 0 0 0 0 1)	(1 1 1 0 1 0 0 1)

for a single-error-correcting code and the same number of shifts $2n$ are required. Table 4.14 can be simplified by taking into account error patterns that can be obtained from cyclic shifts of other patterns. An example of this is the first double error

$$(0\ 0\ 0\ 0\ 0\ 0\ 0\ 0\ 0\ 0\ 0\ 0\ 0\ 1\ 1)$$

which when shifted by one bit towards the right gives

$$(1\ 0\ 0\ 0\ 0\ 0\ 0\ 0\ 0\ 0\ 0\ 0\ 0\ 0\ 1)$$

the last entry in the table. Hence either, but not both, of these patterns can be omitted from the table. The 14 double-error patterns can be arranged into 7 pairs, where within each pair the 2 error patterns differ only by a cyclic shift. By taking this into account the Meggitt decoder needs only to detect 8 error syndromes, namely the single-error syndrome and one syndrome for each of the 7 pairs of double-error patterns. The resulting Meggitt decoder requires a simpler detector but an additional n shifts are needed to ensure that error patterns that have been excluded from the table are decoded.

Problems

4.1 Figure 4.21 shows a linear-feedback shift register with 8 stages. Given that the register is initialized with $(b_0, b_1, b_2, b_3, b_4, b_5, b_6, b_7) = (0\ 0\ 1\ 1\ 0\ 1\ 0\ 1)$ determine the state of the register after:
(a) 5 shifts to the left;
(b) a further 4 shifts to the left;
(c) 18 shifts to the right, from the initial state not from the state after (a) or (b).

4.2 Figure 4.22 shows a linear-feedback shift register with high-order input. Determine the contents of the register for the input (1 1 0 0 1 0).

4.3 A linear-feedback shift register has

$$b_f = b_3$$
$$b_3 = b_2 + b_f$$
$$b_2 = b_1$$
$$b_1 = b_0 + b_f$$
$$b_0 = b_{in} + b_f.$$

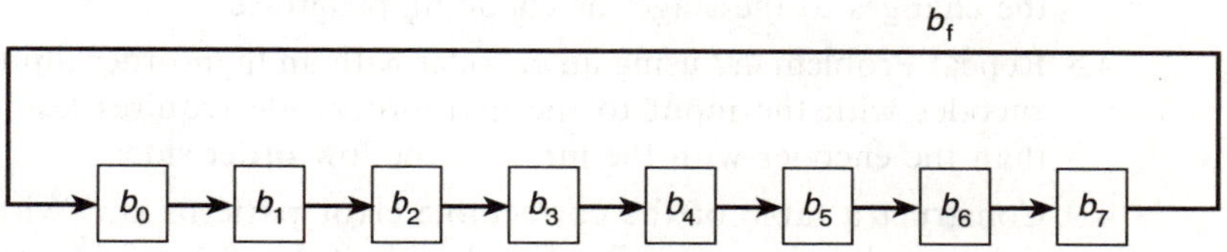

Fig. 4.21 An 8-stage LFSR.

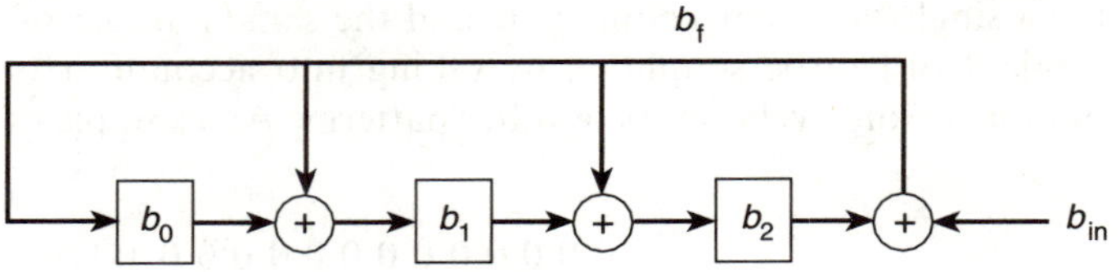

Fig. 4.22 An LFSR with a high-order input.

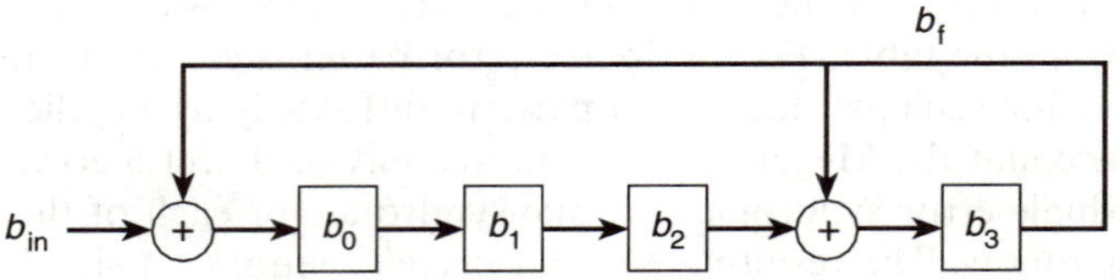

Fig. 4.23 An LFSR for dividing by $x^4 + x^3 + 1$.

Determine the register and the final state of the register given the input (0 1 0 1 1 1).

4.4 Figure 4.23 shows a linear-feedback shift register for dividing by the polynomial $p(x) = x^4 + x^3 + 1$. By considering the operation of the register, determine the remainder and quotient of $x^5 + x^4 + x^2 + 1$ divided by $p(x)$. Check your answer using the Euclidean division algorithm.

4.5 Determine a linear-feedback shift register for dividing by the polynomial $p(x) = x^5 + x^3 + x + 1$. By considering the operation of the register, determine the remainder and quotient of $x^7 + x^2 + 1$ divided by $p(x)$. Check your answer using the 'long hand' method for polynomial division.

4.6 Figure 4.24 shows a systematic encoder for the (15, 7) cyclic code with generator polynomial $g(x) = x^8 + x^4 + x^2 + x + 1$. The input is to the low-order side of the encoder. Given that the encoder input is the information polynomial $i(x) = x^5 + x^2 + 1$ determine the resulting codeword polynomial $c(x)$, showing the changes to the stages as encoding progresses.

4.7 Construct a systematic encoder, with low-order input, for the (15, 11) cyclic code with generator polynomial $g(x) = x^4 + x + 1$. Given the encoder input $i(x) = x^8 + x^6 + x + 1$ determine the resulting codeword polynomial—showing the changes to the stages as encoding progresses.

4.8 Repeat Problem 4.7 using an encoder with an high-order input. Show that the encoder with the input to the high-order side requires less shifts to encode than the encoder with the input to the low-order side.

4.9 Construct a table of the correctable error patterns $e(x)$ with corresponding error syndromes $s(x) = R_{g(x)}[e(x)]$ and $s'(x) = R_{g(x)}[x^{n-k}e(x)]$ for the single-error correcting (15, 11) cyclic code with $g(x) = x^4 + x + 1$.

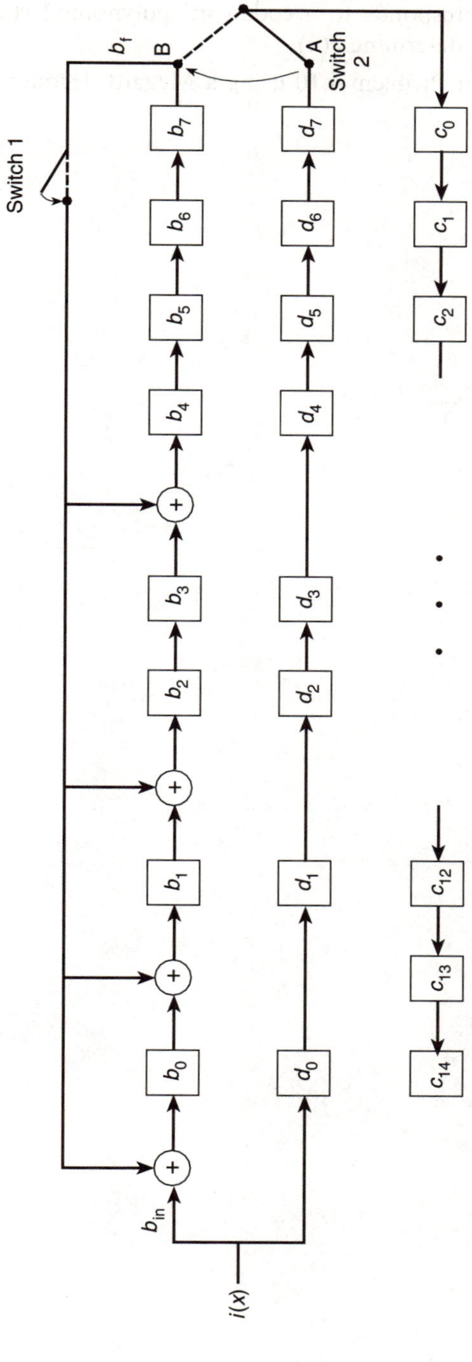

Fig. 4.24 An encoder for the (15, 7) code with $g(x) = x^8 + x^4 + x^2 + x + 1$.

4.10 Construct a Meggitt decoder, with input to the low-order side, for the (15, 11) code given in Problem 4.7. If the decoder input $v(x)=x^6+x^4+x^3+x^2$ corresponds to a codeword polynomial $c(x)$ that has incurred a single error, determine $c(x)$.

4.11 Repeat Problem 4.10 using a Meggitt decoder with input to the high-order side.

Linear algebra

5

In this chapter we consider elementary aspects of linear algebra that are required primarily for an understanding of linear codes in terms of vector spaces, and BCH codes considered in Chapter 7. Once we establish the concepts of vector spaces we reconsider linear codes. The introductory notes on fields given in this chapter are extended considerably in Chapter 6, which considers the mathematics for BCH codes namely that of Galois fields.

5.1 Sets

The mathematical notion of a set differs little from our ordinary idea of a ***set*** as a 'collection of objects'. Objects either belong to the set or do not belong to the set, there is no in between state. A set with n objects is written as

$$\{s_1, s_2, \ldots, s_n\}$$

where the objects $s_1, s_2, \ldots, s_n$ are referred to as the set's *elements*. This is known as a *finite set* as the set contains a finite number of elements. Examples of finite sets are the set of binary digits {0 1} and the set of decimal digits {0 1 2 3 4 5 6 7 8 9}. Given a set with n elements then a set formed by any m (where $m \leq n$) of the elements is a *subset* of the original set. The set {0 1} is a subset of {0 1 2 3 4 5 6 7 8 9}.

A set containing an infinite number of elements is referred to as an *infinite set*, these can often be expressed in terms of the first 3 or 4 elements of the set (or more if necessary). Examples of infinite sets are the set of all nonzero positive integers $\{1, 2, 3, \ldots\}$ and the set of all integers $\{0, \pm 1, \pm 2, \ldots\}$. Not all infinite sets can be expressed in terms of the first few elements of the set, for example the set of all real numbers can clearly not be expressed in terms of a few of its elements.

Mathematical operations between sets can be defined and the reader may be familiar with the idea of the union and the intersection of sets, so giving the set of all elements and the set of common elements respectively. However, it is not operations between sets that we are interested in, but rather operations between elements within a set, and in doing so we create 'higher' mathematical structures known as groups.

5.2 Groups

A *group* is constructed from a set by defining a *group operation* $*$ between elements within the set such that the following conditions are satisfied:

1. The group is *closed* under the operation $*$ in that for any two elements a and b within the set then the element

$$c = a * b$$

also belongs to the set. The group is said to have the property of *closure*.

2. The operation $*$ is *associative*, so that given the elements a, b, and c then

$$a * (b * c) = (a * b) * c. \tag{5.1}$$

In other words the order in which the group operation is carried out is immaterial.

3. There exists a unique *identity element* I within the set such that for any element a

$$a * I = I * a = a. \tag{5.2}$$

4. For each element a there exists a unique *inverse* a' within the set such that

$$a * a' = a' * a = I. \tag{5.3}$$

Furthermore, if

$$a * b = b * a \tag{5.4}$$

for all elements a and b, then the group is said to be a *commutative* group. Conditions 1 to 4 are necessary and sufficient for a set and operation to form a group. Note that a group need not necessarily be commutative. Note also that commutative groups are also called *Abelian* groups (after the Norwegian mathematician Abel). The group operation is normally addition or multiplication so giving *additive groups* or *multiplicative groups* respectively.

Consider the set of binary numbers $\{0, 1\}$ along with modulo-2 addition. The closure property is obeyed because modulo-2 addition always gives 0 or 1 (see Table 1.4). Addition is clearly associative, for example $0+(1+1)=0+0=0$ and $(0+1)+1=1+1=0$. The identity element is 0, and each element is its own inverse as $0+0=0$ and $1+1=0$. Furthermore the order in which elements are added is irrelevant, for example $0+1=1+0$, and therefore modulo-2 addition over the set $\{0, 1\}$ forms a commutative group. Whilst this may appear to be a rather trivial example, this is nevertheless an important group.

Example 5.1

Consider whether ordinary addition forms a group over the sets (a) $\{0, 1, 2, \ldots\}$ and (b) $\{0, \pm 1, \pm 2, \ldots\}$.

(a) The set is closed because the sum of any two positive integers gives another positive integer. Addition is clearly associative, for example adding 7, 4, and 17 gives $7+(4+17)=7+21=28$ and $(7+4)+17=11+17=28$. The identity element is 0 as adding 0 to any integer leaves the integer unaltered. However the nonzero elements in the set do not have an inverse element within the set. For example the inverse of 9 is -9 which does not belong to the set $\{0, 1, 2, \ldots\}$. Therefore addition over the set $\{0, 1, 2, \ldots\}$ fails to generate a group.

(b) As in (a) above, addition over the set $\{0, \pm 1, \pm 2, \ldots\}$ satisfies the closure and associative rules, and again the identity element is 0. However, each element now has an inverse within the set and therefore a group is formed. Furthermore the order in which two elements are added together is immaterial, for example

$13+2=2+13=15$, and so the commutative rule is met. Hence addition over the set $\{0, \pm 1, \pm 2, \ldots\}$ forms a commutative group. □

To subtract two elements within an additive group does not require the definition of a second group operation. Instead subtraction is carried out by replacing the element to be subtracted by its inverse element and then carrying out the group operation, and so

$$a-b=a+(-b)$$

where a and b are elements of an additive group and $-b$ is the inverse of b. In a multiplicative group replacing an element by its inverse defines the process of division and so

$$a \div b = a \times (b')$$

where b' is the multiplicative inverse of b; $a \div b$ is also written as a/b. Subtraction and division can be thought of as inverse group operations defined in terms of the first group operation through $a-b=a+(-b)$ and $a \div b = a \times (b')$ respectively. Note that the associative and commutative properties of a group apply to the group operation and not to the inverse operation. So for example in an additive group commutativity means $7-4=7+(-4)=(-4)+7=-4+7$ and not $7-4=4-7$. In a multiplicative group associativity applies to multiplication and not to division, for example $2 \div (7 \div 4) \neq (2 \div 7) \div 4$.

Example 5.1 shows that addition over the set $\{0, \pm 1, \pm 2, \ldots\}$ forms an additive group. Multiplication over the same set fails to form a group because none of the elements have a multiplicative inverse. Multiplication over the set of nonzero rational numbers (that is numbers that can be expressed as a/b where a and b are nonzero integers) does however form a group. Here the identity element is 1 and given an element $c=a/b$ then its inverse $c'=b/a$. Multiplication is both associative and commutative, and we therefore have a commutative group. An additive group can be constructed by including the 0 element, so that we have the set of rational numbers.

Quite often a subset of elements within a group forms a group under the group operation. In such a case the subset and the group operation are said to form a *subgroup*. The additive group of rational numbers contains all the integers (since an integer is a rational number) and furthermore addition over the set of integers forms a group. Therefore the additive group of integers is a subgroup of the additive group of rational numbers. The set of positive integers is a subset of the set of integers but does not form a subgroup of the additive group of integers.

As with sets, groups can be finite or infinite, and the number of elements within a finite group is known as the *order* of the group. A finite group can be constructed by taking integers *modulo-m*, where n modulo-m is defined as the remainder obtained after dividing n by m. If r is the remainder of n divided by m then we write

$$n = r \text{ modulo-}m$$

which can also be written as

$$n \text{ modulo-}m = r.$$

For example, 5 modulo-4 = 1 and 13 modulo-5 = 3. We have already met modulo-2 arithmetic, in which the only integers allowed are 0 and 1. Taking integers modulo-m

produces the m integers $0, 1, 2, \ldots, m-1$. Note that the remainder is always less than m. If $n=m$ then

$$m \text{ modulo-}m = 0 \tag{5.5}$$

and furthermore if n is a multiple of m so that $n = a \times m$ where a is an integer, then

$$(a \times m) \text{ modulo-}m = 0. \tag{5.6}$$

Given two integers a and b then using *modulo-m addition*

$$(a+b) \text{ modulo-}m = r \tag{5.7}$$

where r is the remainder of $(a+b)$ divided by m. For example under modulo-4 addition $1+5=2$ and $17+6=3$. Under modulo-m addition each nonzero integer a has an additive inverse given by $m-a$, and 0 is its own inverse. Furthermore modulo-m addition is closed, associative and commutative and therefore under modulo-m addition the set $\{0, 1, 2, \ldots, m-1\}$ forms a commutative group.

Example 5.2
Construct the additive group of integers modulo-5 over the set $\{0, 1, 2, 3, 4\}$.

We need to obtain the remainders when integers are added pairwise. Adding 0 to any element leaves the element unaltered, the remaining sums are

$$\begin{aligned}
1+1 &= 2 \text{ modulo-5} \\
1+2 &= 3 \text{ modulo-5} \\
1+3 &= 4 \text{ modulo-5} \\
1+4 &= 0 \text{ modulo-5} \\
2+2 &= 4 \text{ modulo-5} \\
2+3 &= 0 \text{ modulo-5} \\
2+4 &= 1 \text{ modulo-5} \\
3+3 &= 1 \text{ modulo-5} \\
3+4 &= 2 \text{ modulo-5} \\
4+4 &= 3 \text{ modulo-5.}
\end{aligned}$$

There is no need to consider the remaining combinations of pairs of integers because addition is commutative (e.g. $(3+2)$ modulo-5 $=(2+3)$ modulo-5 and so forth). The resulting additive group is shown in Table 5.1. □

Table 5.1
The additive group of integers modulo-5

+	0	1	2	3	4
0	0	1	2	3	4
1	1	2	3	4	0
2	2	3	4	0	1
3	3	4	0	1	2
4	4	0	1	2	3

Example 5.3
Find the inverse of each element in the additive group of integers modulo-5.

The additive inverse of a nonzero integer a is given by $m - a$, therefore under modulo-5 addition the additive inverses of 1, 2, 3, and 4 are 4, 3, 2, and 1 respectively. The inverse of 0 is 0. □

Multiplicative groups can be constructed in a similar way, using *modulo-m multiplication* which gives the remainder of the product of two integers. However, modulo-m multiplication only generates groups for prime values of m and the group is over the set $\{1, 2, \ldots, m-1\}$, the 0 element is excluded from the set because 0 does not have a multiplicative inverse. If m is not a prime number then modulo-m multiplication over the set $\{1, 2, \ldots, m-1\}$ does not produce a group. Hence two groups can be constructed over the set $\{0, 1, 2, \ldots, m-1\}$, an additive group over all the elements in the set and a multiplicative group over the set's nonzero elements.

Example 5.4
Construct the multiplicative group over $\{1, 2, 3, 4\}$ using modulo-5 multiplication.

Taking the product of an element with 1 leaves the element unaltered, the remaining products modulo-5 are

$$2 \times 2 = 4 \text{ modulo-5}$$
$$2 \times 3 = 1 \text{ modulo-5}$$
$$2 \times 4 = 3 \text{ modulo-5}$$
$$3 \times 3 = 4 \text{ modulo-5}$$
$$3 \times 4 = 2 \text{ modulo-5}$$
$$4 \times 4 = 1 \text{ modulo-5.}$$

Table 5.2 shows the resulting group. □

We have seen that the additive inverse of an element a is $m - a$. The multiplicative inverse of a nonzero element a is the element a' such that $a \times a' = 1$. For example, in the modulo-5 multiplicative group the inverse of 2 is 3 because $2 \times 3 = 1$ modulo-5, likewise the inverse of 3 is 2. It is possible for an element to be its own multiplicative inverse, for example $4 \times 4 = 1$ modulo-5 (see Table 5.2) and therefore in the modulo-5 multiplicative group 4 is its own multiplicative inverse.

Table 5.2
The modulo-5 multiplicative group

×	1	2	3	4
1	1	2	3	4
2	2	4	1	3
3	3	1	4	2
4	4	3	2	1

5.3 Fields

To obtain the ordinary arithmetic that we are familiar with the idea of a group needs to be extended to that of a *field*. This is achieved by defining a second operation, within a commutative group, that also forms a commutative group over the same set of elements. Note that both groups must be commutative for a field to be constructed. In a multiplicative group the second operation can be addition, so that elements can be multiplied and added together. Each group operation satisfies the conditions given by eqns 5.1 to 5.4 and furthermore the two group operations must satisfy the *distributive* rule

$$a \times (b + c) = a \times b + a \times c \tag{5.8}$$

where a, b, and c are now referred to as *field elements* or *scalars*. A field has two identity elements; the additive identity and the multiplicative identity. Each field element has an additive inverse and a multiplicative inverse (except 0 which does not have a multiplicative inverse). In summary we can think of a field as:

1. A set of elements.
2. An additive operation which forms a commutative group over the set's elements.
3. A multiplicative operation which forms a commutative group over the set's nonzero elements.
4. The two group operations satisfy the distributive law given by eqn 5.8.

Within a field inverses of the field operations are easily defined (as already discussed for groups) and therefore subtraction and division (by nonzero elements) can be carried out. Hence within a field there is the mechanism for addition, subtraction, multiplication, and division that we encounter in ordinary mathematics. As with sets and group, fields can be finite or infinite, and the number of elements in a field is the order of the field.

As an example of a field consider addition and multiplication over the set of real numbers. Both operations form commutative groups over the set of real numbers and the two operations satisfy the distributive rule. Therefore this gives a field, known as the *real field*. We have seen that addition over the set of integers $\{0, \pm 1, \pm 2, \ldots\}$ forms a group but multiplication fails to do so, and therefore the set of integers under addition and multiplication is not a field. This failure is not because of the multiplication operation, for the product of two integers gives an integer, but rather because the division of an integer by another integer does not always give an integer. Over the set of integers addition, subtraction, and multiplication can be carried out but not division. This gives a mathematical structure called a *ring* that lies between a group and a field. Within a ring, multiplication is possible (which is not possible in an additive group) but division is not possible (which is possible in a field).

Fields can be constructed from modulo-m addition and multiplication. Modulo-2 arithmetic forms an additive group over the set $\{0, 1\}$ and a multiplicative group over the nonzero element of $\{0, 1\}$. It follows therefore that modulo-2 addition and multiplication over the set $\{0, 1\}$ generates a field, known as the *binary field*. Larger fields can be constructed by taking modulo-m addition and multiplication over the set $\{0, 1, \ldots, m-1\}$ where m is a prime number and are known as *prime fields* (recall that m has to be a prime number if modulo-m multiplication is to generate a multiplicative group).

Example 5.5
Construct the prime field under modulo-7 arithmetic.

The field is constructed over the set of integers $\{0, 1, 2, 3, 4, 5, 6\}$. To construct the field we need evaluate the sum modulo-7 and product modulo-7 of integers taken pairwise. Doing this gives Tables 5.3(a) and (b). □

The reader should now realize that the result of a calculation depends on the field within which the calculation is performed. As an example consider $(2+3)\times 3+4$ the result of which is 19, 4 and 5 when the fields are the real field, the modulo-5 field and the modulo-7 field respectively. We have seen that the additive inverse of an element a is $m-a$, and that the multiplicative inverse of a nonzero element a is the element a' such that $a \times a' = 1$ modulo-m. In the prime fields subtraction and division can be carried out by replacing an element by its inverse and respectively carrying out addition and multiplication. For example, in the modulo-7 field dividing by 4 is equivalent to multiplying by 2, as 2 and 4 are multiplicative inverses of each other ($4 \times 2 = 1$ modulo-7).

Example 5.6
Evaluate $((2-3)\times 3)/4$ over the prime field modulo-m when (a) $m=5$ and (b) $m=7$.

(a) In the prime field modulo-5 the additive inverse of 3 is 2 and the multiplicative inverse of 4 is 4. Hence

$$\begin{aligned}((2-3)\times 3)/4 &= (2+2)\times 3\times 4 \text{ modulo-5}\\ &= 4\times 3\times 4 \text{ modulo-5} = 3 \text{ modulo-5} = 3.\end{aligned}$$

(b) Here the additive inverse of 3 is 4 and the multiplicative inverse of 4 is 2. Hence

$$\begin{aligned}((2-3)\times 3)/4 &= (2+4)\times 3\times 2 \text{ modulo-7}\\ &= 6\times 3\times 2 \text{ modulo-7} = 1 \text{ modulo-7} = 1.\end{aligned}$$

□

Table 5.3
The modulo-7 prime field

(a) *Addition*

+	0	1	2	3	4	5	6
0	0	1	2	3	4	5	6
1	1	2	3	4	5	6	0
2	2	3	4	5	6	0	1
3	3	4	5	6	0	1	2
4	4	5	6	0	1	2	3
5	5	6	0	1	2	3	4
6	6	0	1	2	3	4	5

(b) *Multiplication*

×	1	2	3	4	5	6
1	1	2	3	4	5	6
2	2	4	6	1	3	5
3	3	6	2	5	1	4
4	4	1	5	2	6	3
5	5	3	1	6	4	2
6	6	5	4	3	2	1

Linear equations can be defined and solved in prime fields in the same way as in the real field, bearing in mind that variables and constants can only assume values defined within the field and the arithmetic appropriate to the field has to be used in all calculations. Consider for example the simultaneous equations

$$3x + 2y = 2$$
$$4x + 6y = 3$$

defined in the prime field modulo-7. The equations can be solved by first eliminating one of the variables. Multiplying the first equation by 6 gives $4x + 5y = 5$ and subtracting this from the second equation gives $6y - 5y = 3 - 5 = 5$ and therefore $y = 5$. Substituting $y = 5$ into the first equation gives $3x + 3 = 2$ and so $x = 2$. The reader can check that $x = 2$ and $y = 5$ satisfy the second equation.

Example 5.7
Solve the equations

$$2x + 3y = 1$$
$$x + 2y = 2$$

over the prime field modulo-5.

Multiplying $x + 2y = 2$ by 2 gives $2x + 4y = 4$ and subtracting $2x + 3y = 1$ leaves $y = 3$. Substituting $y = 3$ into $x + 2y = 2$ gives $x = 1$. The solution is therefore $x = 1$ and $y = 3$. □

Finite fields are used extensively in the construction and the decoding of codes and in Chapter 6 we consider finite fields in greater depth.

5.4 Vector spaces

Most readers will probably be familiar with the concept of a vector within a plane. Here a two-dimensional x, y coordinate system is used to define the position of any point within the plane. The intersection of the X and Y axis is the origin of the coordinate system and has coordinates $0, 0$. Given a point P with coordinates x, y then the vector $V = (x, y)$ is the straight line starting at the origin and ending at P (see Fig. 5.1a). If vectors are constructed from the origin to all points within the plane then the resulting set of vectors forms a vector space. This is a rather loose definition of a vector space, a formal definition will be given soon. In three dimensions, with axes X, Y, and Z, vectors are denoted by $\boldsymbol{v} = (x, y, z)$ where x, y, and z are now the coordinates of points within the three-dimensional space. Constructing vectors from the origin to all points within the space now gives a 3-dimensional vector space. Whilst it is easy to envisage and give physical interpretation to vectors in 2 or 3 dimensions, this is not so for vectors in 4 or more dimensions. However it is possible to define vectors within 4 or more dimensions. Our interest in vector spaces is because codewords belonging to an (n, k) block code can be interpreted as vectors within a k-dimensional vector space, and in particular vector spaces with high dimensions since k is typically greater than 3.

From a formal point of view a *vector space* is defined as a collection of objects, called *vectors*, together with the operations of *vector addition* and *scalar*

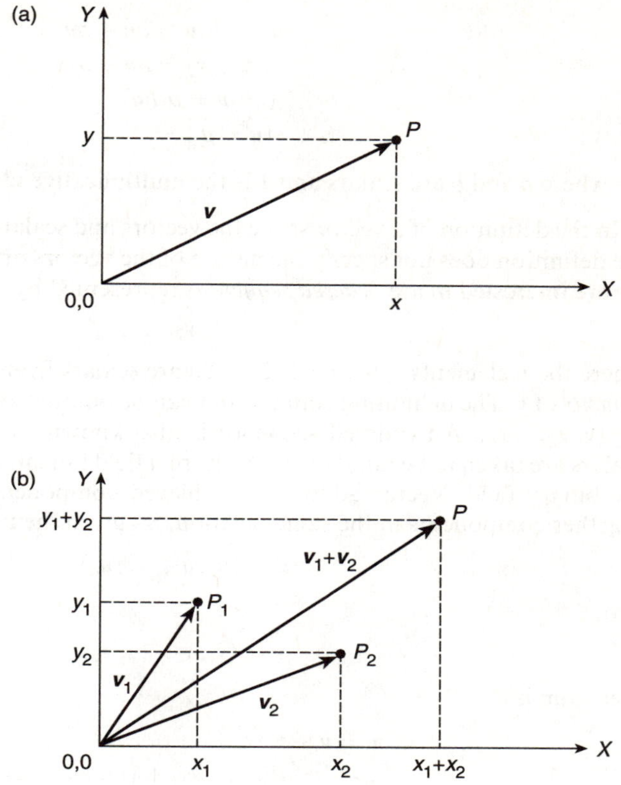

Fig. 5.1 Vectors in a plane.

multiplication which satisfy the conditions:

1. The set of vectors forms an additive commutative group. Hence for any 2 vectors $\boldsymbol{u}$ and $\boldsymbol{v}$ the sum $\boldsymbol{u}+\boldsymbol{v}$ is also a vector (the closure property). Addition is associative and commutative. Each vector has an additive inverse and there exists an additive identity vector $\mathbf{0}$. Hence for any 3 vectors $\boldsymbol{u}$, $\boldsymbol{v}$, and $\boldsymbol{w}$

$$\begin{aligned} \boldsymbol{u}+(\boldsymbol{v}+\boldsymbol{w}) &= (\boldsymbol{u}+\boldsymbol{v})+\boldsymbol{w} \\ \boldsymbol{u}+\boldsymbol{v} &= \boldsymbol{v}+\boldsymbol{u} \\ \boldsymbol{u}+\mathbf{0} &= \boldsymbol{u} \\ \boldsymbol{u}+\boldsymbol{u}' &= \mathbf{0} \end{aligned} \tag{5.9}$$

   where $\boldsymbol{u}'$ is the additive inverse of $\boldsymbol{u}$.

2. Multiplication of vectors by scalars, belonging to some field, is defined. For any vector $\boldsymbol{u}$ and scalar c the *scalar product* $c\boldsymbol{u}$ is a vector. The scalar product is

distributive and associative and given the vectors $\boldsymbol{u}$ and $\boldsymbol{v}$

$$\begin{aligned}(a+b)\boldsymbol{u} &= a\boldsymbol{u} + b\boldsymbol{u}\\ a(\boldsymbol{u}+\boldsymbol{v}) &= a\boldsymbol{u} + a\boldsymbol{v}\\ (ab)\boldsymbol{u} &= a(b\boldsymbol{u})\\ 1\boldsymbol{u} &= \boldsymbol{u}\end{aligned} \tag{5.10}$$

where a and b are scalars and 1 is the multiplicative identity element.

In the definition of a vector space the vectors and scalars can be abstract objects, the definition does not specify the nature of the vectors or scalars. The vectors that we are interested in are *ordered sequences* represented by

$$\boldsymbol{v} = (v_1, v_2, \ldots, v_n)$$

where the n elements v_i for $i = 1, 2, \ldots, n$ are scalars from a field and are the *components* of $\boldsymbol{v}$. The delimiting commas in $\boldsymbol{v}$ can be omitted and we could equally write $\boldsymbol{v} = (v_1\ v_2 \ldots v_n)$. An ordered sequence is also known as an *n-tuple*. For now the scalars are taken to be numbers from the real field but later they will be restricted to the binary field. Vector addition is achieved componentwise, that is by adding together components in the same position. So given the two vectors

$$\boldsymbol{u} = (u_1, u_2, \ldots, u_n)$$

and

$$\boldsymbol{v} = (v_1, v_2, \ldots, v_n)$$

their sum is

$$\begin{aligned}\boldsymbol{w} &= \boldsymbol{u} + \boldsymbol{v}\\ &= (u_1, u_2, \ldots, u_n) + (v_1, v_2, \ldots, v_n)\\ &= (u_1 + v_1, u_2 + v_2, \ldots, u_n + v_n)\\ &= (w_1, w_2, \ldots, w_n)\end{aligned}$$

where $w_i = u_i + v_i$ for $i = 1, 2, \ldots, n$. Note that vectors within the same vector space must have the same number of components. The sum of two 2-dimensional vectors in a plane is illustrated in Fig. 5.1(b). The scalar product is also componentwise, so the product of the vector $\boldsymbol{u}$ and scalar a is

$$\begin{aligned}a\boldsymbol{u} &= a(u_1, u_2, \ldots, u_n)\\ &= (au_1, au_2, \ldots, au_n).\end{aligned}$$

Example 5.8

Given the vectors $\boldsymbol{u} = (2, 7, 1)$ and $\boldsymbol{v} = (10, -3, 8)$ find

(a) $\boldsymbol{u} + \boldsymbol{v}$

(b) $4\boldsymbol{u}$

(c) $2\boldsymbol{u} - 7\boldsymbol{v}$.

(a)

$$\begin{aligned}\boldsymbol{u} + \boldsymbol{v} &= (2, 7, 1) + (10, \ -3, 8)\\ &= (2 + 10, 7 - 3, 1 + 8)\\ &= (12, 4, 9).\end{aligned}$$

(b)

$$4\boldsymbol{u} = 4(2, 7, 1) = (8, 28, 4).$$

(c)

$$\begin{aligned} 2\boldsymbol{u} - 7\boldsymbol{v} &= 2(2, 7, 1) - 7(10, -3, 8) \\ &= (4, 14, 2) - (70, -21, 56) \\ &= (4 - 70, 14 + 21, 2 - 56) \\ &= (-66, 35, -54). \end{aligned}$$

□

Two or more vectors can be added together to form another vector, furthermore multiplying each vector by a scalar and adding the resulting vectors will give a vector. This gives rise to the idea of forming *linear combinations* of vectors, for example given the vectors $\boldsymbol{v}_1, \boldsymbol{v}_2, \ldots, \boldsymbol{v}_m$ then the linear combination

$$\boldsymbol{v} = a_1\boldsymbol{v}_1 + a_2\boldsymbol{v}_2 + \cdots + a_m\boldsymbol{v}_m$$

is also a vector, where $a_1, a_2, \ldots, a_m$ are scalars. A set of vectors $\boldsymbol{v}_1, \boldsymbol{v}_2, \ldots, \boldsymbol{v}_m$ are said to be *linearly dependent* if there exist m scalars $a_1, a_2, \ldots, a_m$ not all of them zero, such that

$$a_1\boldsymbol{v}_1 + a_2\boldsymbol{v}_2 + \cdots + a_m\boldsymbol{v}_m = \boldsymbol{0}. \tag{5.11}$$

Otherwise the vectors are said to be *linearly independent*. If the vectors $\boldsymbol{v}_1, \boldsymbol{v}_2, \ldots, \boldsymbol{v}_m$ are linearly independent then the only scalars that satisfy eqn 5.11 are

$$a_1 = a_2 = \cdots = a_m = 0.$$

If we can find any set of scalars, where not all of them are zero, such that eqn 5.11 is satisfied then the vectors $\boldsymbol{v}_1, \boldsymbol{v}_2, \ldots, \boldsymbol{v}_m$ are linearly dependent. However, if eqn 5.11 is not satisfied for a given set of scalars then the vectors $\boldsymbol{v}_1, \boldsymbol{v}_2, \ldots, \boldsymbol{v}_m$ may still be linearly dependent for there may be some other set of scalars for which eqn 5.11 is satisfied. The example below illustrates this point.

Example 5.9
For each of the three sets of vectors determine whether or not the vectors are linearly dependent

(a) $\boldsymbol{u}_1 = (3, -8, 5)$, $\boldsymbol{u}_2 = (-2, 2, 14)$, and $\boldsymbol{u}_3 = (-1, 6, -19)$
(b) $\boldsymbol{v}_1 = (7, 3, -3)$, $\boldsymbol{v}_2 = (16, 2, -1)$, and $\boldsymbol{v}_3 = (6, -12, 15)$
(c) $\boldsymbol{w}_1 = (-2, 0, 0)$, $\boldsymbol{w}_2 = (0, 7, 0)$, and $\boldsymbol{w}_3 = (0, 0, 9)$.

(a) Taking $a_1 = 1$, $a_2 = 1$, and $a_3 = 1$ gives the linear combination

$$\begin{aligned} a_1\boldsymbol{u}_1 + a_2\boldsymbol{u}_2 + a_3\boldsymbol{u}_3 &= \boldsymbol{u}_1 + \boldsymbol{u}_2 + \boldsymbol{u}_3 \\ &= (3 - 2 - 1, -8 + 2 + 6, 5 + 14 - 19) \\ &= (0, 0, 0) \end{aligned}$$

and therefore the vectors are linearly dependent.

(b) Taking $a_1 = 1$, $a_2 = 1$, and $a_3 = 1$ gives $\boldsymbol{v}_1 + \boldsymbol{v}_2 + \boldsymbol{v}_3 = (29, -7, 11)$ which is nonzero. This though does not necessarily mean that the vectors are independent—there may be other scalars that give a zero linear combination. If we take $a_1 = 2$, $a_2 = -1$, and $a_3 = 1/3$ then

$$\begin{aligned} 2\boldsymbol{v}_1 - \boldsymbol{v}_2 + (1/3)\boldsymbol{v}_3 & \\ &= 2(7,3,-3) - 1(16,2,-1) + 1/3(6,-12,15) \\ &= (14,6,-6) + (-16,-2,1) + (2,-4,5) \\ &= (14-16+2, 6-2-4, -6+1+5) \\ &= (0,0,0) \end{aligned}$$

and the vectors are therefore linearly dependent.

(c) Here any linear combination of $\boldsymbol{w}_1$, $\boldsymbol{w}_2$, and $\boldsymbol{w}_3$ gives a nonzero vector, for example

$$\begin{aligned} 5\boldsymbol{w}_1 - 4\boldsymbol{w}_2 + 10\boldsymbol{w}_3 & \\ &= 5(-2,0,0) - 4(0,7,0) + 10(0,0,9) \\ &= (-10,0,0) + (0,-28,0) + (0,0,90) \\ &= (-10,-28,90). \end{aligned}$$

Consider $a_1\boldsymbol{w}_1 + a_2\boldsymbol{w}_2 + a_3\boldsymbol{w}_3$, this gives

$$\begin{aligned} a_1(-2,0,0) + a_2(0,7,0) + a_3(0,0,9) & \\ &= (-2a_1,0,0) + (0,7a_2,0) + (0,0,9a_3) \\ &= (-2a_1, 7a_2, 9a_3) \end{aligned}$$

and the only way in which $(-2a_1, 7a_2, 9a_3)$ can equal $(0,0,0)$ is if $a_1 = a_2 = a_3 = 0$. Therefore the only scalars that give a zero linear combination of the vectors $\boldsymbol{w}_1$, $\boldsymbol{w}_2$, and $\boldsymbol{w}_3$ are $a_1 = a_2 = a_3 = 0$ and so the vectors are linearly independent. □

The ability to form linear combinations of vectors and the property of linear independence are quite important, for as we shall now see they simplify the construction of vector spaces. Within a vector space there always exists at least one set of linearly independent vectors from which all other vectors can be generated. These are known as *basis vectors* and a set of basis vectors is referred to as a *basis* for the vector space. Each vector within the vector space can be uniquely expressed as a linear combination of the basis vectors. The basis vectors are said to *span* the vector space. The number of vectors within the basis is called the *dimension* of the vector space. In an m-dimensional vector space any set of m linearly independent vectors spans the space and is therefore a basis for the space. As an example lets consider a 3-dimensional space over the real field. Here the vectors

$$\begin{aligned} \boldsymbol{i} &= (1\ 0\ 0) \\ \boldsymbol{j} &= (0\ 1\ 0) \\ \boldsymbol{k} &= (0\ 0\ 1) \end{aligned} \tag{5.12}$$

can be used as basis vectors and all other vectors within the space can be expressed as

$$\boldsymbol{v} = v_x\boldsymbol{i} + v_y\boldsymbol{j} + v_z\boldsymbol{k}$$

where v_x, v_y, and v_z are scalars belonging to the real field and are the projection of $\boldsymbol{v}$ along the X, Y, and Z axes respectively. Substituting eqns 5.12 into $\boldsymbol{v}$ gives

$$\begin{aligned}\boldsymbol{v} &= v_x(1\ 0\ 0) + v_y(0\ 1\ 0) + v_z(0\ 0\ 1)\\ &= (v_x\ 0\ 0) + (0\ v_y\ 0) + (0\ 0\ v_z)\\ &= (v_x\ v_y\ v_z)\end{aligned}$$

which is the standard form for a vector in 3 dimensions. Equations 5.12 can be easily extended to an m-dimensional space so giving

$$\begin{aligned}\boldsymbol{b}_1 &= (1\ 0 \ldots 0)\\ \boldsymbol{b}_2 &= (0\ 1 \ldots 0)\\ &\ \vdots \qquad\qquad \vdots\\ \boldsymbol{b}_m &= (0\ 0 \ldots 1)\end{aligned} \tag{5.13}$$

as a set of basis vectors and are referred to as the *standard basis*.

The vector spaces that we are interested in are those constructed from binary n-tuples, so that vectors take the form of ordered sequences of 0s and 1s. Given the set of 2^n n-tuples it is quite easy to show that under modulo-2 addition a vector space, referred to as $\boldsymbol{V}_n$, is formed. Consider the 16 vectors formed from binary 4-tuples. The vector (0 0 0 0) is the identity element and each vector is its own inverse, for example $(1\ 0\ 1\ 1) + (1\ 0\ 1\ 1) = (1+1, 0+0, 1+1, 1+1) = (0\ 0\ 0\ 0)$. There are only 2 scalars, 0 and 1, in the field and as such we need only to consider multiplication by 0 and 1. Multiplying any vector by 0 always gives (0 0 0 0) and multiplication by 1 leaves a vector unchanged. Vector addition and scalar multiplication obey the required associative, distributive and commutative laws given by eqns 5.9 and 5.10, therefore the set of 4-tuples under modulo-2 addition form a vector space known as $\boldsymbol{V}_4$ (see Table 5.4).

In a vector space subsets of vectors can exist that have all the characteristics of a vector space under vector addition and scalar multiplication, in which case the subset of vectors is known as a *vector subspace* or *subspace*. Consider the four vectors

$$(0\ 0\ 0\ 0), (0\ 1\ 1\ 1), (1\ 0\ 1\ 0), \text{ and } (1\ 1\ 0\ 1).$$

Taking the sum of any two vectors gives one of the four vectors and the set is therefore closed under addition. It is easy to verify that the four vectors satisfy all the conditions necessary to form a vector space and therefore the four vectors form a subspace of $\boldsymbol{V}_4$. This is not the only subspace that can be formed, for example the vectors (0 0 0 0), (0 1 1 0), (0 0 1 0), and (0 1 0 0) are also a vector subspace of $\boldsymbol{V}_4$.

The standard basis of $\boldsymbol{V}_n$ cannot be a basis for a subspace of $\boldsymbol{V}_n$ because the standard basis, like any other basis, generates all the vectors in $\boldsymbol{V}_n$. For example in $\boldsymbol{V}_4$

Table 5.4
The vector space V_4

(0 0 0 0)	(0 1 0 0)	(1 0 0 0)	(1 1 0 0)
(0 0 0 1)	(0 1 0 1)	(1 0 0 1)	(1 1 0 1)
(0 0 1 0)	(0 1 1 0)	(1 0 1 0)	(1 1 1 0)
(0 0 1 1)	(0 1 1 1)	(1 0 1 1)	(1 1 1 1)

the basis vectors (1 0 0 0), (0 1 0 0), (0 0 1 0), and (0 0 0 1) are clearly not basis vectors of the subspace (0 0 0 0), (0 1 1 1), (1 0 1 0), and (1 1 0 1) (they do not even belong to the subspace). A set of linearly independent vectors that is not a basis of a subspace can always be extended as so to form a basis, this is known as *completion* of a basis. If the set is not a basis then there exists at least one vector, in the subspace, that is not a linear combination of the vectors in the set. One of these vectors is appended to the set and the set is checked to see if it now forms a basis. If the set is still not a basis, the process is repeated, so adding vectors one at a time, until the set is a basis. To start the process we only need 2 linearly independent vectors.

Example 5.10

Determine a basis and the dimension of the subspace in V_5 consisting of the vectors

$$\begin{array}{cccc}(0\ 0\ 0\ 0\ 0) & (1\ 1\ 1\ 0\ 0) & (0\ 1\ 0\ 1\ 0) & (1\ 0\ 0\ 0\ 1)\\ (1\ 0\ 1\ 1\ 0) & (0\ 1\ 1\ 0\ 1) & (1\ 1\ 0\ 1\ 1) & (0\ 0\ 1\ 1\ 1).\end{array}$$

Any 2 linearly independent nonzero vectors, say (1 1 1 0 0) and (0 1 0 1 0), can be used as the initial set of vectors. They are linearly independent because if we consider $\boldsymbol{v} = a_1(1\ 1\ 1\ 0\ 0) + a_2(0\ 1\ 0\ 1\ 0)$ then the only values of a_1 and a_2 that give $\boldsymbol{v} = 0$ are $a_1 = a_2 = 0$. Of the remaining 5 nonzero vectors (1 0 1 1 0) cannot be appended to the set, because $(1\ 0\ 1\ 1\ 0) = (1\ 1\ 1\ 0\ 0) + (0\ 1\ 0\ 1\ 0)$ and is therefore linearly dependent on the 2 vectors already in the set. Any one of the vectors (1 0 0 0 1), (0 1 1 0 1), (1 1 0 1 1) or (0 0 1 1 1) can be appended to the set. Taking, say, (0 0 1 1 1) as the third basis vector we find that all the vectors within the subspace can be constructed from a linear combination of the vectors (1 1 1 0 0), (0 1 0 1 0), and (0 0 1 1 1):

$$\begin{aligned}
0\cdot(1\ 1\ 1\ 0\ 0) + 0\cdot(0\ 1\ 0\ 1\ 0) + 0\cdot(0\ 0\ 1\ 1\ 1) &= (0\ 0\ 0\ 0\ 0)\\
0\cdot(1\ 1\ 1\ 0\ 0) + 0\cdot(0\ 1\ 0\ 1\ 0) + 1\cdot(0\ 0\ 1\ 1\ 1) &= (0\ 0\ 1\ 1\ 1)\\
0\cdot(1\ 1\ 1\ 0\ 0) + 1\cdot(0\ 1\ 0\ 1\ 0) + 0\cdot(0\ 0\ 1\ 1\ 1) &= (0\ 1\ 0\ 1\ 0)\\
1\cdot(1\ 1\ 1\ 0\ 0) + 0\cdot(0\ 1\ 0\ 1\ 0) + 0\cdot(0\ 0\ 1\ 1\ 1) &= (1\ 1\ 1\ 0\ 0)\\
0\cdot(1\ 1\ 1\ 0\ 0) + 1\cdot(0\ 1\ 0\ 1\ 0) + 1\cdot(0\ 0\ 1\ 1\ 1) &= (0\ 1\ 1\ 0\ 1)\\
1\cdot(1\ 1\ 1\ 0\ 0) + 1\cdot(0\ 1\ 0\ 1\ 0) + 0\cdot(0\ 0\ 1\ 1\ 1) &= (1\ 0\ 1\ 1\ 0)\\
1\cdot(1\ 1\ 1\ 0\ 0) + 0\cdot(0\ 1\ 0\ 1\ 0) + 1\cdot(0\ 0\ 1\ 1\ 1) &= (1\ 1\ 0\ 1\ 1)\\
1\cdot(1\ 1\ 1\ 0\ 0) + 1\cdot(0\ 1\ 0\ 1\ 0) + 1\cdot(0\ 0\ 1\ 1\ 1) &= (1\ 0\ 0\ 0\ 1)
\end{aligned}$$

A basis of the subspace is therefore (1 1 1 0 0), (0 1 0 1 0), and (0 0 1 1 1), and the dimension of the subspace is 3 because there are 3 basis vectors. Note that the basis is not unique, we could have taken (1 0 0 0 1), (0 1 1 0 1), or (1 1 0 1 1) as the third vector to complete the basis. □

The definition of a vector space requires vector multiplication only by scalars and not by vectors. However vectors can be multiplied together, and there are different ways in which this can be achieved. Consider the two vectors

$$\boldsymbol{u} = (u_1, u_2, \ldots, u_n)$$

and

$$\boldsymbol{v} = (v_1, v_2, \ldots, v_n)$$

the *dot product* (also known as the *inner product*) of $\boldsymbol{u}$ and $\boldsymbol{v}$ is defined as

$$\boldsymbol{u} \cdot \boldsymbol{v} = u_1 v_1 + u_2 v_2 + \cdots + u_n v_n.$$

Note that whilst $\boldsymbol{u}$ and $\boldsymbol{v}$ are vectors the dot product $\boldsymbol{u} \cdot \boldsymbol{v}$ is a scalar belonging to the same field as the components of $\boldsymbol{u}$ and $\boldsymbol{v}$. Two vectors $\boldsymbol{u}$ and $\boldsymbol{v}$ are said to be *orthogonal* to each other if their dot product is zero

$$\boldsymbol{u} \cdot \boldsymbol{v} = 0.$$

Note that orthogonality is a relative property, if $\boldsymbol{u} \cdot \boldsymbol{v} = 0$ then $\boldsymbol{u}$ and $\boldsymbol{v}$ are orthogonal to each other but not necessarily to other vectors. It is possible for a vector in V_n to be orthogonal to itself. Consider, for example, the vector $\boldsymbol{v} = (0\ 1\ 1\ 1\ 1)$ in V_5, the dot product $\boldsymbol{v} \cdot \boldsymbol{v} = (0\ \ 1\ \ 1\ \ 1\ \ 1) \cdot (0\ \ 1\ \ 1\ \ 1\ \ 1) = 0 \cdot 0 + 1 \cdot 1 + 1 \cdot 1 + 1 \cdot 1 + 1 \cdot 1 = 1 + 1 + 1 + 1 = 0$ and therefore $(0\ 1\ 1\ 1\ 1)$ is orthogonal to itself. The property of orthogonality can be used to construct a second vector space from a given vector space. If V is a subspace in V_n then the set of vectors in V_n that are orthogonal to the vectors in V is also a subspace of V_n and is called the *dual space* or *null space* of V. Let V_d be the dual space of V then

$$\boldsymbol{v} \cdot \boldsymbol{v}_\mathrm{d} = 0$$

for all vectors $\boldsymbol{v}$ and $\boldsymbol{v}_\mathrm{d}$ belonging to V and V_d respectively. The property of duality is not an absolute property, but a relative property between 2 spaces. If V_d is the dual space of V, then V is the dual space of V_d. Vectors belonging to the dual space, V_d have the same number of components as vectors belonging to V. If the dimension of V is k then the dimension of its dual space V_d is $n - k$. Over the real field a vector space and its dual space only contain the all-zero vector as a common element. However in V_n, a vector space V can contain vectors that also lie in the dual space of V. Furthermore the dual space can be a subspace of V or contain V as a subspace, or the space V can be it own dual space.

Example 5.11

Show that the vector space V consisting of the vectors $(0\ 0\ 0\ 0)$, $(0\ 0\ 0\ 1)$, $(0\ 0\ 1\ 0)$, and $(0\ 0\ 1\ 1)$, and the space W consisting of $(0\ 0\ 0\ 0)$, $(1\ 0\ 0\ 0)$, $(0\ 1\ 0\ 0)$, and $(1\ 1\ 0\ 0)$, form dual subspaces of V_4.

Taking the dot product of first nonzero vector in V with the nonzero vectors in W gives

$$(0\ 0\ 0\ 1) \cdot (1\ 0\ 0\ 0) = 0 \cdot 1 + 0 \cdot 0 + 0 \cdot 0 + 1 \cdot 0 = 0$$

$$(0\ 0\ 0\ 1) \cdot (0\ 1\ 0\ 0) = 0 \cdot 0 + 0 \cdot 1 + 0 \cdot 0 + 1 \cdot 0 = 0$$

$$(0\ 0\ 0\ 1) \cdot (1\ 1\ 0\ 0) = 0 \cdot 1 + 0 \cdot 1 + 0 \cdot 0 + 1 \cdot 0 = 0.$$

Hence $(0\ 0\ 0\ 1)$ is orthogonal to all the vectors in W. Likewise we can show that $(0\ 0\ 1\ 0)$ and $(0\ 0\ 1\ 1)$ are orthogonal to the vectors in W, and therefore V and W are dual subspaces of V_4. In V the vectors $(0\ 0\ 0\ 1)$ and $(0\ 0\ 1\ 0)$ are linearly independent but the vector $(0\ 0\ 1\ 1) = (0\ 0\ 1\ 0) + (0\ 0\ 0\ 1)$ and so the dimension of V is 2. The dimension of the dual space W is also 2. The reader can verify that no other vector in V_4 lying outside V and W is orthogonal to all the vectors in V or in W. □

Unlike the dot product of two vectors, which gives a scalar, the *cross product* of two vectors gives another vector. For example expressing $\boldsymbol{u}$ and $\boldsymbol{v}$ as the 3-dimensional vectors

$$\boldsymbol{u} = u_x\boldsymbol{i} + u_y\boldsymbol{j} + u_z\boldsymbol{k}$$
$$\boldsymbol{v} = v_x\boldsymbol{i} + v_y\boldsymbol{j} + v_z\boldsymbol{k}$$

then their cross product is

$$\boldsymbol{u} \times \boldsymbol{v} = \begin{vmatrix} \boldsymbol{i} & \boldsymbol{j} & \boldsymbol{k} \\ u_x & u_y & u_z \\ v_x & v_y & v_z \end{vmatrix}$$

and evaluating the determinant on the right-hand side gives

$$\boldsymbol{u} \times \boldsymbol{v} = (u_y v_z - u_z v_y)\boldsymbol{i} + (u_z v_x - u_x v_z)\boldsymbol{j} + (u_x v_y - u_y v_x)\boldsymbol{k}.$$

Note that the end result is a vector. The ability to take the product of vectors gives a mathematical structure above vector spaces namely that of a *linear algebra*. A linear algebra has all the properties of a vector space but further allows vector multiplication subject to

$$\begin{aligned} \boldsymbol{u} * (\boldsymbol{v} * \boldsymbol{w}) &= (\boldsymbol{u} * \boldsymbol{v}) * \boldsymbol{w} \\ \boldsymbol{u} * (\boldsymbol{v} + \boldsymbol{w}) &= \boldsymbol{u} * \boldsymbol{v} + \boldsymbol{u} * \boldsymbol{w} \\ (\boldsymbol{u} + \boldsymbol{v}) * \boldsymbol{w} &= \boldsymbol{u} * \boldsymbol{w} + \boldsymbol{v} * \boldsymbol{w} \\ a(\boldsymbol{u} * \boldsymbol{v}) &= (a\boldsymbol{u}) * \boldsymbol{v} = \boldsymbol{u} * (a\boldsymbol{v}) \end{aligned} \tag{5.14}$$

where $\boldsymbol{u}$, $\boldsymbol{v}$, and $\boldsymbol{w}$ are vectors belonging to the linear algebra, a is a scalar and $*$ is the vector multiplication operation. Both the dot product and the cross product satisfy the distributive rules in eqns 5.14 but neither satisfy the associative rule. The product $\boldsymbol{u} * (\boldsymbol{v} * \boldsymbol{w})$ cannot be constructed when $*$ is the dot product because the first product $\boldsymbol{v} * \boldsymbol{w}$ gives a scalar say c, leaving $\boldsymbol{u} * c$ which is not defined. When $*$ is the cross product then $\boldsymbol{u} * (\boldsymbol{v} * \boldsymbol{w})$ and $(\boldsymbol{u} * \boldsymbol{v}) * \boldsymbol{w}$ give different vectors. Therefore neither the dot product or the cross product can be used to form a linear algebra. We can however generate a linear algebra by defining the vector product

$$\boldsymbol{u} * \boldsymbol{v} = (u_1 v_1, u_2 v_2, \ldots, u_n v_n)$$

where $\boldsymbol{u} = (u_1, u_2, \ldots, u_n)$ and $\boldsymbol{v} = (v_1, v_2, \ldots, v_n)$. It can be easily shown that this definition of vector multiplication satisfies eqns 5.14 and therefore forms a linear algebra.

5.5 Matrices

Here we consider characteristics of matrices relating to vector spaces. Given an n by m matrix $\boldsymbol{A}$ (that is a matrix with n rows and m columns) then $\boldsymbol{A}$ can be written as

$$\boldsymbol{A} = \begin{bmatrix} a_{1,1} & a_{1,2} & \cdots & a_{1,m} \\ a_{2,1} & a_{2,2} & \cdots & a_{2,m} \\ \vdots & & & \vdots \\ a_{n,1} & a_{n,2} & \cdots & a_{n,m} \end{bmatrix}$$

where $a_{i,j} = 0$ or $1, i = 1, \ldots, n$ and $j = 1, \ldots, m$. The ith row in A can be thought of as the vector

$$v_i = (a_{i,1}, a_{i,2}, \ldots, a_{i,m})$$

and is known as a *row vector*. The matrix A can be expressed in terms of its row vectors as

$$A = \begin{bmatrix} v_1 \\ v_2 \\ \vdots \\ v_n \end{bmatrix}.$$

The set of all linear combinations of the row vectors v_i forms a vector space referred to as the *row space* of the matrix A.

Example 5.12
Determine the row space of the matrix

$$A = \begin{bmatrix} 1 & 0 & 0 & 1 & 1 \\ 0 & 1 & 0 & 1 & 0 \\ 0 & 0 & 1 & 0 & 1 \end{bmatrix}.$$

The row vectors are $v_1 = (1\,0\,0\,1\,1)$, $v_2 = (0\,1\,0\,1\,0)$, and $v_3 = (0\,0\,1\,0\,1)$ and taking all the 2^3 linear combinations of v_1, v_2, and v_3 gives

$$\begin{aligned}
0 \cdot v_1 + 0 \cdot v_2 + 0 \cdot v_3 &= (0\,0\,0\,0\,0) \\
0 \cdot v_1 + 0 \cdot v_2 + 1 \cdot v_3 &= v_3 = (0\,0\,1\,0\,1) \\
0 \cdot v_1 + 1 \cdot v_2 + 0 \cdot v_3 &= v_2 = (0\,1\,0\,1\,0) \\
1 \cdot v_1 + 0 \cdot v_2 + 0 \cdot v_3 &= v_1 = (1\,0\,0\,1\,1) \\
0 \cdot v_1 + 1 \cdot v_2 + 1 \cdot v_3 &= v_2 + v_3 = (0\,1\,1\,1\,1) \\
1 \cdot v_1 + 0 \cdot v_2 + 1 \cdot v_3 &= v_1 + v_3 = (1\,0\,1\,1\,0) \\
1 \cdot v_1 + 1 \cdot v_2 + 0 \cdot v_3 &= v_1 + v_2 = (1\,1\,0\,0\,1) \\
1 \cdot v_1 + 1 \cdot v_2 + 1 \cdot v_3 &= v_1 + v_2 + v_3 = (1\,1\,1\,0\,0) \cdot
\end{aligned}$$

The row space of A therefore contains the vectors (0 0 0 0 0), (1 0 0 1 1), (0 1 0 1 0), (0 0 1 0 1), (1 1 0 0 1), (1 0 1 1 0), (0 1 1 1 1), and (1 1 1 0 0). □

We can likewise express a matrix in terms of *column vectors* constructed from the columns of the matrix, and taking all the linear combinations of the column vectors gives the *column space* of a matrix. Hence any matrix A determines 2 vector spaces; the row and column vector spaces of A. The dimensions of the column and row vector spaces, known as the *column* and *row rank* respectively, are the same and are referred to as the *rank* of the matrix.

Interchanging any two rows in a matrix gives the same row space, and furthermore changing a matrix by

(a) interchanging rows
(b) multiplying a row by a nonzero scalar
(c) adding a multiple of one row to another row

produces the same row space. The operations (a), (b), and (c) are known as the *elementary row operations*, and two matrices that differ only by elementary row operations have the same row space. If the elements of a matrix belong to the binary field then the row operation (b) is trivial and (c) simplifies to the addition of one row to another row (since the only nonzero scalar is 1).

Example 5.13

Consider the matrix $\boldsymbol{A}$ in Example 5.12. Adding row 3 to row 1 and then interchanging rows 2 and 3 gives

$$\boldsymbol{A}' = \begin{bmatrix} 1 & 0 & 1 & 1 & 0 \\ 0 & 0 & 1 & 0 & 1 \\ 0 & 1 & 0 & 1 & 0 \end{bmatrix}.$$

Show that the row space generated by $\boldsymbol{A}'$ is the same as that generated by $\boldsymbol{A}$.

The row vectors are now (1 0 1 1 0), (0 0 1 0 1), and (0 1 0 1 0). Taking all the 8 linear combinations of these 3 vectors gives the vectors (0 0 0 0 0), (1 0 1 1 0), (0 0 1 0 1), (0 1 0 1 0), (1 0 0 1 1), (1 1 1 0 0), (0 1 1 1 1), and (1 1 0 0 1). This is the same set of vectors as obtained in Example 5.12 and therefore the row space generated by $\boldsymbol{A}$ and $\boldsymbol{A}'$ are the same. □

5.6 Linear codes as vector spaces

In Chapter 2 a definition of linear codes, based on the property that the sum of two codewords gives another codeword, was given and codewords were loosely referred to as vectors. Here we consider a definition of linear codes based on the concepts of vector spaces.

Given an (n, k) code two sets of words can be defined. The first is the set of 2^n n-bit words which form the vector space V_n. As a vector space V_n has the property that its elements, that is the vectors, form a commutative group, over which vector addition and scalar multiplication can be carried out such that the associative, commutative, and distributive conditions given by eqns 5.9 and 5.10 are satisfied. The second, and smaller, set is formed by the 2^k codewords, which we denote by $\boldsymbol{C}_k$. The requirement for linearity is now that the set of codewords $\boldsymbol{C}_k$ forms a subspace of the vector space V_n, and so an (n, k) code is said to be a linear code if its 2^k codewords form a k-dimensional subspace of the vector space V_n. Recalling that the subspace of a vector space is also a vector space, we see therefore that the codewords of a linear code form a vector space. We have seen that in a k-dimensional vector space V there exists at least one set of k linearly-independent basis vectors which span the vector space, such that the vectors in V can be formed by linear combinations of the basis vectors. For a linear code, the rows of the generator matrix $\boldsymbol{G}$ form a set of basis vectors for the code, and taking all combinations of the rows of $\boldsymbol{G}$ generates the set of codewords.

Example 5.14
The (6, 3) code has generator matrix

$$G = \left[\begin{array}{ccc|ccc} 1 & 0 & 0 & 0 & 1 & 1 \\ 0 & 1 & 0 & 1 & 0 & 1 \\ 0 & 0 & 1 & 1 & 1 & 0 \end{array}\right].$$

Show that the codewords can be generated by taking linear combinations of the rows of $\boldsymbol{G}$.

Taking linear combinations of the 3 rows gives

$$\begin{aligned} \text{row } 1 &= (1\ 0\ 0\ 0\ 1\ 1) \\ \text{row } 2 &= (0\ 1\ 0\ 1\ 0\ 1) \\ \text{row } 3 &= (0\ 0\ 1\ 1\ 1\ 0) \\ \text{row } 1 + \text{row } 2 &= (1\ 1\ 0\ 1\ 1\ 0) \\ \text{row } 1 + \text{row } 3 &= (1\ 0\ 1\ 1\ 0\ 1) \\ \text{row } 2 + \text{row } 3 &= (0\ 1\ 1\ 0\ 1\ 1) \\ \text{row } 1 + \text{row } 2 + \text{row } 3 &= (1\ 1\ 1\ 0\ 0\ 0) \end{aligned}$$

which, along with the all-zero codeword (0 0 0 0 0 0), are the 8 codewords of the (6, 3) linear code (see Table 2.2). □

Recall that for an (n, k) linear code, the generator matrix $\boldsymbol{G}$ has k rows, which we now see are k linearly-independent vectors (row vectors). Denoting the k rows of $\boldsymbol{G}$ by the vectors $\boldsymbol{g}_1, \boldsymbol{g}_2, \ldots, \boldsymbol{g}_k$ we can express $\boldsymbol{G}$ as

$$\boldsymbol{G} = \begin{bmatrix} \boldsymbol{g}_1 \\ \boldsymbol{g}_2 \\ \vdots \\ \boldsymbol{g}_k \end{bmatrix}. \tag{5.15}$$

To check the linear independence of the row vectors we need to consider whether

$$a_1\boldsymbol{g}_1 + a_2\boldsymbol{g}_2 + \cdots + a_k\boldsymbol{g}_k = \boldsymbol{0}$$

where $a_1, a_2, \ldots, a_k$, are 0 or 1. If the row vectors are linearly independent then the only condition that satisfies the above equation is

$$a_1 = a_2 = \cdots = a_k = 0$$

otherwise the rows of $\boldsymbol{G}$ are linearly dependent.

Example 5.15
Consider the (6, 3) code whose generator matrix $\boldsymbol{G}$ is given in Example 5.14. Show that the row vectors of $\boldsymbol{G}$ are linearly independent.

The rows of $\boldsymbol{G}$ give

$$\begin{aligned} g_1 &= (1\ 0\ 0\ 0\ 1\ 1) \\ g_2 &= (0\ 1\ 0\ 1\ 0\ 1) \\ g_3 &= (0\ 0\ 1\ 1\ 1\ 0) \end{aligned}$$

and to check for linear independence we consider values of a_1, a_2, and a_3 for which

$$a_1 g_1 + a_2 g_2 + a_3 g_3 = \mathbf{0}.$$

Taking all 8 combinations of a_1, a_2, and a_3 gives

$$\begin{aligned} 0 \cdot g_1 + 0 \cdot g_2 + 0 \cdot g_3 &= (0\ 0\ 0\ 0\ 0\ 0) \\ 0 \cdot g_1 + 0 \cdot g_2 + 1 \cdot g_3 &= (0\ 0\ 1\ 1\ 1\ 0) \\ 0 \cdot g_1 + 1 \cdot g_2 + 0 \cdot g_3 &= (0\ 1\ 0\ 1\ 0\ 1) \\ 0 \cdot g_1 + 1 \cdot g_2 + 1 \cdot g_3 &= (0\ 1\ 1\ 0\ 1\ 1) \\ 1 \cdot g_1 + 0 \cdot g_2 + 0 \cdot g_3 &= (1\ 0\ 0\ 0\ 1\ 1) \\ 1 \cdot g_1 + 0 \cdot g_2 + 1 \cdot g_3 &= (1\ 0\ 1\ 1\ 0\ 1) \\ 1 \cdot g_1 + 1 \cdot g_2 + 0 \cdot g_3 &= (1\ 1\ 0\ 1\ 1\ 0) \\ 1 \cdot g_1 + 1 \cdot g_2 + 1 \cdot g_3 &= (1\ 1\ 1\ 0\ 0\ 0). \end{aligned}$$

Hence the only values of a_1, a_2, and a_3 that satisfy the condition for linear independence is
$a_1 = a_2 = a_3 = 0$
and therefore the row vectors g_1, g_2, and g_3 are linearly independent. Note that the remaining 7 combinations of a_1, a_2, and a_3 have generated the codewords. □

A codeword $\boldsymbol{c}$ can be expressed as a linear combination of the row vectors of $\boldsymbol{G}$, that is

$$\boldsymbol{c} = a_1 g_1 + a_2 g_2 + \cdots + a_k g_k \tag{5.16}$$

where $a_1, a_2, \ldots, a_k$ are 0 or 1. The codeword for an information word $\boldsymbol{i} = (i_1, i_2, \ldots, i_k)$ is given by $\boldsymbol{c} = \boldsymbol{iG}$. Using the generator matrix, in the form of eqn 5.15, we get

$$\begin{aligned} \boldsymbol{c} &= \boldsymbol{iG} \\ &= (i_1, i_2, \ldots, i_k) \begin{bmatrix} g_2 \\ g_2 \\ \vdots \\ g_k \end{bmatrix} \end{aligned}$$

and so

$$\boldsymbol{c} = i_1 g_1 + i_2 g_2 + \cdots + i_k g_k. \tag{5.17}$$

Comparing eqns 5.16 and 5.17 we see that each information word $\boldsymbol{i}$ generates a different combination of the code's basis vectors and therefore gives a different codeword. The equation $\boldsymbol{c}=\boldsymbol{i}\boldsymbol{G}$ can now be interpreted as generating different combinations of the code's basis vectors, as given in $\boldsymbol{G}$, according to the information word $\boldsymbol{i}$.

The generator matrix of the (7, 4) code can be expressed as

$$\boldsymbol{G} = \begin{bmatrix} \boldsymbol{g}_1 \\ \boldsymbol{g}_2 \\ \boldsymbol{g}_3 \\ \boldsymbol{g}_4 \end{bmatrix}$$

where the row vectors

$$\begin{aligned} \boldsymbol{g}_1 &= (1\ 0\ 0\ 0\ 1\ 0\ 1) \\ \boldsymbol{g}_2 &= (0\ 1\ 0\ 0\ 1\ 1\ 1) \\ \boldsymbol{g}_3 &= (0\ 0\ 1\ 0\ 1\ 1\ 0) \\ \boldsymbol{g}_4 &= (0\ 0\ 0\ 1\ 0\ 1\ 1) \end{aligned}$$

form a basis for the code. Codewords can be expressed as

$$\boldsymbol{c} = i_1\boldsymbol{g}_1 + i_2\boldsymbol{g}_2 + i_3\boldsymbol{g}_3 + i_4\boldsymbol{g}_4$$

where $\boldsymbol{i}=(i_1, i_2, i_3, i_4)$ is an information word. Taking any information word, say, $\boldsymbol{i}=(1\ 0\ 1\ 0)$ gives

$$\begin{aligned} \boldsymbol{c} &= 1 \cdot \boldsymbol{g}_1 + 0 \cdot \boldsymbol{g}_2 + 1 \cdot \boldsymbol{g}_3 + 0 \cdot \boldsymbol{g}_4 \\ &= \boldsymbol{g}_1 + \boldsymbol{g}_3 \\ &= (1\ 0\ 0\ 0\ 1\ 0\ 1) + (0\ 0\ 1\ 0\ 1\ 1\ 0) \\ &= (1\ 0\ 1\ 0\ 0\ 1\ 1) \end{aligned}$$

which is the codeword for $\boldsymbol{i}=(1\ 0\ 1\ 0)$ as given in Table 1.6.

On a practical point, the results of this section show us how to construct codewords from generator matrices without the need of matrix multiplication. Given a generator matrix $\boldsymbol{G}$, the complete set of codewords can be obtained by first writing down the individual rows, followed by every combination of 2 rows added together, then every combination of 3 rows added together and so forth until all rows are added together to give the last codeword. Example 5.14 gave an illustration of this. To construct a specific codeword for a given information word $\boldsymbol{i}$, we add together the rows of $\boldsymbol{G}$ corresponding to the 1s in $\boldsymbol{i}$, as indicated in eqn 5.17. Rows corresponding to the 0 information bits in $\boldsymbol{i}$ are omitted (see also Section 2.2).

Example 5.16

By adding together the appropriate rows of the generator matrix determine codewords for the information words

(a) $\boldsymbol{i}_1=(0\ 1\ 0\ 0)$

(b) $\boldsymbol{i}_2=(1\ 1\ 0\ 1)$

(c) $\boldsymbol{i}_3=(0\ 1\ 1\ 0)$

belonging to the (7, 4) code.

The generator matrix for the (7, 4) code is

$$G = \begin{bmatrix} 1 & 0 & 0 & 0 & 1 & 0 & 1 \\ 0 & 1 & 0 & 0 & 1 & 1 & 1 \\ 0 & 0 & 1 & 0 & 1 & 1 & 0 \\ 0 & 0 & 0 & 1 & 0 & 1 & 1 \end{bmatrix}.$$

(a) For $i_1 = (0\ 1\ 0\ 0)$ we get

$$c = (0\ 1\ 0\ 0) \begin{bmatrix} 1 & 0 & 0 & 0 & 1 & 0 & 1 \\ 0 & 1 & 0 & 0 & 1 & 1 & 1 \\ 0 & 0 & 1 & 0 & 1 & 1 & 0 \\ 0 & 0 & 0 & 1 & 0 & 1 & 1 \end{bmatrix}$$

and clearly we only need to take the second row of G to get $c = (0\ 1\ 0\ 0\ 1\ 1\ 1)$.

(b) For $i_2 = (1\ 1\ 0\ 1)$ we add together all the rows except the third row, which gives $c = (1\ 1\ 0\ 1\ 0\ 0\ 1)$.

(c) Here we add together the middle two rows, so giving $c = (0\ 1\ 1\ 0\ 0\ 0\ 1)$. □

5.7 Dual codes

We have previously considered how the parity-check polynomial of a cyclic code can be used to generate the code's dual code. Here we reconsider dual codes in the context of linear codes and vector spaces. Consider the parity-check matrix H of a linear code. Like the rows of the generator matrix G, the rows of the parity-check matrix are linearly independent and therefore generate a vector space. The vector spaces generated by G and H are dual spaces, so that any vector in one of the spaces is orthogonal to all the vectors in the other space. Recall that the parity-check matrix H, for an (n, k) linear code, is an $n-k$ by n matrix, so H has $n-k$ rows. The vector space generated by H therefore has dimension $n-k$ and contains 2^{n-k} vectors. Compare this to the vector space generated by G which is k-dimensional and contains 2^k vectors. Hence the number of vectors in the vector spaces generated by G and by H will usually differ. The parity-check matrix of a linear code can clearly be used as a generator matrix to generate a linear code. Furthermore as the vector spaces generated by G and H are dual spaces, the linear code generated by H is the dual code of C and is denoted by C_d, where C is the code generated by G. The code C_d is not a dual code in an absolute sense but only relative to the code C. We can equally say that C is the dual code of C_d. If C is an (n, k) linear code, then C_d has blocklength n and information length $n-k$, so that C_d is an $(n, n-k)$ linear code. Note that although the blocklength of the codes are the same, the code C contain 2^k codewords, whilst its dual code C_d contains 2^{n-k} codewords.

Let G_d be the generator matrix of the code that is dual to the (7, 4) code, then $G_d = H$ the parity-check matrix of the (7, 4) code, and so

$$G_d = \begin{bmatrix} 1 & 1 & 1 & 0 & 1 & 0 & 0 \\ 0 & 1 & 1 & 1 & 0 & 1 & 0 \\ 1 & 1 & 0 & 1 & 0 & 0 & 1 \end{bmatrix}. \tag{5.18}$$

Taking all the combinations of the rows of G_d gives the 8 codewords of the dual code as shown in Table 5.5(a) (this is the same set of codewords given in Table 3.6 and is reproduced here for convenience). The order in which the codewords appear in Table 5.5(a) is determined by the order in which the row vectors of G_d are combined and is therefore arbitrary. The codewords appear in a nonsystematic form, but can be rearranged to the systematic form shown in Table 5.5(b). The dual code's blocklength and information length are n and $n-k$ respectively, and therefore the code that is dual to the (7, 4) code is a (7, 3) code (as we have already seen in the context of cyclic codes). Codewords $\boldsymbol{c}$ and $\boldsymbol{c}_d$ from the (7, 4) code and the (7, 3) code respectively are orthogonal, so that the dot product of $\boldsymbol{c}$ and $\boldsymbol{c}_d$ is $\boldsymbol{c}\boldsymbol{c}_d = 0$. For example given the codewords (1 0 1 0 0 1 1) and (0 1 0 0 1 1 1) from the (7, 4) and the (7, 3) codes respectively then

$$\begin{aligned}\boldsymbol{c}\boldsymbol{c}_d &= (1\ 0\ 1\ 0\ 0\ 1\ 1)\cdot(0\ 1\ 0\ 0\ 1\ 1\ 1)\\ &= 1\cdot 0 + 0\cdot 1 + 1\cdot 0 + 0\cdot 0 + 0\cdot 1 + 1\cdot 1 + 1\cdot 1\\ &= 1+1\\ &= 0\end{aligned}$$

and the codewords are therefore orthogonal.

The (7, 3) code should not just be thought of as the dual code of the (7, 4) code, but as a linear code in its own right with generator matrix G_d given by eqn 5.18. The matrix G_d can be easily put into a systematic form to give

$$\boldsymbol{G} = \begin{bmatrix} 1 & 0 & 0 & 1 & 1 & 1 & 0 \\ 0 & 1 & 0 & 0 & 1 & 1 & 1 \\ 0 & 0 & 1 & 1 & 1 & 0 & 1 \end{bmatrix} \tag{5.19}$$

as a systematic generator matrix of the (7, 3) code. Referring to Tables 5.5 and 1.6, we can see that the codewords belonging to the (7, 3) code are also codewords of the (7, 4) code. Hence the (7, 3) linear code is a subspace of its dual space, such a code is known as a *self-orthogonal code*. Each codeword in the (7, 3) code is therefore orthogonal to all the other codewords within the (7, 3) code. A linear code can also

Table 5.5
The (7, 3) code

(a) *Nonsystematic codewords*	(b) *Systematic codewords*
(0 0 0 0 0 0 0)	(0 0 0 | 0 0 0 0)
(1 1 1 0 1 0 0)	(0 0 1 | 1 1 0 1)
(0 1 1 1 0 1 0)	(0 1 0 | 0 1 1 1)
(1 1 0 1 0 0 1)	(0 1 1 | 1 0 1 0)
(1 0 0 1 1 1 0)	(1 0 0 | 1 1 1 0)
(0 0 1 1 1 0 1)	(1 0 1 | 0 0 1 1)
(1 0 1 0 0 1 1)	(1 1 0 | 1 0 0 1)
(0 1 0 0 1 1 1)	(1 1 1 | 0 1 0 0)

be equal to its dual code, in that the two codes have the same set of codewords, in which case the code is known as a *self-dual code*. We have seen that the dual code of an (n, k) code is an $(n, n-k)$ code. Therefore a self-dual code must have $k=n-k$, that is $k=n/2$ and so a self-dual code with blocklength n is an $(n, n/2)$ code. For all even integers n there exists an $(n, n/2)$ self-dual code, for example the $(8, 4)$ code is self-dual. Recall that the dual of a code, with generator matrix $\boldsymbol{G}$, is generated by the code's parity-check matrix $\boldsymbol{H}$, and therefore a self-dual code has $\boldsymbol{H}=\boldsymbol{G}$. Furthermore if $\boldsymbol{G}=[\boldsymbol{I}|\boldsymbol{P}]$ is the generator matrix of a self-dual code, then $\boldsymbol{P}$ is a square matrix that satifies $\boldsymbol{P}\boldsymbol{P}^{\mathrm{T}}=\boldsymbol{I}$.

Problems

5.1 Determine whether the following sets form additive groups:
(a) $S_1=\{\pm 1, \pm 2, \pm 3, \ldots\}$
(b) $S_2=\{0, 1, 2, \ldots\}$
(c) $S_3=\{0, \pm 2, \pm 4, \ldots\}$
(d) $S_4=\{0, \pm 1, \pm 3, \ldots\}$

5.2 Construct the group of integers under modulo-6 addition.

5.3 Determine whether the following sets form multiplicative groups:
(a) The set of real numbers
(b) The set of positive real numbers
(c) The set of negative real numbers.

5.4 Show that modulo-8 multiplication over $\{1, 2, 3, 4, 5, 6, 7\}$ fails to form a multiplicative group.

5.5 Construct the prime field under modulo-5 addition and multiplication. Find the multiplicative inverse and additive inverse for each element within the field. Repeat for the modulo-7 prime field.

5.6 Evaluate the following expressions under modulo-5 arithmetic:
(a) $2 \times 7 + 6$
(b) $(4-8) \times 3 - 2$
(c) $(3+6)/2 - 4/3$.
Repeat using modulo-7 arithmetic.

5.7 Solve the simultaneous equations

$$3x + y = 2$$

$$2x + 3y = 6$$

defined over the prime field modulo-5. Show that the equations do not have a unique solution if they are defined over the prime field modulo-7.

5.8 Show that the set of vectors (0 0 0 0 0), (0 1 0 1 0), (1 1 0 0 1), (1 0 0 1 1), and (1 0 1 1 0) does not form a vector subspace of V_5.

5.9 Find 3 vectors such that when added to the set of vectors in Problem 5.8 the resulting 8 vectors form a vector subspace of V_5. Find a basis for the vector subspace and the dimension of the subspace.

5.10 Determine whether the vectors (0 1 1 0 1 1), (1 1 0 1 1 0), (0 1 1 1 1 0), and (1 1 0 0 1 1) are linearly independent.

5.11 Show that the row space of the matrix

$$G = \begin{bmatrix} 1 & 0 & 0 & 1 & 1 \\ 0 & 1 & 0 & 0 & 1 \\ 0 & 0 & 1 & 1 & 1 \end{bmatrix}$$

gives the codewords of the (5, 3) linear code given in Table 2.3.

5.12 Show that the row vectors of the matrix

$$G = \begin{bmatrix} 1 & 0 & 0 & 1 & 1 & 1 & 0 \\ 0 & 1 & 0 & 0 & 1 & 1 & 1 \\ 0 & 0 & 1 & 1 & 1 & 0 & 1 \end{bmatrix}$$

are linearly independent and that the row space of $\boldsymbol{G}$ gives the codewords of the (7, 3) code given in Table 5.5.

5.13 Show that the (8, 4) code with generator matrix

$$G = \begin{bmatrix} 1 & 0 & 0 & 0 & 1 & 0 & 1 & 1 \\ 0 & 1 & 0 & 0 & 1 & 1 & 1 & 0 \\ 0 & 0 & 1 & 0 & 1 & 1 & 0 & 1 \\ 0 & 0 & 0 & 1 & 0 & 1 & 1 & 1 \end{bmatrix}$$

is self-dual.

5.14 Show that every codeword in the (7, 3) code is orthogonal to every other codeword.

Galois fields

6

We have seen that the mathematical framework for linear codes and cyclic codes are matrices and polynomials respectively. For each type of code, encoding and decoding can be concisely formulated in terms of their respective mathematics. This difference in mathematics signifies more than just mathematical representation, for in going from linear codes to cyclic codes there is an increase in underlying mathematical structure. Moving on, from cyclic codes to the next level of codes with greater mathematical structure are the BCH codes, considered in Chapter 7. The mathematical framework within which the BCH codes are addressed is that of Galois fields and it is these that we consider next.

6.1 Roots of equations

We have already considered fields, in particular finite fields, in Chapter 5. Here we are interested in finite fields constructed from the roots of equations. The motivation for this approach lies in the property of cyclic codes that all codeword polynomials $c(x)$ have the generator polynomial $g(x)$ as a factor. This can be restated by saying that any root of $g(x)=0$ is also a root of $c(x)=0$ and it is the exploitation of roots of equations that takes us to the BCH codes and to the finite fields relevant to the codes.

Consider the *algebraic equation* of the form

$$a_n x^n + a_{n-1}x^{n-1} + \cdots + a_2 x^2 + a_1 x + a_0 = 0 \tag{6.1}$$

where $a_0 \neq 0$. For now the coefficients $a_0, a_1, \ldots, a_n$ are taken to be real numbers but later we consider eqn 6.1 when the coefficients are binary. The simplest algebraic equation is the equation of first degree obtained by setting $n=1$, this is usually written as

$$ax + b = 0$$

where $a=a_1$, $b=a_0$ and there exists one root $x=-b/a$. The second degree (quadratic) equation, commonly expressed as

$$ax^2 + bx + c = 0$$

where now $a=a_2$, $b=a_1$ and $c=a_0$, has 2 roots given by the famous 'formula' for solving quadratic equations

$$x = \frac{-b \pm \sqrt{b^2 - 4ac}}{2a}. \tag{6.2}$$

The roots of equations of third and fourth degree, with 3 and 4 roots respectively, can likewise be obtained algebraically. However, the same is not true for fifth-degree equations for it has been shown that it is not possible to solve the fifth-degree equation using a finite number of algebraic operations, in other words it is impossible to obtain the roots algebraically. This was established in the early 19th century by Abel, a Norwegian mathematician. Furthermore, it is now known that algebraic equations of degree greater than 5 cannot be solved algebraically. This though does not mean that roots for fifth and higher degree equations do not exist, it is just that expressions for giving the roots do not exist. A fifth-degree equation has 5 roots and an equation with degree n, as given by eqn 6.1, has n roots. There is, however, one slight problem regarding the nature of some of the roots. Returning to the quadratic equation $ax^2 + bx + c = 0$ and setting $a = c = 1$ and $b = 0$ gives

$$x^2 + 1 = 0. \tag{6.3}$$

Each root of eqn 6.3 has the property that when multiplied by itself it gives -1. But there are no real numbers that have this property and the only way of obtaining 2 roots for eqn 6.3 is by defining a new *imaginary* or *complex* number j, which satisfies

$$\mathrm{j}^2 + 1 = 0. \tag{6.4}$$

The number j can be combined with 2 real numbers p and q to give the complex number $p + \mathrm{j}q$. Addition and multiplication over the set of all complex numbers obey the rules described in Section 5.3 for a set of elements to form a field, and the resulting field is known as the *complex field*. It is because of the existence of the complex field that all equations of degree n have n roots. The roots to eqn 6.1 are of the form $p + \mathrm{j}q$ with real roots having $q = 0$. If we consider a quadratic equation with $b \neq 0$, for example

$$x^2 - 4x + 13 = 0$$

then using eqn 6.2 we can easily find that there are two complex roots $2 - \mathrm{j}3$ and $2 + \mathrm{j}3$. Complex roots, of an equation with real coefficients, always occur in pairs known as *complex conjugates* or *conjugates*, which take the form $p \pm \mathrm{j}q$. The complex number $p - \mathrm{j}q$ is the conjugate of $p + \mathrm{j}q$, and likewise $p + \mathrm{j}q$ is the conjugate of $p - \mathrm{j}q$. An equation can never have an odd number of complex roots, this would require a complex root without a conjugate. In a cubic equation the roots are either all real or there is one real root and two roots that are complex conjugates. For example, the roots of

$$x^3 - 6x^2 + 13x - 20 = 0$$

are 4, $1 + \mathrm{j}2$ and $1 - \mathrm{j}2$.

The complex field includes all the real numbers as they can be considered as field elements with $q = 0$. We can think of the complex field as expanding or extending the real field that we started with. In this sense, the real field is referred to as a *base field* and the complex field as an *extension field*, the complex field is an extension of the real field. The coefficients $a_0, a_1, \ldots, a_n$ in eqn 6.1 belong to the base field (i.e. the real field) whereas the roots belong to the extension field (i.e. the complex field). The occurrence of complex roots as conjugate pairs is necessary for the coefficients to lie in the real field. To illustrate this, let's try to construct a quadratic equation with 2 roots that are not conjugate pairs. For example let $1+j2$ and $3-j4$ be roots of a quadratic equation, then the quadratic equation will be

$$(x-(1+j2))(x-(3-j4)) = x^2-(4-j2)x+(11+j2)$$

which has coefficients in the complex field. It can be easily shown that if the coefficients of a quadratic equation, with complex roots, lie in the real field then the roots must form a conjugate pair.

Example 6.1
If $p+jq$ is a root of a quadratic equation with real coefficients, show that its conjugate $p-jq$ is the other root.

Let $a+jb$ be the other root, then the quadratic equation is

$$(x-(p+jq))(x-(a+jb)) = 0$$

which gives

$$x^2-x[(a+p)+j(b+q)]+(pa-qb)+j(aq+bp) = 0.$$

If the quadratic equation is to have real coefficients then the complex terms must equal zero, and so

$$b+q = 0$$
$$aq+bp = 0$$

which give $b=-q$ and $a=p$. Hence $a+jb=p-jq$ and so the two roots are conjugates of each other. □

We next consider roots of equations of the form $p(x)=0$ where $p(x)$ is a polynomial with binary coefficients. As we have seen such polynomials are used for encoding and decoding binary cyclic codes. The trivial cases of $x+1=0$ and $x^2+1=0$ have 1 as a root, in the latter case 1 is a double root. The quadratic equation

$$x^2+x+1 = 0 \tag{6.5}$$

presents more of a problem. We cannot use eqn 6.2 for solving this because eqn 6.2 has a 2 on the denominator and $2=0$ when using modulo-2 arithmetic. Since x can only have a value of 0 or 1, we can substitute 0 and 1 directly into eqn 6.5 to see which, if any, is a root. Substituting $x=0$ into eqn 6.5 gives 1 and so 0 is not a root. Likewise $x=1$ gives 1 and therefore neither 0 or 1 are roots of eqn 6.5. As another example consider

$$x^3+x+1 = 0. \tag{6.6}$$

Substituting $x=0$ or $x=1$ into this gives 1, and so again we have a binary polynomial without any binary roots. This is analogous to the situation encountered previously where we considered a real quadratic equation without any real roots, so the complex term j is 'invented' to get around the problem. Here we proceed in the same manner by defining a new term such that it is a root of the polynomial of interest. We will consider eqn 6.6 instead of eqn 6.5 as this proves to be more interesting. We could use j to denote a root of eqn 6.6, but this may cause confusion with the use of j in ordinary complex numbers. Instead it is conventional to use α to represent the newly defined root. Substituting $x=\alpha$ into eqn 6.6 gives

$$\alpha^3 + \alpha + 1 = 0. \tag{6.7}$$

Whilst eqn 6.7 may define the new root α it tells us little else about it and furthermore there are 2 more roots of eqn 6.6 that we need to find. We know that α does not belong to the binary field and to proceed further we need to determine the mathematical structure of the field within which α lies. The root α lies within a finite field known as $GF(2^3)$ which can be generated from eqn 6.7. Once $GF(2^3)$ has been established the other roots can be found.

6.2 The Galois field $GF(2^3)$

The field $GF(2^3)$ can be generated from the newly defined element α given by eqn 6.7. First consider addition and multiplication of α with the binary numbers 0 and 1. The binary numbers 0 and 1 form additive and multiplicative identity elements respectively, so

$$\alpha + 0 = \alpha$$
$$\alpha 1 = \alpha.$$

The additive inverse of α is α itself, as can be easily shown

$$\alpha + \alpha = 1\alpha + 1\alpha = (1+1)\alpha = 0\alpha = 0$$

and so

$$\alpha + \alpha = 0.$$

Furthermore rearranging this gives

$$\alpha = -\alpha$$

and therefore subtraction and addition of α are equivalent. The multiplicative inverse of α is defined as

$$\alpha^{-1} = \frac{1}{\alpha}$$

so that

$$\alpha^{-1}\alpha = \frac{1}{\alpha}\alpha = 1.$$

Table 6.1 summarizes the identity and inverse elements of α.

In eqn 6.7 it is implicit that $\alpha^3 = \alpha\alpha\alpha$ and likewise other powers of α can be defined, for example $\alpha^2 = \alpha\alpha$. Higher powers of α can be determined by rearranging

Table 6.1
Identity and inverse elements of α

Identity elements	Additive	0,	$\alpha+0=\alpha$
	Multiplicative	1,	$\alpha 1=\alpha$
Inverse elements	Additive	α,	$\alpha+\alpha=0$
	Multiplicative	α^{-1}	$\alpha\alpha^{-1}=1$

eqn 6.7 to give $\alpha^3=\alpha+1$ (recall that $-\alpha=\alpha$), repeatedly multiplying by α and substituting $\alpha+1$ for α^3 whenever α^3 appears. Starting first with α^3

$$\begin{aligned}
\alpha^3 &= \alpha+1\\
\alpha^4 &= \alpha\alpha^3 = \alpha(\alpha+1) = \alpha^2+\alpha\\
\alpha^5 &= \alpha\alpha^4 = \alpha(\alpha^2+\alpha) = \alpha^3+\alpha^2 = \alpha^2+\alpha+1\\
\alpha^6 &= \alpha\alpha^5 = \alpha(\alpha^2+\alpha+1) = \alpha^3+\alpha^2+\alpha = \alpha^2+1.
\end{aligned} \tag{6.8}$$

The four elements α^3, α^4, α^5, and α^6 differ from each other and from the four elements 0, 1, α, and α^2. Equations 6.8 are referred to as the *polynomial representations* of the elements α^3, α^4, α^5, and α^6. It may appear that other elements can be constructed by taking further powers of α. This though is not so, for the next power of α gives

$$\alpha^7 = \alpha\alpha^6 = \alpha(\alpha^2+1) = \alpha^3+\alpha = \alpha+1+\alpha = 1$$

and therefore

$$\alpha^7 = 1$$

which is an existing element. Forming further powers of α always generates one of the existing nonzero elements. For example, the next three powers give

$$\begin{aligned}
\alpha^8 &= \alpha\alpha^7 = \alpha 1 = \alpha\\
\alpha^9 &= \alpha\alpha^8 = \alpha\alpha = \alpha^2\\
\alpha^{10} &= \alpha\alpha^9 = \alpha\alpha^2 = \alpha^3.
\end{aligned}$$

A field element with power greater than 6 can be reduced to an element with power of 6 or less by removing factors of α^7. This is equivalent to taking the power modulo-7. For example

$$\begin{aligned}
\alpha^{12} &= \alpha^7\alpha^5 = \alpha^5(12 = 5 \text{ modulo-7})\\
\alpha^{17} &= \alpha^7\alpha^7\alpha^3 = \alpha^3(17 = 3 \text{ modulo-7})
\end{aligned}$$

and so forth. Taking into account 0 and 1 we see that we have constructed a set with the 8 elements

$$0, 1, \alpha, \alpha^2, \alpha^3, \alpha^4, \alpha^5, \text{and } \alpha^6.$$

Along with the operations addition and multiplication the set forms a field, namely the field $GF(2^3)$. Unlike the field of real numbers, which has an infinite number of field elements, $GF(2^3)$ has a finite number of field elements and is therefore a finite field. Finite fields are also referred to as *Galois fields* after the mathematician Evariste Galois (1811–1832). The fields are usually expressed as $GF(p^m)$ where p is the number of elements in the base field, which is referred to as the field's *characteristic*, and m is the degree of the polynomial whose root is used to construct the field. The order of the field is given by $q = p^m$.

Table 6.2 shows the 8 elements of $GF(2^3)$ along with the polynomial representations of α^3, α^4, α^5, and α^6 in terms of 1, α, and α^2. We have already seen that $\alpha + \alpha = 0$, likewise any field element added to itself gives zero. For example

$$\alpha^3 + \alpha^3 = 1\alpha^3 + 1\alpha^3 = \alpha^3(1 + 1) = 0.$$

Two different field elements can be added together by using their polynomial representations. For example, adding α^4 and α^6 gives

$$\alpha^4 + \alpha^6 = (\alpha^2 + \alpha) + (\alpha^2 + 1) = \alpha + 1 = \alpha^3.$$

Example 6.2

Find (a) $\alpha^2 + \alpha$, (b) $\alpha^5 + \alpha + 1$, and (c) $\alpha^6 + \alpha^2 + 1$.

(a) From Table 6.2 we see that $\alpha^2 + \alpha = \alpha^4$.

(b) Here we first rewrite α^5 in terms of its polynomial representation and then cancel out equal field elements

$$\alpha^5 + \alpha + 1 = (\alpha^2 + \alpha + 1) + \alpha + 1 = \alpha^2.$$

(c) We can express α^6 as $\alpha^2 + 1$, and so

$$\alpha^6 + \alpha^2 + 1 = (\alpha^2 + 1) + \alpha^2 + 1 = 0.$$

□

Using the polynomial representation of field elements to add 2 elements together is rather tedious. It is easier to construct a table showing the results of all additions and then to refer to the table when needed. Table 6.3(a) shows addition of elements within $GF(2^3)$.

Table 6.2
The field elements of $GF(2^3)$

0
1
α
α^2
$\alpha^3 = \alpha + 1$
$\alpha^4 = \alpha^2 + \alpha$
$\alpha^5 = \alpha^2 + \alpha + 1$
$\alpha^6 = \alpha^2 + 1$

Table 6.3
Addition and multiplication in $GF(2^3)$

(a) *Addition*

$+$	0	1	α	α^2	α^3	α^4	α^5	α^6
0	0	1	α	α^2	α^3	α^4	α^5	α^6
1	1	0	α^3	α^6	α	α^5	α^4	α^2
α	α	α^3	0	α^4	1	α^2	α^6	α^5
α^2	α^2	α^6	α^4	0	α^5	α	α^3	1
α^3	α^3	α	1	α^5	0	α^6	α^2	α^4
α^4	α^4	α^5	α^2	α	α^6	0	1	α^3
α^5	α^5	α^4	α^6	α^3	α^2	1	0	α
α^6	α^6	α^2	α^5	1	α^4	α^3	α	0

(b) *Multiplication*

$\times$	0	1	α	α^2	α^3	α^4	α^5	α^6
0	0	0	0	0	0	0	0	0
1	0	1	α	α^2	α^3	α^4	α^5	α^6
α	0	α	α^2	α^3	α^4	α^5	α^6	1
α^2	0	α^2	α^3	α^4	α^5	α^6	1	α
α^3	0	α^3	α^4	α^5	α^6	1	α	α^2
α^4	0	α^4	α^5	α^6	1	α	α^2	α^3
α^5	0	α^5	α^6	1	α	α^2	α^3	α^4
α^6	0	α^6	1	α	α^2	α^3	α^4	α^5

Multiplying two field elements together is straightforward, their powers are added together and because $\alpha^7 = 1$ factors of α^7 can be removed. For example, the product of α^3 and α^6 gives

$$\alpha^3\alpha^6 = \alpha^{3+6} = \alpha^9 = \alpha^7\alpha^2 = 1\alpha^2 = \alpha^2.$$

This is the same as taking the sum modulo-7 of the two powers. In the above case we have $3+6=2$ modulo-7 and the resulting field element is α^2 as obtained. Therefore the product of the field elements α^i and α^j in $GF(2^3)$ is

$$\alpha^i\alpha^j = \alpha^{(i+j)\text{modulo-7}}.$$

Although there is no need to refer to a multiplication table when multiplying field elements, one is included for completeness (see Table 6.3b).

Example 6.3
Find (a) $\alpha\alpha^5$, (b) $\alpha^4\alpha^5$, and (c) $\alpha^5\alpha^6\alpha^4$.

(a) $\alpha\alpha^5 = \alpha^6$
(b) $\alpha^4\alpha^5 = \alpha^9 = \alpha^7\alpha^2 = \alpha^2$
(c) $\alpha^5\alpha^6\alpha^4 = \alpha^{15} = \alpha^7\alpha^7\alpha = \alpha.$ □

For any field element α^i in $GF(2^3)$ we have

$$\alpha^i + \alpha^i = 1\alpha^i + 1\alpha^i = \alpha^i(1+1) = 0$$

and therefore each element is its own additive inverse. The multiplicative inverse of α^i is defined as the element α^{-i} such that

$$\alpha^i\alpha^{-i} = 1$$

and is given by

$$\alpha^{-i} = \alpha^{7-i}.$$

Taking the product of α^i with α^{-i} gives $\alpha^i\alpha^{-i}=\alpha^{i+7-i}=\alpha^7=1$ and so α^{-i} satisfies the requirement for a multiplicative inverse. Take, for instance, the multiplicative inverse of α^3

$$\alpha^{-3}=\alpha^{7-3}=\alpha^4$$

and so the multiplicative inverse of α^3 is α^4. Note that the element 1 is its own multiplicative inverse.

One of the requirements of a field is that division by nonzero elements is possible. Given α^i and α^j in $GF(2^3)$, where $\alpha^j\neq 0$, then α^i divided by α^j is

$$\frac{\alpha^i}{\alpha^j}=\alpha^i\alpha^{-j}=\alpha^{(i-j)\text{modulo-7}}$$

where α^{-j} is the multiplicative inverse of α^j. Note that if $i-j<0$ then $(i-j)$ modulo-7 is found by adding 7 to $i-j$. Also if $i=j$ then clearly $\alpha^i/\alpha^j=1$. For example

$$\frac{\alpha^6}{\alpha^2}=\alpha^6\alpha^{-2}=\alpha^{(6-2)\text{modulo-7}}=\alpha^4.$$

The above calculation can also be thought of as follows

$$\frac{\alpha^6}{\alpha^2}=\alpha^6\alpha^{-2}=\alpha^6\alpha^5=\alpha^{11\text{modulo-7}}=\alpha^4$$

where we have now made use of α^5 the multiplicative inverse of α^2, either way the same answer is obtained. Consider α divided by α^5, if we use the multiplicative inverse of α^5, which is α^2, we get

$$\frac{\alpha}{\alpha^5}=\alpha\alpha^{-5}=\alpha\alpha^2=\alpha^3$$

or without using the inverse we can view the calculation as

$$\frac{\alpha}{\alpha^5}=\alpha\alpha^{-5}=\alpha^{(1-5)\text{modulo-7}}=\alpha^{(-4)\text{modulo-7}}=\alpha^3.$$

Example 6.4

Find (a) α^2/α^5, (b) $1/\alpha$, (c) α^3/α, and (d) α/α^3.

(a) The inverse of α^5 is α^2, and so $\alpha^2/\alpha^5=\alpha^2\alpha^2=\alpha^4$. Or we can view this as $\alpha^2/\alpha^5=\alpha^{(2-5)\text{modulo-7}}=\alpha^4$.

(b) $1/\alpha=\alpha^6$

(c) $\alpha^3/\alpha=\alpha^2$

(d) $\alpha/\alpha^3=\alpha^5$. □

We now return to the problem of finding the three roots of the binary equation $x^3+x+1=0$ (see Section 6.1). We have already 'found' 1 root, namely α belonging to $GF(2^3)$, and next we will use a trial-and-error method to test the other elements of $GF(2^3)$ to see if they satisfy $x^3+x+1=0$. Only 5 of the 8 elements need to be considered as α is a root by definition and 0 and 1 are known not to be roots. Starting with $x = \alpha^2$ gives

$$\alpha^6 + \alpha^2 + 1 = 1 + 1 = 0$$

and so α^2 is a root (here we have referred to Table 6.3(a) to get $\alpha^2 + \alpha^6 = 1$). Next try $x = \alpha^3$

$$\alpha^9 + \alpha^3 + 1 = \alpha^7\alpha^2 + \alpha^3 + 1 = \alpha^2 + \alpha = \alpha^4 \neq 0$$

and so α^3 is not a root. Continuing with $x = \alpha^4$ gives

$$\alpha^{12} + \alpha^4 + 1 = \alpha^5 + \alpha^4 + 1 = 1 + 1 = 0$$

and therefore α^4 is the third root. The elements α^5 and α^6 cannot be roots because a cubic equation can only have 3 roots. As a check, substituting α^5 and α^6 into $x^3 + x + 1$ gives α^2 and α respectively, thus confirming that they are not roots. Therefore the three roots of the binary equation

$$x^3 + x + 1 = 0$$

are the field elements α, α^2, and α^4 belonging to the finite field $GF(2^3)$. Hence the original aim of finding the equation's roots has been achieved.

In Section 6.1 we considered equations with real coefficients and complex roots and the idea of base and extension fields were introduced in the context of the real and complex fields. The real field is thought of as the base field containing the equations' coefficients. The extension field contains the base field and extends it to include the complex roots. In the present case the equation of interest $x^3 + x + 1 = 0$ has its coefficients in the binary field and its roots in $GF(2^3)$. The binary field is denoted by $\boldsymbol{GF(2)}$ as it is a finite field with 2 elements. The field $GF(2^3)$ is an extension field of the binary field $GF(2)$. Note that $GF(2^3)$ is not the only extension field of $GF(2)$. The field $GF(2^3)$ has been constructed by determining the roots of the polynomial $p(x) = x^3 + x + 1$. Other extension fields can be generated using different polynomials. However, not all polynomials can generate extension fields, this is considered further in Section 6.5.

6.3 The fields $GF(2^4)$ and $GF(2^5)$

Here we are going to first construct the finite field $GF(2^4)$ and then take a brief look at $GF(2^5)$. $GF(2^4)$ is a field that the reader will encounter in most text books on error control. $GF(2^3)$ was constructed using a cubic polynomial that does not have 0 or 1 as roots. Consider the polynomial

$$p(x) = x^4 + x + 1. \tag{6.9}$$

Neither 0 or 1 are roots of $p(x) = 0$, it can be easily seen that $p(0) = p(1) = 1$. The four roots of eqn 6.9 therefore lie outside the binary field $GF(2)$. If we let α be one of the roots, then $p(\alpha) = 0$ by definition and

$$\alpha^4 + \alpha + 1 = 0. \tag{6.10}$$

Equation 6.10 is used to generate $GF(2^4)$ in the same way as eqn 6.7 was used to generate $GF(2^3)$. The binary elements 0 and 1 are again additive and multiplicative identity elements of α respectively. To determine the elements of $GF(2^4)$ we proceed in the same manner as when constructing $GF(2^3)$, by forming successive powers of α until an existing element is generated. Rearranging 6.10 gives

$$\alpha^4 = \alpha + 1. \tag{6.11}$$

When constructing higher powers of α eqn 6.11 is used to reduce field elements to their lowest power. Starting with α^4 and successively multiplying by α gives

$$\begin{aligned}
\alpha^4 &= \alpha + 1 \\
\alpha^5 &= \alpha\alpha^4 = \alpha(\alpha + 1) = \alpha^2 + \alpha \\
\alpha^6 &= \alpha\alpha^5 = \alpha(\alpha^2 + \alpha) = \alpha^3 + \alpha^2 \\
\alpha^7 &= \alpha\alpha^6 = \alpha(\alpha^3 + \alpha^2) = \alpha^4 + \alpha^3 = \alpha^3 + \alpha + 1.
\end{aligned} \tag{6.12}$$

Note that at this point $\alpha^7 = 1$ in $GF(2^3)$. However, here α^7 differs from all the previous elements and we therefore continue producing higher powers of α until an existing element is obtained

$$\begin{aligned}
\alpha^8 &= \alpha\alpha^7 = \alpha(\alpha^3 + \alpha + 1) = \alpha^4 + \alpha^2 + \alpha = \alpha^2 + 1 \\
\alpha^9 &= \alpha\alpha^8 = \alpha(\alpha^2 + 1) = \alpha^3 + \alpha \\
\alpha^{10} &= \alpha\alpha^9 = \alpha(\alpha^3 + \alpha) = \alpha^4 + \alpha^2 = \alpha^2 + \alpha + 1 \\
\alpha^{11} &= \alpha\alpha^{10} = \alpha(\alpha^2 + \alpha + 1) = \alpha^3 + \alpha^2 + \alpha \\
\alpha^{12} &= \alpha\alpha^{11} = \alpha(\alpha^3 + \alpha^2 + \alpha) = \alpha^3 + \alpha^2 + \alpha + 1.
\end{aligned}$$

All the elements generated so far are different, so the process is continued

$$\begin{aligned}
\alpha^{13} &= \alpha\alpha^{12} = \alpha(\alpha^3 + \alpha^2 + \alpha + 1) = \alpha^3 + \alpha^2 + 1 \\
\alpha^{14} &= \alpha\alpha^{13} = \alpha(\alpha^3 + \alpha^2 + 1) = \alpha^3 + 1
\end{aligned}$$

and finally

$$\alpha^{15} = \alpha\alpha^{14} = \alpha(\alpha^3 + 1) = \alpha^4 + \alpha = 1$$

which is an existing element. Constructing further powers of α will always give existing field elements, for example

$$\begin{aligned}
\alpha^{16} &= \alpha\alpha^{15} = \alpha 1 = \alpha \\
\alpha^{17} &= \alpha\alpha^{16} = \alpha\alpha = \alpha^2.
\end{aligned}$$

The field $GF(2^4)$ therefore has the following 16 elements

$$0, 1, \alpha, \alpha^2, \alpha^3, \alpha^4, \alpha^5, \alpha^6, \alpha^7, \alpha^8, \alpha^9, \alpha^{10}, \alpha^{11}, \alpha^{12}, \alpha^{13}, \alpha^{14}.$$

Table 6.4 lists the elements along with their polynomial representations in terms of 1, α, α^2, and α^3. Field elements of $GF(2^4)$ can be added together by using the polynomial representation of elements given in Table 6.4 or by referring to an addition table (see Table 6.5a).

Table 6.4
The field elements of $GF(2^4)$

0
1
α
α^2
α^3
$\alpha^4=\alpha+1$
$\alpha^5=\alpha^2+\alpha$
$\alpha^6=\alpha^3+\alpha^2$
$\alpha^7=\alpha^3+\alpha+1$
$\alpha^8=\alpha^2+1$
$\alpha^9=\alpha^3+\alpha$
$\alpha^{10}=\alpha^2+\alpha+1$
$\alpha^{11}=\alpha^3+\alpha^2+\alpha$
$\alpha^{12}=\alpha^3+\alpha^2+\alpha+1$
$\alpha^{13}=\alpha^3+\alpha^2+1$
$\alpha^{14}=\alpha^3+1$

Table 6.5(a)
Addition in $GF(2^4)$

$+$	1	α	α^2	α^3	α^4	α^5	α^6	α^7	α^8	α^9	α^{10}	α^{11}	α^{12}	α^{13}	α^{14}
1	0	α^4	α^8	α^{14}	α	α^{10}	α^{13}	α^9	α^2	α^7	α^5	α^{12}	α^{11}	α^6	α^3
α	α^4	0	α^5	α^9	1	α^2	α^{11}	α^{14}	α^{10}	α^3	α^8	α^6	α^{13}	α^{12}	α^7
α^2	α^8	α^5	0	α^6	α^{10}	α	α^3	α^{12}	1	α^{11}	α^4	α^9	α^7	α^{14}	α^{13}
α^3	α^{14}	α^9	α^6	0	α^7	α^{11}	α^2	α^4	α^{13}	α	α^{12}	α^5	α^{10}	α^8	1
α^4	α	1	α^{10}	α^7	0	α^8	α^{12}	α^3	α^5	α^{14}	α^2	α^{13}	α^6	α^{11}	α^9
α^5	α^{10}	α^2	α	α^{11}	α^8	0	α^9	α^{13}	α^4	α^6	1	α^3	α^{14}	α^7	α^{12}
α^6	α^{13}	α^{11}	α^3	α^2	α^{12}	α^9	0	α^{10}	α^{14}	α^5	α^7	α	α^4	1	α^8
α^7	α^9	α^{14}	α^{12}	α^4	α^3	α^{13}	α^{10}	0	α^{11}	1	α^6	α^8	α^2	α^5	α
α^8	α^2	α^{10}	1	α^{13}	α^5	α^4	α^{14}	α^{11}	0	α^{12}	α	α^7	α^9	α^3	α^6
α^9	α^7	α^3	α^{11}	α	α^{14}	α^6	α^5	1	α^{12}	0	α^{13}	α^2	α^8	α^{10}	α^4
α^{10}	α^5	α^8	α^4	α^{12}	α^2	1	α^7	α^6	α	α^{13}	0	α^{14}	α^3	α^9	α^{11}
α^{11}	α^{12}	α^6	α^9	α^5	α^{13}	α^3	α	α^8	α^7	α^2	α^{14}	0	1	α^4	α^{10}
α^{12}	α^{11}	α^{13}	α^7	α^{10}	α^6	α^{14}	α^4	α^2	α^9	α^8	α^3	1	0	α	α^5
α^{13}	α^6	α^{12}	α^{14}	α^8	α^{11}	α^7	1	α^5	α^3	α^{10}	α^9	α^4	α	0	α^2
α^{14}	α^3	α^7	α^{13}	1	α^9	α^{12}	α^8	α	α^6	α^4	α^{11}	α^{10}	α^5	α^2	0

Example 6.5
Find (a) $\alpha^2+\alpha^9$ and (b) $\alpha^7+\alpha^3+\alpha^{11}$ in $GF(2^4)$.

(a) From Table 6.4, $\alpha^9=\alpha^3+\alpha$ and so $\alpha^2+\alpha^9=\alpha^2+\alpha^3+\alpha=\alpha^{11}$.
(b) Again from Table 6.4, $\alpha^7=\alpha^3+\alpha+1$ and $\alpha^{11}=\alpha^3+\alpha^2+\alpha$ and therefore

$$\begin{aligned}\alpha^7+\alpha^3+\alpha^{11} &= \alpha^3+\alpha+1+\alpha^3+\alpha^3+\alpha^2+\alpha\\ &= \alpha^3+\alpha^2+1=\alpha^{13}.\end{aligned}$$

□

When multiplying two field elements together, factors of α^{15} can be taken out and set to unity as $\alpha^{15}=1$. This is equivalent to taking the sum modulo-15 of the exponents. Given two field elements α^i and α^j in $GF(2^4)$ their product is

$$\alpha^i\alpha^j = \alpha^{(i+j)\text{modulo-15}}.$$

Table 6.5(b) show the product of elements in $GF(2^4)$.

Example 6.6
Find (a) $\alpha^2\alpha^9$, (b) $\alpha^{13}\alpha^8$, and (c) $\alpha^7\alpha^{12}\alpha^4$ in $GF(2^4)$.

(a) $\alpha^2\alpha^9=\alpha^{11}$
(b) $\alpha^{13}\alpha^8=\alpha^{21}=\alpha^{15}\alpha^6=1\alpha^6=\alpha^6$
(c) $\alpha^7\alpha^{12}\alpha^4=\alpha^{(7+12+4)\text{modulo-15}}=\alpha^8$.

□

To divide two elements in $GF(2^4)$ the difference modulo-15 in the exponents is required, and so given α^i and α^j, where $\alpha^j\neq 0$, in $GF(2^4)$ then α^i divided by α^j is

$$\frac{\alpha^i}{\alpha^j} = \alpha^{(i-j)\text{modulo-15}}.$$

Table 6.5(b)
Multiplication in $GF(2^4)$

×	1	α	α^2	α^3	α^4	α^5	α^6	α^7	α^8	α^9	α^{10}	α^{11}	α^{12}	α^{13}	α^{14}
1	1	α	α^2	α^3	α^4	α^5	α^6	α^7	α^8	α^9	α^{10}	α^{11}	α^{12}	α^{13}	α^{14}
α	α	α^2	α^3	α^4	α^5	α^6	α^7	α^8	α^9	α^{10}	α^{11}	α^{12}	α^{13}	α^{14}	1
α^2	α^2	α^3	α^4	α^5	α^6	α^7	α^8	α^9	α^{10}	α^{11}	α^{12}	α^{13}	α^{14}	1	α
α^3	α^3	α^4	α^5	α^6	α^7	α^8	α^9	α^{10}	α^{11}	α^{12}	α^{13}	α^{14}	1	α	α^2
α^4	α^4	α^5	α^6	α^7	α^8	α^9	α^{10}	α^{11}	α^{12}	α^{13}	α^{14}	1	α	α^2	α^3
α^5	α^5	α^6	α^7	α^8	α^9	α^{10}	α^{11}	α^{12}	α^{13}	α^{14}	1	α	α^2	α^3	α^4
α^6	α^6	α^7	α^8	α^9	α^{10}	α^{11}	α^{12}	α^{13}	α^{14}	1	α	α^2	α^3	α^4	α^5
α^7	α^7	α^8	α^9	α^{10}	α^{11}	α^{12}	α^{13}	α^{14}	1	α	α^2	α^3	α^4	α^5	α^6
α^8	α^8	α^9	α^{10}	α^{11}	α^{12}	α^{13}	α^{14}	1	α	α^2	α^3	α^4	α^5	α^6	α^7
α^9	α^9	α^{10}	α^{11}	α^{12}	α^{13}	α^{14}	1	α	α^2	α^3	α^4	α^5	α^6	α^7	α^8
α^{10}	α^{10}	α^{11}	α^{12}	α^{13}	α^{14}	1	α	α^2	α^3	α^4	α^5	α^6	α^7	α^8	α^9
α^{11}	α^{11}	α^{12}	α^{13}	α^{14}	1	α	α^2	α^3	α^4	α^5	α^6	α^7	α^8	α^9	α^{10}
α^{12}	α^{12}	α^{13}	α^{14}	1	α	α^2	α^3	α^4	α^5	α^6	α^7	α^8	α^9	α^{10}	α^{11}
α^{13}	α^{13}	α^{14}	1	α	α^2	α^3	α^4	α^5	α^6	α^7	α^8	α^9	α^{10}	α^{11}	α^{12}
α^{14}	α^{14}	1	α	α^2	α^3	α^4	α^5	α^6	α^7	α^8	α^9	α^{10}	α^{11}	α^{12}	α^{13}

Example 6.7

Find (a) α^{13}/α^2, (b) α^3/α^{10}, and (c) α/α^9 in $GF(2^4)$.

(a) $\alpha^{13}/\alpha^2 = \alpha^{(13-2)\text{modulo-}15} = \alpha^{11}$

(b) $\alpha^3/\alpha^{10} = \alpha^{(3-10)\text{modulo-}15} = \alpha^8$

(c) $\alpha/\alpha^9 = \alpha^{(1-9)\text{modulo-}15} = \alpha^7$. □

Each element α^i in $GF(2^4)$ is its own additive inverse and has a multiplicative inverse given by $\alpha^{-i} = \alpha^{15-i}$. Note that elements that are common to $GF(2^3)$ and $GF(2^4)$ do not have the same inverse. For example in $GF(2^4)$ the inverse of α^2 is $\alpha^{-2} = \alpha^{13}$, whereas the inverse of α^2 in $GF(2^3)$ is $\alpha^{-2} = \alpha^5$.

Returning now to the polynomial $p(x) = x^4 + x + 1$ we have established that one of its roots is α belonging to $GF(2^4)$. We can now proceed to find the other 3 roots by using a trial-and-error method in which each field element α^i of $GF(2^4)$ is tested to see if it gives $p(\alpha^i) = 0$. We have already seen that 0 and 1 are not roots of $p(x)$, and so starting with $x = \alpha^2$ we find that

$$p(\alpha^2) = 0$$
$$p(\alpha^3) = \alpha^5$$
$$p(\alpha^4) = 0.$$

So far then, three of the roots are α, α^2 and α^4, recall that the same three field elements belonging to $GF(2^3)$ are roots of $x^3 + x + 1$. Another root is required and so continuing with the search gives

$$p(\alpha^5) = 1$$
$$p(\alpha^6) = \alpha^{10}$$
$$p(\alpha^7) = \alpha^{10}$$
$$p(\alpha^8) = 0$$

and α^8 is therefore the fourth root. The remaining elements need not be tested as there can be only 4 roots, the reader may wish to verify that taking $x = \alpha^9, \alpha^{10}, \alpha^{11}, \alpha^{12}, \alpha^{13}$, and α^{14} gives $p(x) \neq 0$. The roots of the binary equation $x^4 + x + 1 = 0$ are therefore the elements α, α^2, α^4, and α^8 lying in the field $GF(2^4)$.

We now take a brief look at $GF(2^5)$ generated by

$$p(x) = x^5 + x^2 + 1.$$

Setting $p(x) = 0$ and defining an element α such that $p(\alpha) = 0$ gives

$$\alpha^5 = \alpha^2 + 1$$

which can be used to generate the field $GF(2^5)$ with 32 elements and where $\alpha^{31} = 1$ (see Table 6.6). The other 4 roots of $p(x) = x^5 + x^2 + 1 = 0$ are α^2, α^4, α^8, and α^{16} belonging to $GF(2^5)$.

The fields $GF(2^3)$, $GF(2^4)$, and $GF(2^5)$ are just 3 examples of extension fields that can be constructed from the binary field $GF(2)$. Polynomials of degrees 3, 4, and 5 were used to construct $GF(2^3)$, $GF(2^4)$, and $GF(2^5)$ respectively. To construct the

Table 6.6
The field elements of $GF(2^5)$

0	$\alpha^{15}=\alpha^4+\alpha^3+\alpha^2+\alpha+1$
1	$\alpha^{16}=\alpha^4+\alpha^3+\alpha+1$
α	$\alpha^{17}=\alpha^4+\alpha+1$
α^2	$\alpha^{18}=\alpha+1$
α^3	$\alpha^{19}=\alpha^2+\alpha$
α^4	$\alpha^{20}=\alpha^3+\alpha^2$
$\alpha^5=\alpha^2+1$	$\alpha^{21}=\alpha^4+\alpha^3$
$\alpha^6=\alpha^3+\alpha$	$\alpha^{22}=\alpha^4+\alpha^2+1$
$\alpha^7=\alpha^4+\alpha^2$	$\alpha^{23}=\alpha^3+\alpha^2+\alpha+1$
$\alpha^8=\alpha^3+\alpha^2+1$	$\alpha^{24}=\alpha^4+\alpha^3+\alpha^2+\alpha$
$\alpha^9=\alpha^4+\alpha^3+\alpha$	$\alpha^{25}=\alpha^4+\alpha^3+1$
$\alpha^{10}=\alpha^4+1$	$\alpha^{26}=\alpha^4+\alpha^2+\alpha+1$
$\alpha^{11}=\alpha^2+\alpha+1$	$\alpha^{27}=\alpha^3+\alpha+1$
$\alpha^{12}=\alpha^3+\alpha^2+\alpha$	$\alpha^{28}=\alpha^4+\alpha^2+\alpha$
$\alpha^{13}=\alpha^4+\alpha^3+\alpha^2$	$\alpha^{29}=\alpha^3+1$
$\alpha^{14}=\alpha^4+\alpha^3+\alpha^2+1$	$\alpha^{30}=\alpha^4+\alpha$

field $GF(2^m)$ a polynomial of degree m is required. In the following sections we consider some basic properties of extension fields and field elements, along with the characteristics of polynomials that are relevant to the construction of fields.

6.4 Primitive field elements

The nonzero field elements of the Galois fields are generated by taking successive multiples of a single element α. Field elements that can generate all the nonzero elements of a field are said to be *primitive* and α is primitive in $GF(2^3)$, $GF(2^4)$, and $GF(2^5)$. It can be shown that every Galois field has at least one primitive field element. In $GF(2^3)$ α^2 is primitive as can be easily shown. For convenience arbitrary field elements are represented by β. If we let $\beta=\alpha^2$, then constructing successive powers of β gives

$$\beta = \alpha^2$$
$$\beta^2 = (\alpha^2)^2 = \alpha^4$$
$$\beta^3 = (\alpha^2)^3 = \alpha^6$$
$$\beta^4 = (\alpha^2)^4 = \alpha^8 = \alpha \quad \text{(recall that } \alpha^7 = 1 \text{ in } GF(2^3))$$
$$\beta^5 = (\alpha^2)^5 = \alpha^{10} = \alpha^3$$
$$\beta^6 = (\alpha^2)^6 = \alpha^{12} = \alpha^5.$$

So far this has generated six different elements, taking the next power gives

$$\beta^7 = (\alpha^2)^7 = \alpha^{14} = 1$$

and so further multiples of β will produce existing elements

$$\beta^8 = \beta = \alpha^2$$
$$\beta^9 = \beta^2 = \alpha^4$$

and so forth. Hence α^2 can also generate the nonzero elements of $GF(2^3)$ and is therefore a primitive field element of $GF(2^3)$. In fact all the elements (other than 0 and 1) of $GF(2^3)$ are primitive and therefore capable of generating the other nonzero elements.

Example 6.8
Show that α^5 is a primitive element of $GF(2^3)$.

Let $\beta = \alpha^5$ then

$$\beta^2 = \alpha^{10} = \alpha^3$$
$$\beta^3 = \alpha^{15} = \alpha$$
$$\beta^4 = \alpha^{20} = \alpha^6$$
$$\beta^5 = \alpha^{25} = \alpha^4$$
$$\beta^6 = \alpha^{30} = \alpha^2$$
$$\beta^7 = \alpha^{35} = 1.$$

Hence all 7 nonzero elements have been generated and α^5 is therefore primitive in $GF(2^3)$. □

We can likewise show that α^2 is primitive in $GF(2^4)$. However, consider next α^3 in $GF(2^4)$ and let $\beta = \alpha^3$, then

$$\beta = \alpha^3$$
$$\beta^2 = \alpha^6$$
$$\beta^3 = \alpha^9$$
$$\beta^4 = \alpha^{12}.$$

So far this has generated different elements, but the next term gives

$$\beta^5 = \alpha^{15} = 1 \text{(recall that } \alpha^{15} = 1 \text{ in } GF(2^4))$$

and therefore none of the remaining nonzero elements of $GF(2^4)$ can be generated. Continuing to take further powers of β will only generate 1, α^3, α^6, α^9, and α^{12}. For example

$$\beta^6 = \alpha^{18} = \alpha^3$$
$$\beta^7 = \alpha^{21} = \alpha^6$$
$$\beta^8 = \alpha^{24} = \alpha^9$$
$$\beta^9 = \alpha^{27} = \alpha^{12}$$
$$\beta^{10} = \alpha^{30} = 1.$$

Therefore within $GF(2^4)$ the field element α^3 is not primitive. There are other elements within $GF(2^4)$ that are not primitive, for example α^5 can only generate the elements 1 and α^{10}.

Whether or not a field element is primitive can be established by determining the *order* of the element, which for an element β is defined as the smallest positive integer n such that $\beta^n = 1$. This should not be confused with the order of a field, which is the number of elements within the field. In $GF(2^3)$ all the field elements have the same order 7. For example consider α^3 this has an order of 7 because $(\alpha^3)^7 = \alpha^{21} = 1$ and no other smaller power of α^3 gives 1. In $GF(2^4)$ however, not all elements have the same order. For example the order of α^5 is 3, whereas α^2 has an order of 15. The order of an element in $GF(2^m)$ divides $2^m - 1$ and furthermore determines whether or not the element is primitive. In a field $GF(2^m)$ a nonzero field element β is primitive if the order of β is $2^m - 1$. Within $GF(2^3)$ primitive field elements therefore have an order of 7 and primitive elements within $GF(2^4)$ have an order of 15.

Example 6.9
Given that α^{12} and α^7 are field elements of $GF(2^4)$ determine their order, whether or not they are primitive and the field elements generated if they are not primitive.

The smallest power of α^{12} to give unity is 5, as this gives $(\alpha^{12})^5 = \alpha^{60} = 1$. Hence α^{12} is not primitive. The elements generated by α^{12} are $(\alpha^{12})^2 = \alpha^{24} = \alpha^9$, $(\alpha^{12})^3 = \alpha^{36} = \alpha^6$, and $(\alpha^{12})^4 = \alpha^{48} = \alpha^3$. The next power of α^{12} gives $\alpha^{60} = 1$ and therefore α^{12} only generates 1, α^3, α^6, and α^9.

The field element α^7 has order 15 as $(\alpha^7)^{15} = \alpha^{105} = 1$ and no smaller power of α^7 gives unity, it is therefore primitive and generates all the field elements of $GF(2^4)$.

□

6.5 Irreducible and primitive polynomials

The polynomials $x^3 + x + 1$, $x^4 + x + 1$, and $x^5 + x^2 + 1$ used to generate $GF(2^3)$, $GF(2^4)$, and $GF(2^5)$ respectively cannot be factorized. Each polynomial is divisible only by itself and 1, such polynomials are referred to as *irreducible polynomials*. An irreducible polynomial having a primitive field element as a root is called a *primitive polynomial*. We have seen that $x^3 + x + 1$, $x^4 + x + 1$ and $x^5 + x^2 + 1$ have the primitive element α as a root and therefore the polynomials used to generate $GF(2^3)$, $GF(2^4)$, and $GF(2^5)$ are primitive.

For any positive integer m there is at least one irreducible polynomial of degree m. It can be shown that an irreducible polynomial of degree m divides $x^r + 1$ where $r = 2^m - 1$, and this can be used to establish whether or not a polynomial is irreducible. Take for example $x^3 + x + 1$, this should divide $x^7 + 1$ for it to be irreducible. Dividing $x^7 + 1$ by $x^3 + x + 1$ gives the quotient $x^4 + x^2 + x + 1$ and zero remainder, and therefore $x^3 + x + 1$ is irreducible. The reader can likewise show that $x^4 + x + 1$ and $x^5 + x^2 + 1$ are irreducible.

It is not always so easy, however, to establish whether or not an irreducible polynomial is primitive. It can be shown that an irreducible polynomial of degree m is primitive if it divides $x^r + 1$ for no r less than $2^m - 1$. Hence the polynomial must divide $x^r + 1$ but not $x^{r-1} + 1$, $x^{r-2} + 1$ and so forth. Consider again $x^3 + x + 1$, we have seen that it divides $x^7 + 1$ which shows that it is irreducible. To further show that it is primitive we need to show that it does not divide x^6+1, x^5+1 or x^4+1 (there is no need to consider division into x^3+1, x^2+1 or $x+1$ because a polynomial of degree m cannot divide a polynomial of degree $\leq m$). Taking x^6+1 and dividing by x^3+x+1 gives

$$
\begin{array}{rl}
 & x^3+x+1 \\
x^3+x+1 \,\big) & x^6+1 \\
 & x^6+x^4+x^3 \\
 & \overline{-\quad x^4+x^3+1} \\
 & \qquad x^4+x^2+x \\
 & \qquad \overline{-\quad x^3+x^2+x+1} \\
 & \qquad\qquad x^3+x+1 \\
 & \qquad\qquad \overline{-\quad x^2}
\end{array}
$$

resulting in a nonzero remainder. Likewise x^3+x+1 does not divide x^5+1 or x^4+1 and therefore x^3+x+1 is primitive.

The condition given for determining whether an irreducible polynomial is primitive is of limited use. However, a special case arises if 2^m-1 is prime, for an irreducible polynomial of degree m is primitive if 2^m-1 is prime. Care needs to be exercised here, because this special case means that if 2^m-1 is prime then the irreducible polynomial is primitive, but if 2^m-1 is not prime the polynomial may still be primitive. To illustrate this let's consider the polynomials used to generate $GF(2^3)$ and $GF(2^4)$. The field $GF(2^3)$ was generated using x^3+x+1 which has degree $m=3$, therefore $2^m-1=7$ is prime and so x^3+x+1 is primitive (as we have already seen). The polynomial x^4+x+1 used to generate $GF(2^4)$ has $m=4$ and $2^m-1=15$ which is not prime. However, the roots of the polynomial are α, α^2, α^4, and α^8 which are primitive elements of $GF(2^4)$ and therefore x^4+x+1 is primitive (recall that an irreducible polynomial with a primitive root is primitive).

Primitive polynomials are a special type of irreducible polynomials. With regard to generating a finite field it is the irreducible characteristic of a polynomial that is of importance. In order to generate a finite field it is not necessary for a polynomial to be primitive, it must however be irreducible. Primitive polynomials are preferred because it is easier to generate a field from a primitive polynomial than from one that is not primitive. A primitive polynomial has primitive roots and the field can be generated by taking successive powers of any primitive root. If an irreducible polynomial is not primitive then its roots are not primitive and each root generates only a limited number of field elements. To determine the remaining field elements a primitive element must first be found (recall that every finite field has at least one primitive element). To illustrate this consider the polynomial

$$p(x) = x^4+x^3+x^2+x+1. \tag{6.13}$$

The degree of $p(x)$ is $m=4$ and so for $p(x)$ to be irreducible it must divide $x^{15}+1$, which indeed it does. Hence $p(x)$ is irreducible and can be used to generate a field with 15 nonzero elements. Whether or not the polynomial is primitive cannot be determined from its degree because $2^m-1=15$ is not a prime number. To establish whether or not $p(x)$ is primitive we need to determine if it divides into $x^{14}+1$, $x^{13}+1, \ldots,$ or x^5+1. If $p(x)$ divides into any of these polynomials then it is not primitive. It can be shown that $p(x)$ divides into x^5+1 and $p(x)$ is therefore not primitive. To construct the field generated by eqn 6.13 we let α be a root of $p(x)$ then $\alpha^4+\alpha^3+\alpha^2+\alpha+1=0$ and so

$$\alpha^4 = \alpha^3 + \alpha^2 + \alpha + 1.$$

Proceeding as usual to take successive powers of α gives:

$$\begin{aligned}
&\alpha\\
&\alpha^2\\
&\alpha^3\\
&\alpha^4 = \alpha^3 + \alpha^2 + \alpha + 1\\
&\alpha^5 = \alpha\alpha^4 = \alpha(\alpha^3 + \alpha^2 + \alpha + 1) = \alpha^4 + \alpha^3 + \alpha^2 + \alpha = 1.
\end{aligned} \tag{6.14}$$

Hence α has order 5 and is therefore not primitive and fails to generate the 15 nonzero elements of the field. To proceed further we need to find a primitive element. None of the elements α^2, α^3, or α^4 are primitive, taking any one of these will only generate the existing elements. Instead we need to consider some other element. If we let $\beta=\alpha+1$ we find that β is primitive with $\beta, \beta^2, \beta^3, \ldots, \beta^{15}$ giving the 15 nonzero field elements shown in Table 6.7. We have now constructed two finite fields containing 16 elements, the field generated by $x^4+x^3+x^2+x+1$ (Table 6.7) and that generated by x^4+x+1 (Table 6.4). However the two fields are just different representations of the same field $GF(2^4)$, because two finite fields with the same number of elements differ only in the way that the elements are labelled or ordered. Two finite fields with the same number of elements are said to be *isomorphic*, whilst they may have different representations their mathematical structure is the same.

Table 6.7
$GF(2^4)$ generated by $p(x)=x^4+x^3+x^2+x+1$

0
$\beta=\alpha+1$
$\beta^2=\alpha^2+1$
$\beta^3=\alpha^3+\alpha^2+\alpha+1$
$\beta^4=\alpha^3+\alpha^2+\alpha$
$\beta^5=\alpha^3+\alpha^2+1$
$\beta^6=\alpha^3$
$\beta^7=\alpha^2+\alpha+1$
$\beta^8=\alpha^3+1$
$\beta^9=\alpha^2$
$\beta^{10}=\alpha^3+\alpha^2$
$\beta^{11}=\alpha^3+\alpha+1$
$\beta^{12}=\alpha$
$\beta^{13}=\alpha^2+\alpha$
$\beta^{14}=\alpha^3+\alpha$
$\beta^{15}=1$

6.6 Minimal polynomials

We have seen that irreducible polynomials and primitive polynomials are used to construct finite fields. Here we consider minimal polynomials, which are used in the construction of binary codes (see Chapter 7).

Complex roots of equations with real coefficients always occur in pairs of complex conjugates. If $p+jq$ is a root of an equation with real coefficients then its complex conjugate $p-jq$ is also a root. The roots of a polynomial with binary coefficients likewise occur in conjugates, not necessarily in pairs but in groups or *sets of conjugates*. Given that β is a field element of $GF(2^m)$ then the conjugates of β are

$$\beta, \beta^2, \beta^4, \beta^8, \ldots, \beta^{2^{r-1}}$$

where r is the smallest integer such that $\beta^{2^r} = \beta$. For example consider the conjugates of α^5 in $GF(2^4)$

$$(\alpha^5)^2 = \alpha^{10}$$
$$(\alpha^5)^4 = \alpha^{20} = \alpha^5$$

therefore in $GF(2^4)$ α^5 has only one conjugate, α^{10}. The conjugates of α^7 in $GF(2^4)$ are

$$(\alpha^7)^2 = \alpha^{14}$$
$$(\alpha^7)^4 = \alpha^{28} = \alpha^{13}$$
$$(\alpha^7)^8 = \alpha^{56} = \alpha^{11}$$
$$(\alpha^7)^{16} = \alpha^{112} = \alpha^7$$

and therefore α^7 has the conjugates α^{11}, α^{13}, and α^{14}.

Example 6.10
Determine the conjugates of α^3 in $GF(2^3)$ and in $GF(2^4)$.

In $GF(2^4)$ we have:

$$(\alpha^3)^2 = \alpha^6$$
$$(\alpha^3)^4 = \alpha^{12}$$
$$(\alpha^3)^8 = \alpha^{24} = \alpha^9$$
$$(\alpha^3)^{16} = \alpha^{48} = \alpha^3$$

and therefore the conjugates of α^3 are α^6, α^9 and α^{12}. Whereas in $GF(2^3)$ the conjugates of α^3 are:

$$(\alpha^3)^2 = \alpha^6$$
$$(\alpha^3)^4 = \alpha^{12} = \alpha^5$$
$$(\alpha^3)^8 = \alpha^{24} = \alpha^3.$$

Note that the set of conjugates of α^3 in $GF(2^3)$ is different from that in $GF(2^4)$.

□

Table 6.8
Conjugate elements in $GF(2^4)$ and in $GF(2^3)$

(a) $GF(2^4)$ Conjugates	*Order*	(b) $GF(2^3)$ Conjugates	*Order*
1	1	1	1
$\alpha, \alpha^2, \alpha^4, \alpha^8$	15	$\alpha, \alpha^2, \alpha^4$	7
$\alpha^3, \alpha^6, \alpha^9, \alpha^{12}$	5	$\alpha^3, \alpha^5, \alpha^6$	7
α^5, α^{10}	3		
$\alpha^7, \alpha^{11}, \alpha^{13}, \alpha^{14}$	15		

Table 6.8 shows the sets of conjugate elements in $GF(2^3)$ and $GF(2^4)$ along with the order of elements in the same conjugate set. Note that conjugate elements have the same order and therefore if an element β of $GF(2^m)$ is primitive then its conjugates are also primitive. Recall that the order of an element divides $2^m - 1$ and therefore if $2^m - 1$ is prime the field elements will have order $2^m - 1$ and be primitive. If $2^m - 1$ is not prime, then some elements will be nonprimitive with order less than $2^m - 1$ but there will be at least 1 primitive element. In $GF(2^3)$ $2^3 - 1 = 7$ is prime and therefore the field elements have order 7 and are primitive. $GF(2^4)$ has $2^4 - 1 = 15$ which is not prime and therefore the field has some nonprimitive elements with order less than 15.

One of the properties of conjugates is that they provide a mechanism for going from an extension field to its base field. Consider the pair of complex conjugates $z = p + jq$ and $z^* = p - jq$, their product gives the real number

$$zz^* = p^2 + q^2.$$

Taking the product of the two factors $(x - z)$ and $(x - z^*)$ likewise gives a real expression

$$(x - z)(x - z^*) = x^2 - 2px + p^2 + q^2.$$

In finite fields sets of conjugate elements perform the same function. Consider α^7, belonging to $GF(2^4)$, and its conjugates α^{11}, α^{13}, and α^{14}. Let

$$m(x) = (x + \alpha^7)(x + \alpha^{11})(x + \alpha^{13})(x + \alpha^{14})$$

then

$$\begin{aligned} m(x) &= (x^2 + x(\alpha^7 + \alpha^{11}) + \alpha^{18})(x^2 + x(\alpha^{13} + \alpha^{14}) + \alpha^{27}) \\ &= (x^2 + \alpha^8 x + \alpha^3)(x^2 + \alpha^2 x + \alpha^{12}) \\ &= x^4 + x^3(\alpha^2 + \alpha^8) + x^2(\alpha^{12} + \alpha^{10} + \alpha^3) + x(\alpha^{20} + \alpha^5) + \alpha^{15} \\ &= x^4 + x^3 + 1 \end{aligned}$$

which is a polynomial in the base field $GF(2)$. The polynomial $m(x)$ is referred to as the *minimal polynomial* of α^7, α^{11}, α^{13}, and α^{14}. It is the binary polynomial of smallest degree that has α^7, α^{11}, α^{13}, and α^{14} as roots. Let $m_i(x)$ denote the minimal polynomial of α^i, then $m_i(x)$ is defined to be the smallest degree polynomial in $GF(2)$

that has α^i as a root, and so

$$m_i(\alpha^i) = 0. \tag{6.15}$$

The minimal polynomial $m_i(x)$ is also the minimal polynomial of the conjugates of α^i and therefore

$$m_7(x) = m_{11}(x) = m_{13}(x) = m_{14}(x) = x^4 + x^3 + 1$$

where $m_7(x), m_{11}(x), m_{13}(x)$, and $m_{14}(x)$ are the minimal polynomials of $\alpha^7, \alpha^{11}, \alpha^{13}$, and α^{14} respectively.

To determine the minimal polynomial $m(x)$ of an element β, a factor $(x + \beta^*)$ is required for each conjugate β^* of β. This ensures that the conjugate β^* is a root of $m(x)$. The minimal polynomial is then given by the product of all such factors, so that

$$m(x) = (x + \beta)(x + \beta^2)(x + \beta^4)\dots(x + \beta^{2^{r-1}}) \tag{6.16}$$

where r is the smallest integer such that $\beta^{2^r} = \beta$. In $GF(2^4)$ the conjugates of α are α^2, α^4, α^8 and the minimal polynomial of α is therefore

$$\begin{aligned} m_1(x) &= (x + \alpha)(x + \alpha^2)(x + \alpha^4)(x + \alpha^8) \\ &= (x^2 + \alpha^5 x + \alpha^3)(x^2 + \alpha^5 x + \alpha^{12}) \\ &= x^4 + x + 1. \end{aligned}$$

The minimal polynomials of α^2, α^4, and α^8 are all equal to $m_1(x)$

$$m_2(x) = m_4(x) = m_8(x) = m_1(x) = x^4 + x + 1.$$

Table 6.9 gives the minimal polynomials of field elements in $GF(2^3)$ and $GF(2^4)$.

Table 6.9
Minimal polynomials in $GF(2^3)$ and $GF(2^4)$

Field elements	*Minimal polynomials*
(a) $GF(2^3)$	
0	x
1	$x+1$
$\alpha, \alpha^2, \alpha^4$	x^3+x+1
$\alpha^3, \alpha^5, \alpha^6$	x^3+x^2+1
(b) $GF(2^4)$	
0	x
1	$x+1$
$\alpha, \alpha^2, \alpha^4, \alpha^8$	x^4+x+1
α^5, α^{10}	x^2+x+1
$\alpha^3, \alpha^6, \alpha^9, \alpha^{12}$	$x^4+x^3+x^2+x+1$
$\alpha^7, \alpha^{11}, \alpha^{13}, \alpha^{14}$	x^4+x^3+1

Example 6.11
Find the minimal polynomial of α^5 in $GF(2^3)$.

The conjugates of α^5 are $(\alpha^5)^2 = \alpha^3$ and $(\alpha^5)^4 = \alpha^6$. The minimal polynomial of α^5 is therefore

$$\begin{aligned} m_5(x) &= (x+\alpha^5)(x+\alpha^3)(x+\alpha^6) \\ &= (x^2+\alpha^2 x+\alpha)(x+\alpha^6) \\ &= x^3 + x^2(\alpha^6+\alpha^2) + x(\alpha^8+\alpha) + \alpha^7 \\ &= x^3 + x^2 + 1. \end{aligned}$$

This is also the minimal polynomial of α^3 and α^6. □

Consider the field element α^i and its conjugate α^{2i}, conjugate elements have the same minimal polynomial and therefore $m_{2i} = m_i$. Furthermore given that any even integer can be expressed as $2i$ where i is a smaller odd integer, we see therefore that the minimal polynomial of an even power of a field element is always equal to the minimal polynomial of some odd lower power of the field element.

6.7 Solution of equations in $GF(2^4)$ and $GF(2^3)$

We have seen that addition and multiplication of field elements can be carried out in finite fields. The finite fields though are not restricted to just addition and multiplication of field elements, for much of the mathematics that can be performed in the real and complex fields can also be performed in finite fields. Indeed it is often easier to carry out a mathematical operation in a finite field than in the real or complex fields due to the finite number of elements. Here we consider various characteristics of Galois fields, with particular reference to the solution of equations within the fields.

Consider first the linear equation

$$\alpha^3 x + \alpha^{11} = 0$$

defined in $GF(2^4)$. This differs from the equations previously considered in that its coefficients belong to an extension field. In the previous sections the coefficients of equations were real or binary, only the roots of equations were in extension fields. Nevertheless there is no reason why we cannot construct an equation whose coefficients belong to an extension field. We can easily solve the above equation, taking α^{11} over to the right-hand side gives $\alpha^3 x = \alpha^{11}$ and dividing through by α^3 gives $x = \alpha^{11}/\alpha^3 = \alpha^8$.

Every element in the field $GF(2^m)$ has a square root within $GF(2^m)$. Consider α^4 in $GF(2^4)$, its square root is $\sqrt{\alpha^4} = \alpha^2$. Likewise in $GF(2^3)$ the square root of α^4 is α^2. However the square root of α^5 in $GF(2^4)$ is not so obvious. To deal with this we multiply α^5 by α^{15} and then take the square root, so giving

$$\sqrt{\alpha^5} = \sqrt{(\alpha^5\alpha^{15})} = \sqrt{\alpha^{20}} = \alpha^{10}.$$

Note, however, that in $GF(2^3)$ the square root of α^5 is

$$\sqrt{\alpha^5} = \sqrt{(\alpha^5\alpha^7)} = \sqrt{\alpha^{12}} = \alpha^6.$$

Hence the square root of α^5 in $GF(2^3)$ differs from its square root in $GF(2^4)$.

Next consider the roots of the quadratic equation

$$x^2 + \alpha^{12}x + \alpha^9 = 0$$

defined over $GF(2^4)$. Here again we have an equation whose coefficients belong to an extension field. We can factorize the above equation by using the standard approach of establishing two terms whose sum and product give the required coefficients. If β_1 and β_2 are the required roots then

$$(x + \beta_1)(x + \beta_2) = 0$$

and expanding this gives

$$x^2 + x(\beta_1 + \beta_2) + \beta_1\beta_2 = 0$$

and therefore we need to find the field elements β_1 and β_2 that satisfy

$$\beta_1 + \beta_2 = \alpha^{12}$$
$$\beta_1\beta_2 = \alpha^9.$$

Referring to Table 6.5 we see that the field elements α^2 and α^7 meet this requirement, since

$$\alpha^2\alpha^7 = \alpha^9$$
$$\alpha^2 + \alpha^7 = \alpha^{12}$$

and therefore

$$(x + \alpha^2)(x + \alpha^7) = x^2 + \alpha^{12}x + \alpha^9 = 0$$

so giving α^2 and α^7 as the roots. The solution of equations in a finite field can be achieved by a trial-and-error method in which field elements are systematically tested to see if they are roots. Such an approach of searching for roots is referred to as a *Chien search*. For example consider the roots of $p(x) = x^3 + \alpha^9x^2 + \alpha^6x + \alpha^2$ over $GF(2^4)$. Starting with $x = 0$ gives

$$p(0) = \alpha^2$$
$$p(1) = \alpha^4$$
$$p(\alpha) = \alpha^{14}$$
$$p(\alpha^2) = 0$$

and so $x = \alpha^2$ is one of the roots.Continuing with the search shows that the other roots are α^7 and α^8. Note that within an extension field polynomials can exist that do not have roots within the field but lie within some other field. Consider for example

$$p(x) = x^2 + \alpha^2x + \alpha^{10}$$

in $GF(2^4)$. A search fails to find any roots and therefore $p(x)$ is irreducible over $GF(2^4)$.

A useful property of the field $GF(2^m)$ is that the square of a series of terms added together is equal to sum of the individual terms squared. Consider x_1+x_2 squared

$$(x_1+x_2)^2 = x_1^2 + x_1x_2 + x_2x_1 + x_2^2$$

and because $x_1x_2+x_2x_1=2x_1x_2=0$ we see that

$$(x_1+x_2)^2 = x_1^2 + x_2^2.$$

Squaring this again gives

$$\{(x_1+x_2)^2\}^2 = \{(x_1^2+x_2^2)\}^2 = x_1^4 + x_2^4$$

and so

$$(x_1+x_2)^4 = x_1^4 + x_2^4.$$

This can be extended to all powers 2^i, where i is a positive integer, of (x_1+x_2) so giving

$$(x_1+x_2)^{2^i} = x_1^{2^i} + x_2^{2^i}. \tag{6.17}$$

For example in $GF(2^4)$

$$\begin{aligned}(x+\alpha^7)^8 &= x^8 + (\alpha^7)^8\\ &= x^8 + \alpha^{56}\\ &= x^8 + \alpha^{11}.\end{aligned}$$

Care must be taken not to incorrectly apply eqn 6.17, for instance $(x_1+x_2)^6 \neq (x_1^6+x_2^6)$. Equation 6.17 can, though, still be used to expand such an expression

$$\begin{aligned}(x_1+x_2)^6 &= (x_1+x_2)^4(x_1+x_2)^2\\ &= (x_1^4+x_2^4)(x_1^2+x_2^2)\\ &= x_1^6 + x_2^2x_1^4 + x_2^4x_1^2 + x_2^6.\end{aligned}$$

Example 6.12

Expand (a) $(x+\alpha^4)^2$ in $GF(2^3)$ and (b) $(x+\alpha^3)^5\,(x+\alpha^{10})$ in $GF(2^4)$.

(a) In $GF(2^3)$ we have $(x+\alpha^4)^2 = x^2+\alpha^8 = x^2+\alpha$.

(b) In $GF(2^4)$

$$\begin{aligned}(x+\alpha^3)^5(x+\alpha^{10}) &= (x+\alpha^3)^4(x+\alpha^3)(x+\alpha^{10})\\ &= (x^4+\alpha^{12})(x^2+\alpha^{12}x+\alpha^{13})\\ &= x^6+\alpha^{12}x^5+\alpha^{13}x^4+\alpha^{12}x^2+\alpha^9x+\alpha^{10}.\end{aligned}$$ □

Equation 6.17 can be applied to a series of r terms, given $x_1+x_2+\cdots+x_r$ in $GF(2^m)$ then

$$(x_1+x_2+\cdots+x_r)^{2^i} = x_1^{2^i} + x_2^{2^i} + \cdots + x_r^{2^i} \tag{6.18}$$

where i is a positive integer.

Matrices and determinants of field elements can be constructed and are subject to the same algebraic rules as when constructed with real or complex numbers, obviously though using the additive and multiplicative rules of the finite field within which the field elements exist. Over a field $GF(2^m)$ the matrix

$$\boldsymbol{A} = \begin{bmatrix} a_{11} & a_{12} & \dots & a_{1n} \\ a_{21} & a_{22} & \dots & a_{2n} \\ \vdots & & & \vdots \\ a_{r1} & a_{r2} & \dots & a_{rn} \end{bmatrix}$$

can be defined where r and n are positive integers, and where a_{ij} are field elements with $i = 1, 2, \dots, r$ and $j = 1, 2, \dots, n$. Consider the 3 by 3 matrix defined in $GF(2^4)$

$$\boldsymbol{A}_1 = \begin{bmatrix} \alpha^2 & \alpha & \alpha^{13} \\ 0 & \alpha^{10} & \alpha \\ \alpha^7 & \alpha^3 & 1 \end{bmatrix}.$$

The determinant of $\boldsymbol{A}_1$ is

$$\begin{aligned} \det \boldsymbol{A}_1 &= \alpha^2 \begin{vmatrix} \alpha^{10} & \alpha \\ \alpha^3 & 1 \end{vmatrix} + \alpha \begin{vmatrix} 0 & \alpha \\ \alpha^7 & 1 \end{vmatrix} + \alpha^{13} \begin{vmatrix} 0 & \alpha^{10} \\ \alpha^7 & \alpha^3 \end{vmatrix} \\ &= \alpha^2(\alpha^{10} + \alpha^4) + \alpha(0 + \alpha^8) + \alpha^{13}(0 + \alpha^{17}) \\ &= \alpha^2\alpha^2 + \alpha\alpha^8 + \alpha^{13}\alpha^2 = \alpha^4 + \alpha^9 + 1 = \alpha^3. \end{aligned}$$

Note that $\det \boldsymbol{A}_1$ is a field element of $GF(2^4)$. As $\det \boldsymbol{A}_1$ is nonzero we can determine the inverse of $\boldsymbol{A}_1$. The inverse $\boldsymbol{A}^{-1}$ of a square matrix $\boldsymbol{A}$ is given by $\operatorname{adj} \boldsymbol{A}/\det \boldsymbol{A}$ where $\operatorname{adj} \boldsymbol{A}$ is the adjoint of $\boldsymbol{A}$ and $\det \boldsymbol{A} \neq 0$. The adjoint of a matrix is the transpose of the matrix formed from the cofactors of $\boldsymbol{A}$ and so

$$\operatorname{adj} \boldsymbol{A} = \begin{bmatrix} A_{11} & A_{21} & \dots & A_{r1} \\ A_{12} & A_{22} & \dots & A_{r2} \\ \vdots & & & \vdots \\ A_{1r} & A_{2r} & \dots & A_{rr} \end{bmatrix}$$

where the cofactor A_{ij} is the determinant constructed by excluding the row and column that a_{ij} lies in. A plus or minus sign is normally attached to the cofactor, however as the base field is $GF(2)$ there is no need for this. To find the inverse of $\boldsymbol{A}_1$ we first determine the cofactors

$$A_{11} = \begin{vmatrix} \alpha^{10} & \alpha \\ \alpha^3 & 1 \end{vmatrix} = \alpha^{10}1 + \alpha\alpha^3 = \alpha^{10} + \alpha^4 = \alpha^2$$

$$A_{21} = \begin{vmatrix} \alpha & \alpha^{13} \\ \alpha^3 & 1 \end{vmatrix} = \alpha 1 + \alpha^{13}\alpha^3 = \alpha + \alpha = 0$$

and continuing to evaluate the other 7 cofactors we get

$$\text{adj } \boldsymbol{A}_1 = \begin{bmatrix} A_{11} & A_{21} & A_{31} \\ A_{12} & A_{22} & A_{32} \\ A_{13} & A_{23} & A_{33} \end{bmatrix} = \begin{bmatrix} \alpha^2 & 0 & 1 \\ \alpha^8 & \alpha & \alpha^3 \\ \alpha^2 & \alpha^4 & \alpha^{12} \end{bmatrix}.$$

The inverse of $\boldsymbol{A}_1$, is therefore

$$\begin{aligned} \boldsymbol{A}_1^{-1} &= \text{adj } \boldsymbol{A}_1/\det \boldsymbol{A}_1 \\ &= (1/\alpha^3)\begin{bmatrix} \alpha^2 & 0 & 1 \\ \alpha^8 & \alpha & \alpha^3 \\ \alpha^2 & \alpha^4 & \alpha^{12} \end{bmatrix} \\ &= \begin{bmatrix} \alpha^2/\alpha^3 & 0 & 1/\alpha^3 \\ \alpha^8/\alpha^3 & \alpha/\alpha^3 & \alpha^3/\alpha^3 \\ \alpha^2/\alpha^3 & \alpha^4/\alpha^3 & \alpha^{12}/\alpha^3 \end{bmatrix} \end{aligned}$$

which gives

$$\boldsymbol{A}_1^{-1} = \begin{bmatrix} \alpha^{14} & 0 & \alpha^{12} \\ \alpha^5 & \alpha^{13} & 1 \\ \alpha^{14} & \alpha & \alpha^9 \end{bmatrix}.$$

The reader can check that $\boldsymbol{A}_1\boldsymbol{A}_1^{-1} = \begin{bmatrix} 1 & 0 & 0 \\ 0 & 1 & 0 \\ 0 & 0 & 1 \end{bmatrix} = \boldsymbol{I}.$

Example 6.13

Determine the inverse of the matrix

$$\boldsymbol{A} = \begin{bmatrix} \alpha^3 & \alpha^5 \\ 1 & \alpha \end{bmatrix}$$

in $GF(2^3)$ and $GF(2^4)$.

In $GF(2^3)$ the determinant of $\boldsymbol{A}$ is

$$\det \boldsymbol{A} = \begin{vmatrix} \alpha^3 & \alpha^5 \\ 1 & \alpha \end{vmatrix} = \alpha^3\alpha + \alpha^5 1 = \alpha^4 + \alpha^5 = 1.$$

As $\det \boldsymbol{A} \neq 0$ the inverse therefore exists. The cofactors of $\boldsymbol{A}$ are $A_{11} = \alpha$, $A_{12} = 1$, $A_{21} = \alpha^5$ and $A_{22} = \alpha^3$ and so the adjoint of $\boldsymbol{A}$ is

$$\text{adj } \boldsymbol{A} = \begin{bmatrix} A_{11} & A_{21} \\ A_{12} & A_{22} \end{bmatrix} = \begin{bmatrix} \alpha & \alpha^5 \\ 1 & \alpha^3 \end{bmatrix}.$$

The inverse of $\boldsymbol{A}$ is therefore

$$\boldsymbol{A}^{-1} = (\text{adj } \boldsymbol{A})/\det \boldsymbol{A} = \begin{bmatrix} \alpha & \alpha^5 \\ 1 & \alpha^3 \end{bmatrix}.$$

We can check that

$$AA^{-1} = \begin{bmatrix} \alpha^3 & \alpha^5 \\ 1 & \alpha \end{bmatrix} \begin{bmatrix} \alpha & \alpha^5 \\ 1 & \alpha^3 \end{bmatrix}$$

$$= \begin{bmatrix} \alpha^4 + \alpha^5 & \alpha^8 + \alpha^8 \\ \alpha + \alpha & \alpha^5 + \alpha^4 \end{bmatrix}$$

$$= \begin{bmatrix} 1 & 0 \\ 0 & 1 \end{bmatrix}$$

as required.

In $GF(2^4)$ we get

$$\det A = \alpha^8$$

$$\operatorname{adj} A = \begin{bmatrix} \alpha & \alpha^5 \\ 1 & \alpha^3 \end{bmatrix}$$

which gives

$$A^{-1} = \begin{bmatrix} \alpha^8 & \alpha^{12} \\ \alpha^7 & \alpha^{10} \end{bmatrix}.$$

□

Linear equations can be defined over finite fields and solved using standard methods. For example, take

$$\alpha x + \alpha^5 y = \alpha^3$$
$$x + \alpha^7 y = \alpha^{11}$$

defined over $GF(2^4)$. These can be solved in the normal manner of first multiplying one of the equations by a suitable number and then subtracting the equations, thus eliminating one variable. Here multiplying $x + \alpha^7 y = \alpha^{11}$ by α and then adding it to $\alpha x + \alpha^5 y = \alpha^3$ eliminates the x variable so leaving

$$(\alpha^5 + \alpha^8)y = \alpha^3 + \alpha^{12}$$

which is easily solved to give $y = \alpha^6$. Substituting this into either of the simultaneous equations gives $x = \alpha^4$.

Matrix inversion techniques can also be used to solve linear equations in finite fields, adopting this approach the above linear equations can be written as

$$\begin{bmatrix} \alpha & \alpha^5 \\ 1 & \alpha^7 \end{bmatrix} \begin{bmatrix} x \\ y \end{bmatrix} = \begin{bmatrix} \alpha^3 \\ \alpha^{11} \end{bmatrix}$$

with the solution given by

$$\begin{bmatrix} x \\ y \end{bmatrix} = \begin{bmatrix} \alpha & \alpha^5 \\ 1 & \alpha^7 \end{bmatrix}^{-1} \begin{bmatrix} \alpha^3 \\ \alpha^{11} \end{bmatrix}.$$

Here we get

$$\operatorname{adj} \begin{bmatrix} \alpha & \alpha^5 \\ 1 & \alpha^7 \end{bmatrix} = \begin{bmatrix} \alpha^7 & \alpha^5 \\ 1 & \alpha \end{bmatrix}$$

and

$$\begin{vmatrix} \alpha & \alpha^5 \\ 1 & \alpha^7 \end{vmatrix} = \alpha\alpha^7 + 1\alpha^5 = \alpha^4$$

so that the required inverse matrix is

$$\begin{bmatrix} \alpha & \alpha^5 \\ 1 & \alpha^7 \end{bmatrix}^{-1} = \frac{1}{\alpha^4}\begin{bmatrix} \alpha^7 & \alpha^5 \\ 1 & \alpha \end{bmatrix}$$
$$= \begin{bmatrix} \alpha^7/\alpha^4 & \alpha^5/\alpha^4 \\ 1/\alpha^4 & \alpha/\alpha^4 \end{bmatrix}$$
$$= \begin{bmatrix} \alpha^3 & \alpha \\ \alpha^{11} & \alpha^{12} \end{bmatrix}.$$

Therefore the solution is

$$\begin{bmatrix} x \\ y \end{bmatrix} = \begin{bmatrix} \alpha^3 & \alpha \\ \alpha^{11} & \alpha^{12} \end{bmatrix}\begin{bmatrix} \alpha^3 \\ \alpha^{11} \end{bmatrix}$$
$$= \begin{bmatrix} \alpha^3\alpha^3 + \alpha\alpha^{11} \\ \alpha^{11}\alpha^3 + \alpha^{12}\alpha^{11} \end{bmatrix}$$
$$= \begin{bmatrix} \alpha^4 \\ \alpha^6 \end{bmatrix}$$

which gives $x = \alpha^4$, $y = \alpha^6$ as obtained previously. The next example considers a set of three linear equations.

Example 6.14

Using matrix inversion, determine the solution of the following set of linear equations over $GF(2^4)$:

$$\alpha^3 x_1 + \alpha x_2 + x_3 = \alpha^5$$
$$\alpha^2 x_1 + \alpha^6 x_2 + x_3 = \alpha^6$$
$$\alpha^{14} x_1 + \alpha^7 x_2 + \alpha^7 x_3 = 1.$$

Representing the equations in matrix form gives

$$\boldsymbol{Ax} = \boldsymbol{c}$$

where

$$\boldsymbol{x} = \begin{bmatrix} x_1 \\ x_2 \\ x_3 \end{bmatrix} \quad \text{and} \quad \boldsymbol{c} = \begin{bmatrix} \alpha^5 \\ \alpha^6 \\ 1 \end{bmatrix}.$$

are column vectors, and $\boldsymbol{A}$ is the matrix

$$\boldsymbol{A} = \begin{bmatrix} \alpha^3 & \alpha & 1 \\ \alpha^2 & \alpha^6 & 1 \\ \alpha^{14} & \alpha^7 & \alpha^7 \end{bmatrix}.$$

The determinant of $\boldsymbol{A}$ is

$$\begin{aligned}\det \boldsymbol{A} &= \alpha^3\begin{vmatrix}\alpha^6 & 1\\ \alpha^7 & \alpha^7\end{vmatrix} + \alpha\begin{vmatrix}\alpha^2 & 1\\ \alpha^{14} & \alpha^7\end{vmatrix} + 1\begin{vmatrix}\alpha^2 & \alpha^6\\ \alpha^{14} & \alpha^7\end{vmatrix}\\ &= \alpha^3(\alpha^{13}+\alpha^7) + \alpha(\alpha^9+\alpha^{14}) + 1(\alpha^9+\alpha^5)\\ &= \alpha^8+\alpha^5+\alpha^6 = \alpha^{12}.\end{aligned}$$

The adjoint of $\boldsymbol{A}$ is

$$\operatorname{adj} \boldsymbol{A} = \begin{bmatrix}\alpha^5 & \alpha^{11} & \alpha^{11}\\ \alpha^4 & \alpha^{11} & \alpha^6\\ \alpha^6 & \alpha^5 & \alpha\end{bmatrix}$$

and using $\boldsymbol{A}^{-1} = \operatorname{adj} \boldsymbol{A}/\det \boldsymbol{A}$ gives

$$\boldsymbol{A}^{-1} = \begin{bmatrix}\alpha^8 & \alpha^{14} & \alpha^{14}\\ \alpha^7 & \alpha^{14} & \alpha^9\\ \alpha^9 & \alpha^8 & \alpha^4\end{bmatrix}.$$

Multiplying $\boldsymbol{Ax} = \boldsymbol{c}$ by $\boldsymbol{A}^{-1}$ gives $\boldsymbol{x} = \boldsymbol{A}^{-1}\boldsymbol{c}$ and therefore

$$\begin{aligned}\begin{bmatrix}x_1\\ x_2\\ x_3\end{bmatrix} &= \begin{bmatrix}\alpha^8 & \alpha^{14} & \alpha^{14}\\ \alpha^7 & \alpha^{14} & \alpha^9\\ \alpha^9 & \alpha^8 & \alpha^4\end{bmatrix}\begin{bmatrix}\alpha^5\\ \alpha^6\\ 1\end{bmatrix}\\ &= \begin{bmatrix}\alpha^{13}+\alpha^5+\alpha^{14}\\ \alpha^{12}+\alpha^5+\alpha^9\\ \alpha^{14}+\alpha^{14}+\alpha^4\end{bmatrix}\\ &= \begin{bmatrix}\alpha\\ \alpha^4\\ \alpha^4\end{bmatrix}\end{aligned}$$

which gives $x_1 = \alpha$, $x_2 = \alpha^4$, and $x_3 = \alpha^4$. □

Problems

6.1 Given that α is a field element of $GF(2^3)$ evaluate
(a) $(\alpha^2\alpha^{-5}+1)\,(\alpha^4+\alpha)$
(b) $\sqrt{(\alpha^4\alpha^5+\sqrt{\alpha})}$.
Repeat when α is an element of $GF(2^4)$.

6.2 Determine whether the polynomials

$$\begin{aligned}p_1(x) &= x^4+x^3+x+1\\ p_2(x) &= x^2+x+1\\ p_3(x) &= x^3+x^2+1\end{aligned}$$

over $GF(2)$ are (a) irreducible and (b) primitive.

6.3 Given the polynomial $p(x)=x^2+x+1$ over $GF(2)$, construct the field $GF(2^2)$ and therefore find the roots of $p(x)=0$.

6.4 Show that the field elements of $GF(2^3)$ are primitive (except for 0 and 1). Determine whether the elements α^7 and α^{12} in $GF(2^4)$ are primitive.

6.5 Given that β is a root of the irreducible polynomial $p(x)=x^3+x^2+1$ over $GF(2)$, construct the field $GF(2^3)$ using $p(x)$. Note that the field $GF(2^3)$ constructed using x^3+x^2+1 is the same as that constructed using x^3+x+1, they differ only in the way in which elements are labelled.

6.6 Determine the conjugate sets for the field elements of $GF(2^5)$ (see Table 6.6). Show that the minimal polynomials of α and α^3 in $GF(2^5)$ are

$$m_1(x) = x^5 + x^2 + 1$$

$$m_3(x) = x^5 + x^4 + x^3 + x^2 + 1$$

respectively.

6.7 Find the roots of $x^3+\alpha^8x^2+\alpha^{12}x+\alpha=0$ defined over $GF(2^4)$.

6.8 Find the determinant of the matrix

$$\boldsymbol{A} = \begin{bmatrix} 1 & \alpha^4 & \alpha^3 \\ \alpha^2 & 0 & \alpha \\ \alpha^4 & \alpha & \alpha^5 \end{bmatrix}$$

over $GF(2^3)$ and $GF(2^4)$.

6.9 Determine the inverse of the matrix

$$\boldsymbol{A} = \begin{bmatrix} \alpha^{12} & 0 & \alpha \\ 1 & \alpha^8 & \alpha^{14} \\ \alpha^2 & \alpha^{11} & \alpha^5 \end{bmatrix}$$

over $GF(2^3)$ and $GF(2^4)$.

6.10 Solve the linear equations

$$x + \alpha^4 y = \alpha^5$$

$$\alpha^5 x + \alpha^2 y = \alpha^3$$

defined over $GF(2^4)$. Show that the equations do not have a unique solution when defined over $GF(2^3)$.

6.11 Solve the linear equations

$$\alpha^{10}x + \alpha^4 y + \alpha z = \alpha^9$$

$$\alpha^5 x + \alpha^{12} y + z = \alpha^4$$

$$\alpha^2 x + \alpha^4 y + \alpha^7 z = \alpha^7$$

defined over $GF(2^3)$. Repeat over $GF(2^4)$.

Bose–Chaudhuri–Hocquenghem codes

7

Having considered the properties of finite fields we are now in a good position to move on, from cyclic codes, to the next level of codes namely the *Bose–Chaudhuri–Hocquenghem* (BCH) codes. Cyclic codes were introduced from the point of view of their cyclic property. Later we saw that one of the properties of cyclic codes is that codewords have their generator polynomial as a factor and the roots of a code's generator polynomial are therefore roots of the codewords. Here we first reconsider cyclic codes in terms of roots in an extension field and then we see that by using a well-defined set of roots we can construct BCH codes. The BCH codes are a subset of cyclic codes, they are a powerful class of multiple-error correcting codes with well understood mathematical properties. Binary and nonbinary BCH codes exist, and in particular the *Reed–Solomon codes* are a popular nonbinary class of BCH codes which find many applications. BCH codes are generally considered to be the most important class of codes; a study of error-control codes is incomplete without considering BCH codes.

7.1 Cyclic codes revisited

We have seen that a codeword polynomial $c(x)$ of an (n, k) cyclic code can always be written as

$$c(x) = f(x)g(x) \tag{7.1}$$

where for a nonsystematic code $f(x)$ is the information polynomial $i(x)$, whilst for a systematic code $f(x)$ is the quotient $q(x)$ obtained by dividing $i(x)x^{n-k}$ by $g(x)$. From eqn 7.1, it is clear that any root of $g(x)$, is also a root of $c(x)$, for if we let β be a root of $g(x)$, then $g(\beta) = 0$ and

$$c(\beta) = f(\beta)g(\beta) = 0.$$

Hence in a cyclic code the roots of the generator polynomial $g(x)$ are also roots of the codeword polynomials. Consider the codeword polynomial

$$c(x) = x^5 + x^2 + x + 1$$

belonging to the $(7, 4)$ cyclic code generated by $g(x) = x^3 + x + 1$. The 3 roots of $x^3 + x + 1$ are the field elements α, α^2, and α^4 belonging to $GF(2^3)$. Substituting $x = \alpha$ into $c(x)$ gives

$$c(\alpha) = \alpha^5 + \alpha^2 + \alpha + 1 = 0$$

where the calculations are carried out in $GF(2^3)$. Likewise if we let $x = \alpha^2$ then

$$c(\alpha^2) = \alpha^{10} + \alpha^4 + \alpha^2 + 1 = 0$$

and $x = \alpha^4$ gives

$$c(\alpha^4) = \alpha^{20} + \alpha^8 + \alpha^4 + 1 = 0.$$

The reader may wish to verify that any of the other codeword polynomials of the (7, 4) code also have α, α^2, and α^4 as roots.

Next consider the (15, 11) cyclic code with the generator polynomial $g(x) = x^4 + x + 1$ and

$$c(x) = x^6 + x^4 + x^3 + x^2 + x + 1$$

as one of its codeword polynomials. The roots of $g(x)$ are $\alpha, \alpha^2, \alpha^4$, and α^8 in $GF(2^4)$, and substituting these into $c(x)$ gives

$$\begin{aligned} c(\alpha) &= \alpha^6 + \alpha^4 + \alpha^3 + \alpha^2 + \alpha + 1 = 0 \\ c(\alpha^2) &= \alpha^{12} + \alpha^8 + \alpha^6 + \alpha^4 + \alpha^2 + 1 = 0 \\ c(\alpha^4) &= \alpha^9 + \alpha + \alpha^{12} + \alpha^8 + \alpha^4 + 1 = 0 \\ c(\alpha^8) &= \alpha^3 + \alpha^2 + \alpha^9 + \alpha + \alpha^8 + 1 = 0 \end{aligned}$$

and so, yet again, the roots of the generator polynomial are roots of the codeword polynomial.

The polynomials $x^3 + x + 1$ and $x^4 + x + 1$ serve 2 functions, they are generator polynomials of cyclic codes and, because they are irreducible, they are used to construct finite fields. A generator polynomial need not be irreducible in which case it can not be used to construct a finite field. For example, the generator polynomial for the (15, 7) cyclic code

$$g(x) = x^8 + x^7 + x^6 + x^4 + 1$$

can be factorized as

$$g(x) = (x^4 + x + 1)(x^4 + x^3 + x^2 + x + 1)$$

and is therefore reducible in $GF(2)$ and cannot be used to construct a finite field. The 8 roots of $g(x)$ are field elements in $GF(2^4)$ with minimal polynomials $x^4 + x + 1$ or $x^4 + x^3 + x^2 + x + 1$ but $g(x)$ cannot be used to construct $GF(2^4)$.

The generator polynomials $x^3 + x + 1$ and $x^4 + x + 1$ can be expressed as

$$x^3 + x + 1 = (x + \alpha)(x + \alpha^2)(x + \alpha^4)$$

and

$$x^4 + x + 1 = (x + \alpha)(x + \alpha^2)(x + \alpha^4)(x + \alpha^8)$$

and we think of each generator polynomial as being specified by a chosen set of field elements. So $x^3 + x + 1$ is specified by α, α^2 and α^4 in $GF(2^3)$, and $x^4 + x + 1$ by $\alpha, \alpha^2, \alpha^4$, and α^8 in $GF(2^4)$. Note that α, α^2, and α^4 form a conjugate set in $GF(2^3)$ with minimal polynomial $x^3 + x + 1$, and that in $GF(2^4)$ α, α^2, α^4, and α^8 form a conjugate set with minimal polynomial $x^4 + x + 1$ (see Tables 6.8 and 6.9).

To construct a generator polynomial $g(x)$ from an arbitrary set of r field elements $\beta_1, \beta_2, \ldots, \beta_r$ we need to find the polynomial of least degree that has $\beta_1, \beta_2, \ldots, \beta_r$ as its roots. The polynomial

$$g(x) = (x + \beta_1)(x + \beta_2) \cdots (x + \beta_r) \tag{7.2}$$

satisfies this requirement, but may not be a binary polynomial. For example taking $\beta_1 = \alpha$ and $\beta_2 = \alpha^2$ in $GF(2^3)$ gives $g(x) = (x + \alpha)(x + \alpha^2) = x^2 + \alpha^4 x + \alpha^3$ which has coefficients in $GF(2^3)$. Replacing each factor $(x + \beta_i)$ by $m_i(x)$, the minimal polynomial of β_i, gives

$$g(x) = m_1(x) m_2(x) \cdots m_r(x) \tag{7.3}$$

which is now a binary polynomial with roots $\beta_1, \beta_2, \ldots, \beta_r$. If any of the elements $\beta_1, \beta_2, \ldots, \beta_r$ are conjugates of each other then eqn 7.3 will contain multiple factors of the conjugates' minimal polynomial. Taking the *Least Common Multiple* (LCM) of $g(x)$ will exclude all such common multiples and give

$$g(x) = \mathrm{LCM}[m_1(x), m_2(x), \ldots, m_r(x)] \tag{7.4}$$

as the binary polynomial of least degree with roots $\beta_1, \beta_2, \ldots, \beta_r$. In $GF(2^m)$ the product of 2 or more minimal polynomials divides $x^{q-1} + 1$, where $q = 2^m$, and therefore $g(x)$ as given by eqn 7.4 is a generator polynomial for a cyclic code.

Let's reconsider the (7, 4) code whose generator polynomial $x^3 + x + 1$ has roots α, α^2, and α^4 in $GF(2^3)$. According to eqn 7.4 the generator polynomial specified by α, α^2, and α^4 is

$$g(x) = \mathrm{LCM}[m_1(x), m_2(x), m_4(x)]$$

where $m_1(x)$, $m_2(x)$, and $m_4(x)$ are the minimal polynomials of α, α^2 and α^4 respectively. However in $GF(2^3)$ $m_1(x) = m_2(x) = m_4(x) = x^3 + x + 1$ and therefore we get

$$g(x) = \mathrm{LCM}[m_1(x), m_1(x), m_1(x)] = m_1(x) = x^3 + x + 1$$

as required.

7.2 Definition and construction of binary BCH codes

We have seen how to construct a generator polynomial, of a cyclic code, with an arbitrary set of field elements as its roots. The BCH codes are a subset of cyclic codes whose generator polynomials have roots carefully specified so as to give good error-correcting capability. A t-error-correcting cyclic code with generator polynomial $g(x)$ is a binary BCH code if and only if $g(x)$ is the least-degree polynomial over $GF(2)$ that has

$$\beta, \beta^2, \beta^3, \ldots, \beta^{2t}$$

as roots, where β is an element of $GF(2^m)$. It can be shown that with this selection of roots the resulting code is capable of correcting t errors. If the field element β is primitive then the codes are known as *primitive BCH codes* and have a blocklength of $n = 2^m - 1$. The BCH codes considered here are primitive, unless stated otherwise, with

$$\alpha, \alpha^2, \alpha^3, \ldots, \alpha^{2t}$$

as the $2t$ consecutive roots. Using eqn 7.4 we can see therefore that the generator polynomial $g(x)$ of a t-error-correcting binary BCH code is given by

$$g(x) = \text{LCM}[m_1(x), m_2(x), m_3(x), \ldots, m_{2t}(x)] \tag{7.5}$$

where $m_i(x)$ is the minimal polynomial of α^i and α is an element of $GF(2^m)$. As the minimal polynomial of an even power of a field element is always equal to the minimal polynomial of some odd and lower power of the element, the minimal polynomials with even i can be omitted from eqn 7.5 and therefore

$$g(x) = \text{LCM}[m_1(x), m_3(x), m_5(x), \ldots, m_{2t-1}(x)]. \tag{7.6}$$

The blocklength of a primitive BCH code constructed over $GF(2^m)$ is $n = 2^m - 1$. BCH codes are cyclic codes and the degree r of the generator polynomial of an (n, k) cyclic code is $n - k$. Hence the information length k of a BCH code is $k = 2^m - 1 - r$.

As an example let's construct a double-error-correcting BCH code over $GF(2^4)$. Here $t = 2$ and taking $\beta = \alpha$, where α is a primitive element of $GF(2^4)$, gives

$$g(x) = \text{LCM}[m_1(x), m_2(x), m_3(x), m_4(x)]$$

where $m_1(x)$, $m_2(x)$, $m_3(x)$ and $m_4(x)$ are the minimal polynomials of α, α^2, α^3, and α^4 respectively. In $GF(2^4)$ the minimal polynomials

$$\begin{aligned} m_1(x) &= x^4 + x + 1 \\ m_2(x) &= m_1(x) \\ m_3(x) &= x^4 + x^3 + x^2 + x + 1 \\ m_4(x) &= m_1(x) \end{aligned}$$

and so $m_2(x)$ and $m_4(x)$ can be excluded from $g(x)$, therefore giving

$$\begin{aligned} g(x) &= m_1(x)m_3(x) \\ &= (x^4 + x + 1)(x^4 + x^3 + x^2 + x + 1) \\ &= x^8 + x^7 + x^6 + x^4 + 1. \end{aligned}$$

As α is a primitive element of $GF(2^4)$, the blocklength of the constructed code is $n = 2^m - 1 = 15$. The degree of $g(x)$ is $r = 8$ and the information length is $k = n - r = 7$. We have therefore constructed the double-error-correcting (15, 7) BCH code with $g(x) = x^8 + x^7 + x^6 + x^4 + 1$.

A single-error-correcting code with the same blocklength can be constructed over the same field. Let $t = 1$ then

$$g(x) = \text{LCM}[m_1(x), m_2(x)].$$

Over $GF(2^4)$ the minimal polynomial $m_2(x) = m_1(x) = x^4 + x + 1$ and therefore

$$g(x) = m_1(x) = x^4 + x + 1.$$

The blocklength is $n = 2^m - 1 = 15$ and the information length is $k = n - r = 11$ because the degree of $g(x)$ is $r = 4$. This is therefore the single-error-correcting (15, 11) BCH code with $g(x) = x^4 + x + 1$.

Example 7.1

Construct a triple-error-correcting BCH code with blocklength $n = 31$ over $GF(2^5)$.

Let $t = 3$ and α be a primitive element of $GF(2^5)$. The generator polynomial is

$$g(x) = \text{LCM}[m_1(x), m_2(x), m_3(x), m_4(x), m_5(x), m_6(x)].$$

In $GF(2^5)$

$$\begin{aligned} m_1(x) &= x^5 + x^2 + 1 \\ m_2(x) &= m_1(x) \\ m_3(x) &= x^5 + x^4 + x^3 + x^2 + 1 \\ m_4(x) &= m_2(x) \\ m_5(x) &= x^5 + x^4 + x^2 + x + 1 \\ m_6(x) &= m_3(x) \end{aligned}$$

and $g(x)$ therefore reduces to

$$\begin{aligned} g(x) &= m_1(x)m_3(x)m_5(x) \\ &= (x^5 + x^2 + 1)(x^5 + x^4 + x^3 + x^2 + 1)(x^5 + x^4 + x^2 + x + 1) \\ &= x^{15} + x^{11} + x^{10} + x^9 + x^8 + x^7 + x^5 + x^3 + x^2 + x + 1. \end{aligned}$$

The blocklength and information length are $n = 2^5 - 1 = 31$ and $k = 31 - 15 = 16$ respectively. This is therefore the (31, 16) triple-error-correcting binary BCH code.

□

A t-error-correcting BCH code has a guaranteed minimum distance of $d = 2t + 1$. However the minimum distance $d_{\min}$ of the code may be greater than d, therefore giving the code an error-control capability greater than that designed. As such the minimum distance $d = 2t + 1$ is known as the *designed distance* of the code, and for any BCH code $d_{\min} \geq d$.

BCH codes are cyclic and linear, and so once a code's generator polynomial $g(x)$ is constructed, encoding can be carried out in the usual manner using $g(x)$ or the generator matrix $\boldsymbol{G}$ constructed from $g(x)$. It is at the decoding stage that techniques specific to BCH codes are used.

7.3 Error syndromes in finite fields

At the decoding stage of an error-correcting code, decisions are made on the basis of error syndromes that depend on the presence of errors. Codewords, whether represented by vectors in linear codes or polynomials in cyclic codes, are constructed so as to give a zero contribution to the error syndromes. In a linear code, codewords $\boldsymbol{c}$ satisfy $\boldsymbol{cH}^{\mathrm{T}} = 0$ and the error syndrome of a word $\boldsymbol{v}$ to be decoded is $\boldsymbol{s} = \boldsymbol{vH}^{\mathrm{T}}$ (where $\boldsymbol{H}$ is the parity-check matrix). In cyclic codes, codeword polynomials $c(x)$ satisfy $R_{g(x)}[c(x)] = 0$ and the error syndrome of a polynomial $v(x)$ to be decoded is $s(x) = R_{g(x)}[v(x)]$. With BCH codes, error syndromes are also defined, this time they are field elements in an extension field and again codewords do not contribute to the error syndromes.

Consider a t-error-correcting BCH code with $\alpha, \alpha^2, \ldots, \alpha^{2t}$ as the roots of its generator polynomial $g(x)$. The roots of $g(x)$ are also the roots of the codeword polynomials $c(x)$, and therefore

$$c(\alpha^i) = 0 \tag{7.7}$$

for $i = 1, 2, \ldots, 2t$. Equation 7.7 provides a means for testing whether a polynomial $v(x)$ is a codeword of a BCH code. A polynomial $v(x)$ is a codeword if and only if $\alpha, \alpha^2, \alpha^3, \ldots, \alpha^{2t}$ are roots of $v(x)$. Now consider a codeword $c(x)$ which incurs an error pattern $e(x)$, so giving

$$v(x) = c(x) + e(x)$$

as the polynomial to be decoded. Substituting $x = \alpha$ gives

$$v(\alpha) = c(\alpha) + e(\alpha).$$

However, from eqn 7.7, $c(\alpha) = 0$ and so

$$v(\alpha) = e(\alpha).$$

Hence $v(x)$ evaluated at $x = \alpha$ depends solely on the error pattern $e(x)$ and can therefore be used as an error syndrome of $v(x)$. Evaluating $v(x)$ at any of the field elements α^i gives

$$v(\alpha^i) = c(\alpha^i) + e(\alpha^i)$$

which again reduces to

$$v(\alpha^i) = e(\alpha^i)$$

for $i = 1, 2, \ldots, 2t$. Hence from $v(x)$ we can obtain $2t$ error syndromes and we define the ith error syndrome of $v(x)$ as

$$S_i = v(\alpha^i) \tag{7.8}$$

where $i = 1, 2, \ldots, 2t$.

The error syndromes $S_1, S_2, \ldots, S_{2t}$ are elements of the field $GF(2^m)$ containing α. Table 7.1 gives a comparison of syndrome definitions in linear, cyclic and BCH codes, along with the condition that codewords satisfy.

If $v(x)$ is error free then $v(x) = c(x)$ and

$$S_i = v(\alpha^i) = c(\alpha^i) = 0$$

Table 7.1
Error syndromes in linear, cyclic, and BCH codes

	Linear	*Cyclic*	*BCH*
Codewords	$\boldsymbol{cH}^{\mathrm{T}} = 0$	$R_{g(x)}[c(x)] = 0$	$c(\alpha^i) = 0$
Error syndromes	$\boldsymbol{s} = \boldsymbol{vH}^{\mathrm{T}}$	$s(x) = R_{g(x)}[v(x)]$	$S_i = v(\alpha^i)$

for all $2t$ error syndromes. For example, consider the codeword

$$c(x) = x^9 + x^6 + x^5 + x^4 + x + 1$$

belonging to double-error-correcting (15, 7) BCH code. Let $v(x) = c(x)$ and $t = 2$, then over $GF(2^4)$ we get

$$\begin{aligned} S_1 &= v(\alpha) = \alpha^9 + \alpha^6 + \alpha^5 + \alpha^4 + \alpha + 1 = 0 \\ S_2 &= v(\alpha^2) = \alpha^3 + \alpha^{12} + \alpha^{10} + \alpha^8 + \alpha^2 + 1 = 0 \\ S_3 &= v(\alpha^3) = \alpha^{12} + \alpha^3 + 1 + \alpha^{12} + \alpha^3 + 1 = 0 \\ S_4 &= v(\alpha^4) = \alpha^6 + \alpha^9 + \alpha^5 + \alpha + \alpha^4 + 1 = 0. \end{aligned}$$

If $v(x)$ contains errors, then some or all of the error syndromes will be nonzero. Introducing an error, say $e(x) = x^7 + x^4$, to $c(x)$ gives

$$v(x) = c(x) + e(x) = x^9 + x^7 + x^6 + x^5 + x + 1$$

and recalculating the error syndromes now gives

$$\begin{aligned} S_1 &= v(\alpha) = \alpha^9 + \alpha^7 + \alpha^6 + \alpha^5 + \alpha + 1 = \alpha^3 \\ S_2 &= v(\alpha^2) = \alpha^3 + \alpha^{14} + \alpha^{12} + \alpha^{10} + \alpha^2 + 1 = \alpha^6 \\ S_3 &= v(\alpha^3) = \alpha^{12} + \alpha^6 + \alpha^3 + 1 + \alpha^3 + 1 = \alpha^4 \\ S_4 &= v(\alpha^4) = \alpha^6 + \alpha^{13} + \alpha^9 + \alpha^5 + \alpha^4 + 1 = \alpha^{12}. \end{aligned}$$

If the number of errors does not exceed the error-correction limit of a BCH code then, as we shall see, the error pattern $e(x)$ can be determined from $S_1, S_2, \ldots, S_{2t}$.

Calculating error syndromes in a finite field can be simplified by using eqn 6.18, which shows that for $x_1, x_2, \ldots, x_n$ in $GF(2^m)$ we can write

$$x_1^2 + x^2 + \cdots + x_n^2 = (x_1 + x_2 + \cdots + x_n)^2.$$

For example, here the error syndrome S_2, given above can be expressed as

$$\begin{aligned} S_2 &= (\alpha^2)^9 + (\alpha^2)^7 + (\alpha^2)^6 + (\alpha^2)^5 + \alpha^2 + 1 \\ &= (\alpha^9)^2 + (\alpha^7)^2 + (\alpha^6)^2 + (\alpha^5)^2 + (\alpha)^2 + 1 \\ &= (\alpha^9 + \alpha^7 + \alpha^6 + \alpha^5 + \alpha + 1)^2 \end{aligned}$$

and as

$$S_1 = \alpha^9 + \alpha^7 + \alpha^6 + \alpha^5 + \alpha + 1$$

we see therefore that

$$S_2 = (\alpha^9 + \alpha^7 + \alpha^6 + \alpha^5 + \alpha + 1)^2 = S_1^2.$$

Likewise $S_4 = S_2^2$ and clearly the error syndrome S_{2i} is given by

$$S_{2i} = S_i^2. \tag{7.9}$$

When calculating error syndromes we therefore need only to evaluate $S_i = v(\alpha^i)$ for odd values of i and then use $S_{2i} = S_i^2$ to obtain the error syndromes for even values of i. However, note that this only applies to binary codes, eqn 7.9 cannot be used to determine the error syndromes of nonbinary codes.

The decoder has no *a priori* knowledge of the error pattern, the only information that the decoder has is the polynomial $v(x)$ (or word $\boldsymbol{v}$) and the error syndromes that it calculates from $v(x)$. An error pattern with μ errors can be represented as

$$e(x) = x^{p_1} + x^{p_2} + \cdots + x^{p_\mu} \tag{7.10}$$

where the *error positions* $p_1, p_2, \ldots, p_\mu$ give the locations of the errors in the corresponding error vector $\boldsymbol{e}$. For example in an 8-bit word the 3-bit error pattern $\boldsymbol{e} = (0\ 0\ 1\ 0\ 1\ 0\ 0\ 1)$ gives $p_1 = 5$, $p_2 = 3$, $p_3 = 0$ and so

$$e(x) = x^{p_1} + x^{p_2} + x^{p_3} = x^5 + x^3 + 1.$$

A codeword $c(x)$ incurring μ errors gives

$$v(x) = c(x) + e(x) = c(x) + (x^{p_1} + x^{p_2} + \cdots + x^{p_\mu})$$

as the word to be decoded. The decoder for a t-error correcting code evaluates $v(x)$ at $x = \alpha, \alpha^2, \ldots, \alpha^{2t}$ to obtain the error syndromes

$$\begin{aligned}
S_1 &= v(\alpha) = c(\alpha) + e(\alpha) = \alpha^{p_1} + \alpha^{p_2} + \cdots + \alpha^{p_\mu} \\
S_2 &= v(\alpha^2) = c(\alpha^2) + e(\alpha^2) = \alpha^{2p_1} + \alpha^{2p_2} + \cdots + \alpha^{2p_\mu} \\
S_3 &= v(\alpha^3) = c(\alpha^3) + e(\alpha^3) = \alpha^{3p_1} + \alpha^{3p_2} + \cdots + \alpha^{3p_\mu} \\
&\vdots \\
S_{2t} &= v(\alpha^{2t}) = c(\alpha^{2t}) + e(\alpha^{2t}) = \alpha^{2tp_1} + \alpha^{2tp_2} + \cdots + \alpha^{2tp_\mu}.
\end{aligned} \tag{7.11}$$

For clarity the right-hand side of eqn 7.11 is usually expressed in terms of *error-location numbers* X_i where

$$X_i = \alpha^{p_i}.$$

Note that the exponents of the error-location numbers give the error positions. The error-location numbers $X_1, X_2, \ldots, X_\mu$ are nonzero field elements in $GF(2^m)$ and provide a convenient representation of the unknown error positions $p_1, p_2, \ldots, p_\mu$.

Equation 7.11 can now be expressed as

$$\begin{aligned} S_1 &= X_1 + X_2 + X_3 + \cdots + X_\mu \\ S_2 &= X_1^2 + X_2^2 + X_3^2 + \cdots + X_\mu^2 \\ S_3 &= X_1^3 + X_2^3 + X_3^3 + \cdots + X_\mu^3 \\ &\vdots \\ S_{2t} &= X_1^{2t} + X_2^{2t} + X_3^{2t} + \cdots + X_\mu^{2t}. \end{aligned} \tag{7.12}$$

Equations 7.12 contain μ unknown variables $X_1, X_2, X_3, \ldots, X_\mu$ and $2t$ known terms $S_1, S_2, S_3, \ldots, S_{2t}$ and are referred to as the *syndrome equations*. Earlier we saw that $S_{2i} = S_i^2$ and therefore of the $2t$ syndrome equations only t equations are independent. Hence this is a set of t simultaneous equations with μ unknowns, and if $\mu \leq t$ then a unique solution exists. In other words, if the number of errors falls within the error-correction capability of the code, then the error-location numbers can be determined. The exponents of the error-location numbers are then taken as the error positions. However, there is a problem in determining the solutions of eqns 7.12. The syndrome equations are nonlinear and therefore cannot be solved using standard linear techniques such as matrix inversion. Instead, indirect methods are used involving the transformation of the syndrome equations into a form that can be readily solved. The solution of the syndrome equations lies at the heart of the decoding of BCH codes, any method that can solve the syndrome equations can be considered to be a decoding technique for the BCH codes. The most important method for decoding BCH codes is the *Peterson–Gorenstein–Zierler decoder* which is capable of dealing with multiple errors. However we first examine the simpler problems of decoding single-error-correcting and double-error-correcting BCH codes without the use of the Peterson–Gorenstein–Zierler decoder.

7.4 Decoding SEC and DEC binary BCH codes

Decoding a single-error-correcting (SEC) BCH code is quite straightforward. For a single-error-correcting code $t = 1$ and we assume that a single error (i.e. the maximum number of correctable errors) has occurred so that $\mu = 1$. Substituting $t = \mu = 1$ into the syndrome equations (eqns 7.12) gives

$$\begin{aligned} S_1 &= X_1 \\ S_2 &= X_1^2 \end{aligned} \tag{7.13}$$

and therefore the error-location number is directly given by

$$X_1 = S_1.$$

For example, consider the $(7,4)$ code with codeword $c(x)=x^5+x^2+x+1$ and let's assume that $c(x)$ incurs the single error $e(x)=x^5$, then the word to be decoded is

$$v(x) = c(x) + e(x) = x^2 + x + 1.$$

The $(7,4)$ code is a single-error-correcting code constructed over $GF(2^3)$ and so the error syndromes are evaluated over $GF(2^3)$. In $GF(2^3)$

$$S_1 = v(\alpha) = \alpha^2 + \alpha + 1 = \alpha^5$$

and from the syndrome equations, given by eqns 7.13, we get $X_1=\alpha^5$. The exponents of the error-location numbers give the error positions in the error pattern and therefore $X_1=\alpha^5$ gives $e(x)=x^5$ as the decoder's estimate of the error pattern. The resulting codeword is

$$c(x) = v(x) + e(x) = x^5 + x^2 + x + 1$$

which we know is correct. Decoding will always be correct providing 2 or more errors do not occur. If the error syndromes are zero, then the decoder assumes that the received word is the correct codeword. Note that a syndrome table has not been used, but we have used a table for addition in $GF(2^3)$. Note also that S_2 is not required in the decoding process and therefore need not be computed.

Consider next decoding double-error-correcting (DEC) BCH codes. We again assume the occurrence of the maximum number of correctable errors, so that $\mu=t=2$ and the syndrome equations (eqns 7.12) reduce to

$$\begin{aligned} S_1 &= X_1 + X_2 \\ S_2 &= X_1^2 + X_2^2 \\ S_3 &= X_1^3 + X_2^3 \\ S_4 &= X_1^4 + X_2^4. \end{aligned} \tag{7.14}$$

The second and fourth of these equations, involving S_2 and S_4 respectively, are dependent on the first equation and therefore solutions for X_1 and X_2 can be obtained from

$$\begin{aligned} S_1 &= X_1 + X_2 \\ S_3 &= X_1^3 + X_2^3 \end{aligned}$$

as these are two independent equations with two unknowns. The two equations are nonlinear and cannot be solved using matrix inversion, instead we proceed as follows. Consider $(X_1+X_2)^3$:

$$\begin{aligned} (X_1 + X_2)^3 &= (X_1 + X_2)^2(X_1 + X_2) \\ &= (X_1^2 + X_2^2)(X_1 + X_2) \\ &= X_1^3 + X_2^3 + X_1X_2(X_1 + X_2). \end{aligned}$$

Substituting $X_1+X_2=S_1$ and $X_1^3 + X_1^3 = S_3$ into the above gives

$$S_1^3 = S_3 + S_1X_1X_2$$

and replacing X_2 by $X_2 = X_1 + S_1$ gives

$$S_1^3 = S_3 + S_1 X_1 (X_1 + S_1).$$

Rearranging and dividing through by S_1 gives the quadratic equation

$$X_1^2 + S_1 X_1 + \frac{(S_1^3 + S_3)}{S_1} = 0 \tag{7.15}$$

and the solution of this gives X_1, with the other solution giving X_2. If we let $X_1 = X_2 + S_1$, instead of $X_2 = X_1 + S_1$, then

$$X_2^2 + S_1 X_2 + \frac{(S_1^3 + S_3)}{S_1} = 0 \tag{7.16}$$

is obtained instead of eqn 7.15 and again the two roots give X_1 and X_2. Whether eqn 7.15 or 7.16 is used to determine X_1 and X_2 is quite arbitrary and we can therefore write

$$x^2 + S_1 x + \frac{(S_1^3 + S_3)}{S_1} = 0 \tag{7.17}$$

where X_1 and X_2 are the two roots. The roots of eqn 7.17 can be obtained by using a Chien search, that is by systematically testing to see if field elements satisfy the equation, or by establishing the two factors (see Section 6.7). As an example, we consider the codeword

$$c(x) = x^{11} + x^8 + x^7 + x^6 + x^3 + x^2$$

belonging to the double-error-correcting (15, 7) BCH code. Introducing an error pattern, say $e(x) = x^{10} + x^2$ gives

$$v(x) = x^{11} + x^{10} + x^8 + x^7 + x^6 + x^3$$

and over $GF(2^4)$ the error syndromes S_1 and S_3 are

$$S_1 = v(\alpha) = \alpha^{11} + \alpha^{10} + \alpha^8 + \alpha^7 + \alpha^6 + \alpha^3 = \alpha^4$$

$$S_3 = v(\alpha^3) = \alpha^3 + 1 + \alpha^9 + \alpha^6 + \alpha^3 + \alpha^9 = \alpha^{13}.$$

Evaluating $(S_1^3 + S_3)/S_1$ gives

$$\frac{(S_1^3 + S_3)}{S_1} = \frac{(\alpha^{12} + \alpha^{13})}{\alpha^4} = \frac{\alpha}{\alpha^4} = \alpha^{12}$$

and substituting this into eqn 7.17, along with $S_1 = \alpha^4$, gives

$$x^2 + \alpha^4 x + \alpha^{12} = 0.$$

The two roots of this quadratic equation give the required error-location numbers. Let $p(x) = x^2 + \alpha^4 x + \alpha^{12}$, then using a Chien search we systematically test the nonzero field elements of $GF(2^4)$ to see if they are roots of $p(x)$. Starting with $x=1$

$$p(1) = \alpha^{13}$$
$$p(\alpha) = \alpha^{13}$$
$$p(\alpha^2) = 0$$
$$p(\alpha^3) = \alpha^3.$$

So far α^2 is one solution. We could continue searching for the second root but it is easier to use the first expression in eqns 7.14, namely $S_1 = X_1 + X_2$, which gives

$$X_2 = S_1 + X_1 = \alpha^4 + \alpha^2 = \alpha^{10}$$

as the other solution. The reader can verify that $p(\alpha^{10})=0$. The error-location numbers are therefore $X_1=\alpha^2$ and $X_2=\alpha^{10}$, the exponents of the field elements X_1 and X_2 correspond to the errors x^2 and x^{10} respectively. We have therefore correctly determined the error polynomial $e(x)=x^{10}+x^2$ present in $v(x)$.

In the event of a single error occurring there will be only one nonzero error-location number and so $X_1=S_1$, $X_2=0$, $S_3 = X_1^3 = S_1^3$ and eqn 7.17 reduces to

$$x + S_1 = 0.$$

The error-location number is therefore directly given by S_1 (i.e. it is the same as decoding a single-error-correcting code). For example consider again the (15, 7) double-error-correcting BCH code with codeword $c(x)=x^{11}+x^8+x^7+x^6+x^3+x^2$, but this time incurring the single error $e(x)=x^4$. Here $v(x)=x^{11}+x^8+x^7+x^6+x^4+x^3+x^2$ giving the error syndromes

$$S_1 = v(\alpha) = \alpha^4$$
$$S_3 = v(\alpha^3) = \alpha^{12}$$

over $GF(2^4)$. Evaluating $(S_1^3 + S_3)/S_1$ gives

$$\frac{(S_1^3 + S_3)}{S_1} = \frac{(\alpha^4)^3 + \alpha^{12}}{\alpha^4} = \frac{\alpha^{12} + \alpha^{12}}{\alpha^4} = 0.$$

The constant term in eqn 7.17 is therefore 0 and so eqn 7.17 reduces to

$$x + S_1 = 0$$

as required. Therefore the error-location number $X=S_1=\alpha^4$ which gives the error pattern x^4 (which we know is correct).

Example 7.2

Given that the codewords $c_1(x)$ and $c_2(x)$, belonging to the double-error-correcting (15, 7) code constructed over $GF(2^4)$, incur 2 and 1 errors so giving

(a) $v_1(x)=x^{11}+x^9+x^8+x^6+x^5+x+1$

(b) $v_2(x)=x^{12}+x^{11}+x^{10}+x^9+x^7+x^5+x$

respectively, determine $c_1(x)$ and $c_2(x)$.

(a) The error syndromes are

$$S_1 = v_1(\alpha) = \alpha^3$$
$$S_3 = v_1(\alpha^3) = \alpha^{13}$$

over $GF(2^4)$. Substituting these into eqn 7.17 gives

$$x^2 + \alpha^3 x + \alpha^7 = 0.$$

By inspection we can see that $\alpha^{10} + \alpha^{12} = \alpha^3$ and $\alpha^{10}\alpha^{12} = \alpha^7$ over $GF(2^4)$ and so

$$x^2 + \alpha^3 x + \alpha^7 = (x + \alpha^{10})(x + \alpha^{12}) = 0.$$

The roots of $x^2 + \alpha^3 x + \alpha^7 = 0$ are therefore α^{10} and α^{12}, giving the error-location numbers $X_1 = \alpha^{10}$ and $X_2 = \alpha^{12}$. Hence

$$e(x) = x^{10} + x^{12}$$

and so

$$c_1(x) = x^{12} + x^{11} + x^{10} + x^9 + x^8 + x^6 + x^5 + x + 1.$$

(b) Here we have $S_1 = \alpha^4$ and $S_3 = \alpha^{12}$. Therefore $S_1^3 + S_3 = \alpha^{12} + \alpha^{12} = 0$ and eqn 7.17 reduces to $x + \alpha^4 = 0$. The error-location number is $X_1 = \alpha^4$ giving an error pattern $e(x) = x^4$ and codeword $c_2(x) = x^{12} + x^{11} + x^{10} + x^9 + x^7 + x^5 + x^4 + x$. □

To summarize, when decoding a double-error-correcting BCH code the occurrence of two errors results in a quadratic equation whose two roots give two error-location numbers. However, in the event of a single error occurring, the quadratic equation reduces to a linear equation and the error-location number is given by S_1 (as for a single-error-correcting code). In the event of three or more errors occurring a decoding error will occur if eqn 7.17 has 1 or 2 solutions. If eqn 7.17 has no solution then an uncorrectable error pattern will have been detected. If the error pattern is identical to a codeword, then the syndromes are zero and again a decoding error occurs.

Example 7.3

Consider the (15, 7) double-error-correcting BCH code and codeword $c(x) = x^8 + x^7 + x^6 + x^4 + 1$. Determine the outcome of a decoder when $c(x)$ incurs the error patterns

(a) $e(x) = x^7 + x^2 + 1$
(b) $e(x) = x^{11} + x^9 + x^6 + x^4$.

(a) The polynomial to be decoded is

$$v(x) = c(x) + e(x) = x^8 + x^6 + x^4 + x^2$$

giving error syndromes

$$S_1 = v(\alpha) = \alpha^8 + \alpha^6 + \alpha^4 + \alpha^2 = \alpha^{11}$$
$$S_3 = v(\alpha^3) = \alpha^9 + \alpha^3 + \alpha^{12} + \alpha^6 = 1$$

over $GF(2^4)$. Substituting S_1 and S_3 in eqn 7.17 gives

$$x^2 + \alpha^{11} x + \alpha^3 = 0.$$

It can be shown, by inspection or by testing the field elements, that none of the elements in $GF(2^4)$ are solutions of $x^2+\alpha^{11}x+\alpha^3=0$. No error-location numbers can therefore be obtained and the decoder concludes that an uncorrectable error pattern has been detected, i.e. a decoding failure occurs.

(b) Here $v(x)=x^{11}+x^9+x^8+x^7+1$ and so

$$S_1 = v(\alpha) = \alpha^7$$
$$S_3 = v(\alpha^3) = 0.$$

Substituting S_1 and S_3 into eqn 7.17 gives

$$x^2 + \alpha^7 x + \alpha^{14} = 0$$

and by inspection it can be seen that $\alpha^2+\alpha^{12}=\alpha^7$ and $\alpha^2\alpha^{12}=\alpha^{14}$ over $GF(2^4)$. Hence

$$x^2 + \alpha^7 x + \alpha^{14} = (x + \alpha^2)(x + \alpha^{12})$$

giving α^2 and α^{12} as the required roots and error-location numbers. The decoder therefore concludes that the double error pattern $e(x)=x^2+x^{12}$ occurred and adding this to $v(x)$ gives the codeword

$$c(x) = x^{12} + x^{11} + x^9 + x^8 + x^7 + x^2 + 1.$$

This is the wrong codeword and so a decoding error has occurred. □

7.5 The error-location polynomial

The method described in Section 7.4 for decoding single error-correcting and double-error-correcting BCH codes can be extended to deal with multiple-error-correcting BCH codes. For a t-error-correcting code a polynomial of degree t or less can be defined whose coefficients are functions of the error syndromes. The occurrence of $\mu \le t$ errors gives a polynomial of degree μ whose μ roots are the reciprocal of the required error-location numbers. Consider again eqn 7.17 and let

$$\begin{aligned}\sigma_1 &= S_1 \\ \sigma_2 &= (S_1^3 + S_3)/S_1\end{aligned} \tag{7.18}$$

then eqn 7.17 becomes

$$x^2 + \sigma_1 x + \sigma_2 = 0.$$

This is the polynomial of highest degree that we need to consider when decoding double-error-correcting codes. In the event of 1 error occurring we get $\sigma_2=0$ and therefore $x+\sigma_1=0$, which gives $X=\sigma_1=S_1$ as the required error-location number. For a t-error-correcting code we need to consider polynomials of the form

$$x^\mu + \sigma_1 x^{\mu-1} + \sigma_2 x^{\mu-2} + \cdots + \sigma_{\mu-1} x + \sigma_\mu = 0$$

where $\mu \le t$ and where the polynomial coefficients are again functions of the error syndromes and the μ roots give the μ error-location numbers. If we replace x by its reciprocal $1/x$ and then multiply through by x^{μ} we get

$$\sigma_0 + \sigma_1 x + \sigma_2 x^2 + \cdots + \sigma_{\mu-1} x^{\mu-1} + \sigma_\mu x^\mu = 0$$

where $\sigma_0 = 1$. We now define the *error-location polynomial*

$$\sigma(x) = \sigma_0 + \sigma_1 x + \sigma_2 x^2 + \cdots + \sigma_{\mu-1} x^{\mu-1} + \sigma_\mu x^\mu \tag{7.19}$$

which is a polynomial whose roots are the reciprocal of the error-location numbers. The error-location polynomial can be defined so that its roots are error-location numbers, and whilst it is easier to think of roots as representing error-location numbers, it is however convenient and conventional to use error-location polynomials whose reciprocal roots are error-location numbers. Note that for a double-error-correcting code the error-location polynomial is

$$\sigma(x) = 1 + \sigma_1 x + \sigma_2 x^2 \tag{7.20}$$

where σ_1 and σ_2 are again given by eqns 7.18.

Example 7.4
Given that $v(x) = x^9 + x^8 + x^6 + x^4 + 1$ represents a codeword $c(x)$, of the double-error-correcting (15, 7) code, that has incurred 2 errors determine $c(x)$.

Over $GF(2^4)$ we get

$$S_1 = v(\alpha) = 1$$
$$S_3 = v(\alpha^3) = \alpha^4$$

and substituting these in to eqns 7.18 gives

$$\sigma_1 = 1$$
$$\sigma_2 = \alpha.$$

The error-location polynomial is therefore

$$\sigma(x) = 1 + \sigma_1 x + \sigma_2 x^2 = 1 + x + \alpha x^2.$$

Using a Chien search we find that the roots of $\sigma(x)$ are α^6 and α^8 over $GF(2^4)$. The error-location numbers are therefore

$$X_1 = 1/\alpha^6 = \alpha^9$$
$$X_2 = 1/\alpha^8 = \alpha^7$$

which give the error polynomial $e(x) = x^9 + x^7$ and codeword polynomial

$$c(x) = v(x) + e(x) = x^8 + x^7 + x^6 + x^4 + 1.$$ □

If the roots of the error-location polynomial are the field elements $\beta_1, \beta_2, \ldots, \beta_\mu$ then the error-location numbers are

$$X_1 = 1/\beta_1$$
$$X_2 = 1/\beta_2$$
$$\vdots$$
$$X_\mu = 1/\beta_\mu$$

and the error-location polynomial can be expressed as

$$\sigma(x) = (xX_1 + 1)(xX_2 + 1)\cdots(xX_\mu + 1). \tag{7.21}$$

Determining the error-location polynomial is the most difficult part of decoding a BCH code. The coefficients $\sigma_0, \sigma_1, \sigma_2, \ldots, \sigma_\mu$ of the error-location polynomial $\sigma(x)$, have to be determined from the known error syndromes. To achieve this eqn 7.21 is expanded and its coefficients are compared with those of eqn 7.19, and in doing so we find that

$$\begin{aligned}
\sigma_0 &= 1\\
\sigma_1 &= X_1 + X_2 + X_3 + \cdots + X_{\mu-1} + X_\mu\\
\sigma_2 &= X_1X_2 + X_2X_3 + X_3X_4 + \cdots + X_{\mu-1}X_\mu\\
&\vdots\\
\sigma_\mu &= X_1X_2X_3\ldots X_{\mu-1}X_\mu.
\end{aligned} \tag{7.22}$$

The coefficients of $\sigma(x)$ as given above are said to be *elementary symmetric functions* of the error-location numbers. We now have two sets of equations involving the error-location numbers of a t-error-correcting code:

(1) equations 7.12 relating the error-location numbers to the error syndromes;
(2) equations 7.22 relating the error-location numbers to the polynomial coefficients.

From these two sets of equations we can eliminate the error-location numbers to obtain expressions involving only the error syndromes and the coefficients of the error-location polynomial. It can be shown that for the first μ error syndromes

$$\begin{aligned}
S_1 &= \sigma_1\\
S_2 &= \sigma_1S_1 + 2\sigma_2\\
S_3 &= \sigma_1S_2 + \sigma_2S_1 + 3\sigma_3\\
&\vdots\\
S_\mu &= \sigma_1S_{\mu-1} + \sigma_2S_{\mu-2} + \cdots + \sigma_{\mu-1}S_1 + \mu\sigma_\mu.
\end{aligned} \tag{7.23}$$

Note that the last term $i\sigma_i$ in each expression in the eqns 7.23 is 0 for even values of i and σ_i for odd values of i. The remaining error syndromes are given by

$$\begin{aligned}
S_{\mu+1} &= \sigma_1S_\mu + \sigma_2S_{\mu-1} + \cdots + \sigma_{\mu-1}S_2 + \sigma_\mu S_1\\
S_{\mu+2} &= \sigma_1S_{\mu+1} + \sigma_2S_\mu + \cdots + \sigma_{\mu-1}S_3 + \sigma_\mu S_2\\
S_{\mu+3} &= \sigma_1S_{\mu+2} + \sigma_2S_{\mu+1} + \cdots + \sigma_{\mu-1}S_4 + \sigma_\mu S_3\\
&\vdots\\
S_{2\mu} &= \sigma_1S_{2\mu-1} + \sigma_2S_{2\mu-2} + \cdots + \sigma_{\mu-1}S_{\mu+1} + \sigma_\mu S_\mu.
\end{aligned} \tag{7.24}$$

Equations 7.23 and 7.24 are a set of linear equations, referred to as *Newton's identities*, from which the coefficients of $\sigma(x)$ can be determined. Although they are a single set of equations, they fall naturally into two groups, the first μ equations given by eqns 7.23, and the remaining μ equations given by eqns 7.24. For clarity, they can be expressed in the matrix forms

$$\begin{bmatrix} S_1 \\ S_2 \\ S_3 \\ \vdots \\ S_\mu \end{bmatrix} = \begin{bmatrix} 1 & 0 & 0 & \cdots & 0 & 0 \\ S_1 & 2 & 0 & \cdots & 0 & 0 \\ S_2 & S_1 & 3 & \cdots & 0 & 0 \\ \vdots & & & & & \vdots \\ S_{\mu-1} & S_{\mu-2} & S_{\mu-3} & \cdots & S_1 & \mu \end{bmatrix} \begin{bmatrix} \sigma_1 \\ \sigma_2 \\ \sigma_3 \\ \vdots \\ \sigma_\mu \end{bmatrix} \tag{7.25}$$

and

$$\begin{bmatrix} S_{\mu+1} \\ S_{\mu+2} \\ S_{\mu+3} \\ \vdots \\ S_{2\mu} \end{bmatrix} = \begin{bmatrix} S_\mu & S_{\mu-1} & S_{\mu-2} & \cdots & S_2 & S_1 \\ S_{\mu+1} & S_\mu & S_{\mu-1} & \cdots & S_3 & S_2 \\ S_{\mu+2} & S_{\mu+1} & S_\mu & \cdots & S_4 & S_3 \\ \vdots & & & & & \vdots \\ S_{2\mu-1} & S_{2\mu-2} & S_{2\mu-3} & \cdots & S_{\mu+1} & S_\mu \end{bmatrix} \begin{bmatrix} \sigma_1 \\ \sigma_2 \\ \sigma_3 \\ \vdots \\ \sigma_\mu \end{bmatrix} \tag{7.26}$$

In Section 7.6 we look at the Peterson–Gorenstein–Zierler decoder, which uses eqn 7.26 as the basis for a decoder for multiple-error-correcting BCH codes. Later Berlekamp's algorithm is considered, this is a fast algorithm that uses eqns 7.23 and 7.24. For the remaining part of this section, we look at how the coefficients of the error-location polynomial of a binary code can be obtained algebraically from eqns 7.23–7.26.

For a binary code eqns 7.23–7.26 can be simplified by taking into account the relationship $S_{2i} = S_i^2$. Consider eqns 7.23, using $S_2 = S_1^2$ we find that the second equation $S_2 = \sigma_1 S_1 + 2\sigma_2$ reduces to the first equation $S_1 = \sigma_1$ and can therefore be excluded. Likewise all the equations for S_i with even values of i can be excluded from eqns 7.23–7.26. Furthermore we can combine eqns 7.25 and 7.26 to get

$$\begin{bmatrix} S_1 \\ S_3 \\ S_5 \\ \vdots \\ S_{2\mu-1} \end{bmatrix} = \begin{bmatrix} 1 & 0 & 0 & \cdots & 0 & 0 \\ S_2 & S_1 & 1 & \cdots & 0 & 0 \\ S_4 & S_3 & S_2 & \cdots & 0 & 0 \\ \vdots & & & & & \vdots \\ S_{2\mu-2} & S_{2\mu-3} & S_{2\mu-4} & \cdots & S_\mu & S_{\mu-1} \end{bmatrix} \begin{bmatrix} \sigma_1 \\ \sigma_2 \\ \sigma_3 \\ \vdots \\ \sigma_\mu \end{bmatrix} \tag{7.27}$$

For a double-error-correcting code the maximum number of correctable errors is 2, and setting $\mu = 2$ in eqn 7.27 gives

$$\begin{bmatrix} S_1 \\ S_3 \end{bmatrix} = \begin{bmatrix} 1 & 0 \\ S_2 & S_1 \end{bmatrix} \begin{bmatrix} \sigma_1 \\ \sigma_2 \end{bmatrix}$$

from which we get

$$S_1 = \sigma_1$$
$$S_3 = S_2\sigma_1 + S_1\sigma_2.$$

Rearranging these 2 expressions gives

$$\sigma_1 = S_1$$
$$\sigma_2 = (S_1^3 + S_3)/S_1$$

which agree with the coefficients of $\sigma(x)$ given previously (eqns 7.18).

Next we consider the slightly more difficult example of a triple-error-correcting code. Here the error-location polynomial is

$$\sigma(x) = 1 + \sigma_1 x + \sigma_2 x^2 + \sigma_3 x^3$$

and as the maximum number of correctable errors is $\mu = 3$ eqn 7.27 reduces to

$$\begin{bmatrix} S_1 \\ S_3 \\ S_5 \end{bmatrix} = \begin{bmatrix} 1 & 0 & 0 \\ S_2 & S_1 & 1 \\ S_4 & S_3 & S_2 \end{bmatrix} \begin{bmatrix} \sigma_1 \\ \sigma_2 \\ \sigma_3 \end{bmatrix}$$

which when expanded gives

$$\begin{aligned} S_1 &= \sigma_1 \\ S_3 &= S_2\sigma_1 + S_1\sigma_2 + \sigma_3 \\ S_5 &= S_4\sigma_1 + S_3\sigma_2 + S_2\sigma_3. \end{aligned} \tag{7.28}$$

Multiplying the middle equation by S_2 and adding it to the last equation eliminates σ_1 and σ_3, so allowing σ_2 to be determined

$$\begin{aligned} S_2S_3 + S_5 &= (S_2^2 + S_4)\sigma_1 + (S_2S_1 + S_3)\sigma_2 + (S_2 + S_2)\sigma_3 \\ &= (S_1^4 + S_1^4)\sigma_1 + (S_1^2S_1 + S_3)\sigma_2 \\ &= (S_1^3 + S_3)\sigma_2 \end{aligned}$$

and so

$$\sigma_2 = \frac{(S_2S_3 + S_5)}{(S_1^3 + S_3)}.$$

Now rearranging the middle expression of eqn 7.28 gives

$$\sigma_3 = S_3 + S_2\sigma_1 + S_1\sigma_2$$

and we could leave σ_3 like this since σ_1 and σ_2 are known. However, for completeness, we can substitute σ_1 and σ_2 as given above into σ_3 to get

$$\sigma_3 = (S_3 + S_1^3) + \frac{S_1(S_2S_3 + S_5)}{(S_1^3 + S_3)}.$$

Therefore for a triple-error-correcting code the coefficients of the error-location polynomial are

$$\begin{aligned} \sigma_1 &= S_1 \\ \sigma_2 &= \frac{(S_2S_3 + S_5)}{(S_1^3 + S_3)} \\ \sigma_3 &= (S_3 + S_1^3) + \frac{S_1(S_2S_3 + S_5)}{(S_1^3 + S_3)}. \end{aligned} \tag{7.29}$$

Example 7.5
Given a triple-error-correcting code and error syndromes $S_1 = \alpha^3$, $S_3 = \alpha^8$ and $S_5 = 1$ over $GF(2^4)$, determine the error-location polynomial $\sigma(x)$.

Using eqns 7.29 gives

$$\sigma_1 = S_1 = \alpha^3$$
$$\sigma_2 = \frac{(S_2 S_3 + S_5)}{(S_1^3 + S_3)} = \frac{(\alpha^6 \alpha^8 + 1)}{(\alpha^9 + \alpha^8)} = \frac{\alpha^3}{\alpha^{12}} = \alpha^6$$
$$\sigma_3 = (S_3 + S_1^3) + \frac{S_1(S_2 S_3 + S_5)}{(S_1^3 + S_3)} = (\alpha^9 + \alpha^8) + \frac{(\alpha^9 \alpha^8 + \alpha^3 1)}{(\alpha^9 + \alpha^8)} = \alpha^8.$$

The error-location-polynomial is therefore

$$\sigma(x) = 1 + \sigma_1 x + \sigma_2 x^2 + \sigma_3 x^3 = 1 + \alpha^3 x + \alpha^6 x^2 + \alpha^8 x^3. \qquad \square$$

The method described in this section can be applied to any t-error-correcting code. However beyond $t = 4$ or 5 the resulting equations, relating the coefficients of the error-location polynomial to the error syndromes, become rather complicated and so this approach becomes impractical. Instead the Peterson–Gorenstein–Zierler decoder forms the basis for multiple-error correction.

7.6 The Peterson–Gorenstein–Zierler decoder

We are now in a position where we can consider an algorithm that typifies the decoding of BCH codes, namely the *Peterson–Gorenstein–Zierler decoder*. This is a general purpose decoder that can be used for decoding any t-error-correcting BCH code. It is based on the error-location polynomial, and as we shall see the decoder brings together into a single algorithm the various ideas considered in the previous sections.

The error syndromes $S_{\mu+1}, S_{\mu+2}, \ldots, S_{2\mu}$ are related to the coefficients of the error-location polynomial by eqn 7.26. By convention the order of the columns of the matrix in eqn 7.26 are reversed, along with the rows of the column vector containing the polynomial coefficients. This gives

$$\begin{bmatrix} S_{\mu+1} \\ S_{\mu+2} \\ S_{\mu+3} \\ \vdots \\ S_{2\mu} \end{bmatrix} = \begin{bmatrix} S_1 & S_2 & S_3 & \cdots & S_{\mu-1} & S_\mu \\ S_2 & S_3 & S_4 & \cdots & S_\mu & S_{\mu+1} \\ S_3 & S_4 & S_5 & \cdots & S_{\mu+1} & S_{\mu+2} \\ \vdots & & & & & \vdots \\ S_\mu & S_{\mu+1} & S_{\mu+2} & \cdots & S_{2\mu-2} & S_{2\mu-1} \end{bmatrix} \begin{bmatrix} \sigma_\mu \\ \sigma_{\mu-1} \\ \sigma_{\mu-2} \\ \vdots \\ \sigma_1 \end{bmatrix} \qquad (7.30)$$

which can be expressed as

$$\boldsymbol{S} = \boldsymbol{M\sigma}$$

where

$$\boldsymbol{M} = \begin{bmatrix} S_1 & S_2 & S_3 & \dots & S_{\mu-1} & S_\mu \\ S_2 & S_3 & S_4 & \dots & S_\mu & S_{\mu+1} \\ S_3 & S_4 & S_5 & \dots & S_{\mu+1} & S_{\mu+2} \\ \vdots & & & & & \vdots \\ S_\mu & S_{\mu+1} & S_{\mu+2} & \dots & S_{2\mu-2} & S_{2\mu-1} \end{bmatrix} \tag{7.31}$$

$$\boldsymbol{S} = \begin{bmatrix} S_{\mu+1} \\ S_{\mu+2} \\ S_{\mu+3} \\ \vdots \\ S_{2\mu} \end{bmatrix} \tag{7.32}$$

$$\boldsymbol{\sigma} = \begin{bmatrix} \sigma_\mu \\ \sigma_{\mu-1} \\ \sigma_{\mu-2} \\ \vdots \\ \sigma_1 \end{bmatrix} \tag{7.33}$$

Assuming that $\boldsymbol{M}$ is nonsingular, so that its inverse $\boldsymbol{M}^{-1}$ exists, and multiplying $\boldsymbol{S} = \boldsymbol{M}\boldsymbol{\sigma}$ through by $\boldsymbol{M}^{-1}$ gives

$$\boldsymbol{M}^{-1}\boldsymbol{S} = (\boldsymbol{M}^{-1}\boldsymbol{M})\boldsymbol{\sigma} = \boldsymbol{\sigma}$$

and therefore the coefficients of the error-location polynomial are given by

$$\boldsymbol{\sigma} = \boldsymbol{M}^{-1}\boldsymbol{S}. \tag{7.34}$$

To evaluate eqn 7.34, $\boldsymbol{M}$ has to be nonsingular and we need to know the number of errors μ that have occurred. It can be shown that the matrix

$$\boldsymbol{M} = \begin{bmatrix} S_1 & S_2 & S_3 & \dots & S_{i-1} & S_i \\ S_2 & S_3 & S_4 & \dots & S_i & S_{i+1} \\ S_3 & S_4 & S_5 & \dots & S_{i+1} & S_{i+2} \\ \vdots & & & & & \vdots \\ S_i & S_{i+1} & S_{i+2} & \dots & S_{2i-2} & S_{2i-1} \end{bmatrix}$$

is nonsingular if $i = \mu$, but singular if $i > \mu$. For example consider a code that can correct 5 errors but only 3 errors actually occur, so $t = 5$ and $\mu = 3$. Taking $i = 5$ and constructing

$$\boldsymbol{M} = \begin{bmatrix} S_1 & S_2 & S_3 & S_4 & S_5 \\ S_2 & S_3 & S_4 & S_5 & S_6 \\ S_3 & S_4 & S_5 & S_6 & S_7 \\ S_4 & S_5 & S_6 & S_7 & S_8 \\ S_5 & S_6 & S_7 & S_8 & S_9 \end{bmatrix}$$

will give $\det(\boldsymbol{M})=0$, and so $\boldsymbol{M}$ is singular. Likewise the 4 by 4 matrix $\boldsymbol{M}$ of error syndromes, obtained when $i=4$ will also be singular. However, when i equals the number of errors that have occurred, i.e. $i=\mu=3$, the 3 by 3 matrix $\boldsymbol{M}$ is non-singular. $\boldsymbol{M}^{-1}$ can then be determined and the polynomial coefficients can be found using $\boldsymbol{\sigma}=\boldsymbol{M}^{-1}\boldsymbol{S}$.

For a BCH code with error-correction limit t, decoding proceeds as follows. The decoder first assumes that the maximum number of correctable errors have occurred, $i=t$, and constructs $\boldsymbol{M}$ and determines $\det(\boldsymbol{M})$. If $\boldsymbol{M}$ is nonsingular, then $\boldsymbol{M}^{-1}$ and $\boldsymbol{\sigma}=\boldsymbol{M}^{-1}\boldsymbol{S}$ can be found. If $\boldsymbol{M}$ is singular the decoder assumes that t errors did not occur and repeats the calculations on the assumption of 1 less error, i.e. $i=t-1$. The decoder continues in an iterative manner, decreasing i by 1 each time $\boldsymbol{M}$ is found to be singular. On obtaining $\det(\boldsymbol{M})\neq 0$, the value of i is taken to be the number of errors that occurred and $\boldsymbol{\sigma}$ is determined. The main steps of the Peterson–Gorenstein–Zierler decoder are:

1. Calculate the error syndromes $S_1, S_2, \ldots, S_{2t}$ from $v(x)$.
2. Assume the maximum number of errors, $i=t$.
3. Construct the matrix $\boldsymbol{M}$.
4. Find the determinant of $\boldsymbol{M}$ and check if $\det(\boldsymbol{M})=0$. If $\det(\boldsymbol{M})=0$ reduce i by 1 and go back to step 3, otherwise continue to Step 5.
5. Determine $\boldsymbol{M}^{-1}$ and construct $\boldsymbol{S}$.
6. Find the polynomial coefficients using $\boldsymbol{\sigma}=\boldsymbol{M}^{-1}\boldsymbol{S}$ and construct $\sigma(x)$ from σ.
7. Determine the roots of $\sigma(x)$ and take their reciprocals. The error-location numbers are given by the reciprocal roots.

As an example of how the decoder works, let's consider the triple-error-correcting (15, 5) BCH code with

$$v(x) = x^8 + x^5 + x^2 + x + 1$$

where $v(x)$ corresponds to a codeword $c(x)$ with 2 errors. For clarity, the example is referenced to the seven steps given above.

Step 1 In $GF(2^4)$ the error syndromes are

$$\begin{aligned}
S_1 &= \alpha^8 + \alpha^5 + \alpha^2 + \alpha + 1 = \alpha^2 \\
S_2 &= S_1^2 = \alpha^4 \\
S_3 &= \alpha^9 + \alpha^{15} + \alpha^6 + \alpha^3 + 1 = \alpha^{11} \\
S_4 &= S_2^2 = \alpha^8 \\
S_5 &= \alpha^{40} + \alpha^{25} + \alpha^{10} + \alpha^5 + 1 = 0 \\
S_6 &= S_3^2 = \alpha^7.
\end{aligned}$$

Step 2 Assume the maximum number of errors, $i=3$.

Step 3 The matrix $\boldsymbol{M}$ is

$$\boldsymbol{M} = \begin{bmatrix} S_1 & S_2 & S_3 \\ S_2 & S_3 & S_4 \\ S_3 & S_4 & S_5 \end{bmatrix} = \begin{bmatrix} \alpha^2 & \alpha^4 & \alpha^{11} \\ \alpha^4 & \alpha^{11} & \alpha^8 \\ \alpha^{11} & \alpha^8 & 0 \end{bmatrix}$$

Step 4 The determinant of $\boldsymbol{M}$ is

$$\det(\boldsymbol{M}) = \begin{vmatrix} \alpha^2 & \alpha^4 & \alpha^{11} \\ \alpha^4 & \alpha^{11} & \alpha^8 \\ \alpha^{11} & \alpha^8 & 0 \end{vmatrix}$$

$$= \alpha^2 \begin{vmatrix} \alpha^{11} & \alpha^8 \\ \alpha^8 & 0 \end{vmatrix} + \alpha^4 \begin{vmatrix} \alpha^4 & \alpha^8 \\ \alpha^{11} & 0 \end{vmatrix} + \alpha^{11} \begin{vmatrix} \alpha^4 & \alpha^{11} \\ \alpha^{11} & \alpha^8 \end{vmatrix}$$

$$= \alpha^2\alpha + \alpha^4\alpha^4 + \alpha^{11}\alpha^2 = \alpha^3 + \alpha^8 + \alpha^{13} = 0.$$

The matrix $\boldsymbol{M}$ is therefore singular, so i is reduced by 1 to give $i = 2$ and steps 3 and 4 are repeated.

Step 3—repeated We now have

$$\boldsymbol{M} = \begin{bmatrix} S_1 & S_2 \\ S_2 & S_3 \end{bmatrix} = \begin{bmatrix} \alpha^2 & \alpha^4 \\ \alpha^4 & \alpha^{11} \end{bmatrix}.$$

Step 4—repeated The revised value of $\det(\boldsymbol{M})$ is

$$\det(\boldsymbol{M}) = \begin{vmatrix} \alpha^2 & \alpha^4 \\ \alpha^4 & \alpha^{11} \end{vmatrix} = \alpha^{13} + \alpha^8 = \alpha^3.$$

Now $\boldsymbol{M}$ is nonsingular so we can move to Step 5 and find $\boldsymbol{M}^{-1}$.

Step 5 Using $\boldsymbol{M}^{-1} = \text{adj}(\boldsymbol{M})/\det(\boldsymbol{M})$ we get

$$\text{adj}(\boldsymbol{M}) = \begin{bmatrix} \alpha^{11} & \alpha^4 \\ \alpha^4 & \alpha^2 \end{bmatrix}$$

and so

$$\boldsymbol{M}^{-1} = \frac{1}{\alpha^3} \begin{bmatrix} \alpha^{11} & \alpha^4 \\ \alpha^4 & \alpha^2 \end{bmatrix} = \begin{bmatrix} \alpha^8 & \alpha \\ \alpha & \alpha^{14} \end{bmatrix}.$$

Also

$$\boldsymbol{S} = \begin{bmatrix} S_3 \\ S_4 \end{bmatrix} = \begin{bmatrix} \alpha^{11} \\ \alpha^8 \end{bmatrix}.$$

Step 6 Using $\sigma = \boldsymbol{M}^{-1}\boldsymbol{S}$ gives

$$\begin{bmatrix} \sigma_2 \\ \sigma_1 \end{bmatrix} = \begin{bmatrix} \alpha^8 & \alpha \\ \alpha & \alpha^{14} \end{bmatrix} \begin{bmatrix} \alpha^{11} \\ \alpha^8 \end{bmatrix} = \begin{bmatrix} \alpha^{14} \\ \alpha^2 \end{bmatrix}$$

and so $\sigma_2 = \alpha^{14}$, $\sigma_1 = \alpha^2$ and the error-location polynomial is therefore

$$\sigma(x) = 1 + \sigma_1 x + \sigma_2 x^2 = 1 + \alpha^2 x + \alpha^{14} x^2.$$

Step 7 Using a Chien over $GF(2^4)$ we find that the roots of $\sigma(x)$ are α^5 and α^{11}, the reciprocals of which give the error-location numbers α^{10} and α^4 respectively. The error pattern is therefore $x^{10}+x^4$ and so the required codeword polynomial is

$$c(x) = v(x) + x^{10} + x^4 = x^{10} + x^8 + x^5 + x^4 + x^2 + x + 1.$$

In the example below the (15, 5) triple-error-correcting BCH code is again considered, but with a codeword that has incurred 3 errors.

Example 7.6

A codeword $c(x)$, of the (15, 5) triple-error-correcting BCH code, incurs errors so as to give $v(x)=x^{13}+x^{10}+x^8+x^4+x+1$. Find the number of errors that $c(x)$ has incurred, the error pattern and $c(x)$.

Over $GF(2^4)$ the error syndromes are

$$\begin{aligned}
S_1 &= v(\alpha) = \alpha^{13} + \alpha^{10} + \alpha^8 + \alpha^4 + \alpha + 1 = \alpha^{12}\\
S_2 &= S_1^2 = \alpha^9\\
S_3 &= v(\alpha^3) = \alpha^9 + 1 + \alpha^9 + \alpha^{12} + \alpha^3 + 1 = \alpha^{10}\\
S_4 &= S_2^2 = \alpha^3\\
S_5 &= \alpha^5 + \alpha^5 + \alpha^{10} + \alpha^5 + \alpha^5 + 1 = \alpha^5\\
S_6 &= S_3^2 = \alpha^5.
\end{aligned}$$

Assuming the maximum number of errors, $i=3$, we construct the matrix

$$\boldsymbol{M} = \begin{bmatrix} \alpha^{12} & \alpha^9 & \alpha^{10} \\ \alpha^9 & \alpha^{10} & \alpha^3 \\ \alpha^{10} & \alpha^3 & \alpha^5 \end{bmatrix}.$$

The determinant of $\boldsymbol{M}$ is

$$\det(\boldsymbol{M}) = \alpha^{12}(1+\alpha^6) + \alpha^9(\alpha^{14}+\alpha^{13}) + \alpha^{10}(\alpha^{12}+\alpha^5) = \alpha^4.$$

As $\boldsymbol{M}$ is nonsingular we assume that 3 errors have occurred. The inverse of M is

$$\boldsymbol{M}^{-1} = \frac{\operatorname{adj}(\boldsymbol{M})}{\det(\boldsymbol{M})} = (1/\alpha^4)\begin{bmatrix} \alpha^{13} & \alpha^2 & \alpha^{14} \\ \alpha^2 & \alpha & \alpha \\ \alpha^{14} & \alpha & \alpha^4 \end{bmatrix} = \begin{bmatrix} \alpha^9 & \alpha^{13} & \alpha^{10} \\ \alpha^{13} & \alpha^{12} & \alpha^{12} \\ \alpha^{10} & \alpha^{12} & 1 \end{bmatrix}$$

and

$$\boldsymbol{S} = \begin{bmatrix} \alpha^3 \\ \alpha^5 \\ \alpha^5 \end{bmatrix}.$$

Using $\sigma = \boldsymbol{M}^{-1}\boldsymbol{S}$ gives

$$\begin{bmatrix} \sigma_3 \\ \sigma_2 \\ \sigma_1 \end{bmatrix} = \begin{bmatrix} \alpha^9 & \alpha^{13} & \alpha^{10} \\ \alpha^{13} & \alpha^{12} & \alpha^{12} \\ \alpha^{10} & \alpha^{12} & 1 \end{bmatrix} \begin{bmatrix} \alpha^3 \\ \alpha^5 \\ \alpha^5 \end{bmatrix} = \begin{bmatrix} \alpha^5 \\ \alpha \\ \alpha^{12} \end{bmatrix}$$

and so $\sigma_3 = \alpha^5$, $\sigma_2 = \alpha$, $\sigma_1 = \alpha^{12}$ giving the error-location polynomial

$$\sigma(x) = 1 + \alpha^{12}x + \alpha x^2 + \alpha^5 x^3.$$

Searching $GF(2^4)$ for the roots of $\sigma(x)$ gives α^2, α^{10}, and α^{13} as roots and taking their reciprocals gives α^{13}, α^5, and α^2 as the error-location numbers respectively. The error pattern is therefore $e(x) = x^{13} + x^5 + x^2$ and the codeword polynomial $c(x) = v(x) + e(x) = x^{10} + x^8 + x^5 + x^4 + x^2 + x + 1$. □

The Peterson–Gorenstein–Zierler decoder forms the basis of decoding algorithms for BCH codes. It is a relatively simple decoder, as the reader should find after working through a few examples. The matrix inversion does however present a problem to the decoder. For a large error-correction limit t, evaluating the resulting determinants can be computationally slow and inefficient. Furthermore, as we shall see later, a second matrix inversion is required when dealing with non-binary codes, and so aggravating the problem. To develop fast decoding algorithms we need to avoid the matrix inversions, this is considered in Sections 7.8 and 7.9.

7.7 Reed–Solomon codes

The codes considered so far have all been binary codes, and we now turn our attention to non-binary codes. At first the idea of a non-binary code may seem rather strange or of little practical use. Information processing, transmission, and storage is usually thought of in terms of a binary representation. Bits are manipulated either individually or in blocks of convenient length, for example as 8-bit words. However, an 8-bit word can be thought of as a single *non-binary symbol* with 256 different values, irrespective of its underlying structure (i.e. the fact that it is really a collection of 8 bits and not a single symbol). Likewise any sequence of r bits can be viewed as a single non-binary symbol that has one of 2^r values. Furthermore symbols need not necessarily be restricted to 2^r values but can be defined for any positive integer.

Non-binary codes are concerned with the detection and correction of errors in symbols. The construction of non-binary codes, along with encoding and decoding techniques, follows directly from that of binary codes. The main difference arises in the need to determine the magnitude of errors and not just the error locations. In binary codes error magnitudes are 1 and it is only necessary to determine the position of errors. Once located error correction is achieved by simply inverting the erroneous bits. With a non-binary code we first locate the position of the errors and then determine the magnitude of the errors.

Binary codes can be viewed as codes whose symbols have 2 values, 0 and 1, that is the code's symbols lie in $GF(2)$. A non-binary code has its symbols in the field $GF(q)$ where q is a prime number or any power of a prime number. A non-binary (n, k) linear code will have codewords of the form $\boldsymbol{c} = (c_{n-1}, c_{n-2}, \ldots, c_2, c_1, c_0)$ where the codeword components lie in $GF(q)$ and there exists at least one set of k codewords from which all the other codewords can be obtained by linear combinations of the k codewords. A non-binary (n, k) cyclic code can be constructed from a polynomial $g(x)$ of degree $n - k$, where $g(x)$ has its coefficients in $GF(q)$ and divides x^n-1. Note that for binary codes $x^n - 1 = x^n + 1$ and so $g(x)$ can be said to divide the latter if it is to generate a cyclic code.

A t-error-correcting nonbinary BCH code of blocklength $n = q^m - 1$ is a (n, k) cyclic code whose generator polynomial $g(x)$ has its coefficients in $GF(q)$ and roots

$$\beta, \beta^2, \ldots, \beta^{2t}$$

in $GF(q^m)$ an extension field of $GF(q)$. Recall that the generator polynomial of a t-error-correcting binary BCH code is given by the least common multiple LCM of the minimal polynomials $m_i(x)$, over $GF(2)$, of $2t$ consecutive field elements. To construct a non-binary BCH code minimal polynomials over $GF(q)$ are required and the generator polynomial of the code is given by

$$g(x) = \text{LCM}[m_1(x), m_2(x), \ldots, m_{2t}(x)]$$

where now $m_i(x)$ is the minimal polynomial over $GF(q)$ of β^i. If $q = 2$ then the minimal polynomials are binary and we obtain the binary BCH codes.

The most important class of non-binary BCH codes are the *Reed–Solomon codes*, which differ from other non-binary codes in that the base field and extension field are taken to be the same. Both the symbols and the generator polynomial roots lie in the field $GF(q)$ and define a Reed–Solomon code with blocklength $n = q - 1$. Here we consider Reed–Solomon codes where $q = 2^m$ and so symbols and roots lie in $GF(2^m)$. A t-error-correcting Reed–Solomon code is a cyclic code whose generator polynomial is the least-degree polynomial that has $\beta, \beta^2, \beta^3, \ldots, \beta^{2t}$ as roots, where β belongs to $GF(2^m)$. The minimal polynomial over $GF(2^m)$ of an element β in $GF(2^m)$ is the factor

$$m_\beta = x + \beta$$

as this is clearly the least-degree polynomial that has β as a root. The generator polynomial of a Reed–Solomon code is therefore

$$g(x) = (x + \beta)(x + \beta^2)(x + \beta^3) \cdots (x + \beta^{2t}). \tag{7.35}$$

Note that there is also no need to take the least common multiple of the factors as all the factors are distinct.

Consider a double-error-correcting Reed–Solomon code over $GF(2^4)$, taking $\beta=\alpha$ gives

$$g(x) = (x+\alpha)(x+\alpha^2)(x+\alpha^3)(x+\alpha^4)$$

and expanding this gives the generator polynomial

$$g(x) = x^4 + \alpha^{13}x^3 + \alpha^6 x^2 + \alpha^3 x + \alpha^{10}. \tag{7.36}$$

Note that the coefficients of $g(x)$ are no longer binary. The code's blocklength is $n=2^4-1=15$, and as the degree of $g(x)$ is $r=4$, then using $n-k=r$ gives $k=11$ (recall that the degree of the generator polynomial of an (n, k) cyclic code is $n-k$). This is therefore the generator polynomial of a double-error-correcting (15, 11) Reed–Solomon code.

Example 7.7
Construct a single-error-correcting Reed–Solomon code with blocklength 7.

The code is constructed over $GF(2^3)$ as this gives a code with blocklength $n=2^3-1=7$. Substituting $t=1$ and $\beta=\alpha$ in eqn 7.35 gives

$$g(x) = (x+\alpha)(x+\alpha^2) = x^2 + \alpha^4 x + \alpha^3$$

where α is a primitive element in $GF(2^3)$. The information length k is given by $k=n-r$, where $r=2$ is the degree of $g(x)$ and $n=7$, and so $k=5$. This is therefore a single-error-correcting (7, 5) Reed–Solomon code. □

For encoding purposes the Reed–Solomon codes can be treated as cyclic codes or a generator matrix can be constructed from the generator polynomial and the codes can then be treated as linear codes. For example consider the (7, 5) Reed–Solomon code with generator polynomial

$$g(x) = x^2 + \alpha^4 x + \alpha^3 \tag{7.37}$$

and let's construct the systematic codeword for the information word, say, $\boldsymbol{i}=(1\ 0\ \alpha\ \alpha^5\ \alpha^2)$ where α is an element of $GF(2^3)$. The information polynomial corresponding to $\boldsymbol{i}$ is

$$i(x) = x^4 + \alpha x^2 + \alpha^5 x + \alpha^2$$

and multiplying this by $x^{n-k}=x^2$ gives

$$x^2 i(x) = x^6 + \alpha x^4 + \alpha^5 x^3 + \alpha^2 x^2.$$

Recall that to construct systematic codewords we require the remainder of $x^{n-k}i(x)$ divided by $g(x)$. When dividing two non-binary polynomials, care has to be taken to ensure that at each step the coefficients of the highest power of x are the same. The division is a bit more awkward than that of dividing two binary polynomials, however the principle is the same. Dividing $x^2 i(x)$ by $g(x)$ we get

$$
\begin{array}{rl}
 & x^4+\alpha^4x^3+\alpha^3x^2+\alpha^5x+\alpha^6 \\
x^2+\alpha^4x+\alpha^3 \,) & x^6+\alpha x^4+\alpha^5x^3+\alpha^2x^2 \\
 & x^6+\alpha^4x^5+\alpha^3x^4 \\
 & -\quad \alpha^4x^5+x^4+\alpha^5x^3+\alpha^2x^2 \\
 & \alpha^4x^5+\alpha x^4+x^3 \\
 & -\quad \alpha^3x^4+\alpha^4x^3+\alpha^2x^2 \\
 & \alpha^3x^4+x^3+\alpha^6x^2 \\
 & -\quad \alpha^5x^3+x^2 \\
 & \alpha^5x^3+\alpha^2x^2+\alpha x \\
 & -\quad \alpha^6x^2+\alpha x \\
 & \alpha^6x^2+\alpha^3x+\alpha^2 \\
 & -\quad x+\alpha^2
\end{array}
$$

and the remainder is therefore $r(x)=x+\alpha^2$. Adding $r(x)$ to $x^2 i(x)$ gives the codeword polynomial

$$c(x) = x^2 i(x) + r(x) = x^6 + \alpha x^4 + \alpha^5 x^3 + \alpha^2 x^2 + x + \alpha^2$$

which gives the codeword $\boldsymbol{c} = (1\ 0\ \alpha\ \alpha^5\ \alpha^2\ 1\ \alpha^2)$.

Example 7.8

Construct the (15, 13) single-error-correcting Reed–Solomon code and determine the systematic codeword corresponding to $\boldsymbol{i} = (0\ 0\ \alpha\ 0\ 0\ 1\ \alpha^7\ \alpha^2\ 0\ 0\ 1\ \alpha\ \alpha^2)$ where α is a primitive element of $GF(2^4)$.

The generator polynomial is

$$g(x) = (x+\alpha)(x+\alpha^2) = x^2 + \alpha^5 x + \alpha^3.$$

Note that the degree of $g(x)$ is 2, which is consistent with the code's (n, k) parameters, $n - k = 15 - 13 = 2$. The information polynomial corresponding to $\boldsymbol{i}$ is

$$i(x) = \alpha x^{10} + x^7 + \alpha^7 x^6 + \alpha^2 x^5 + x^2 + \alpha x + \alpha^2$$

and dividing $x^2 i(x)$ by $g(x)$ gives the quotient and remainder

$$
\begin{aligned}
q(x) &= \alpha x^{10} + \alpha^6 x^9 + \alpha^{13} x^8 + \alpha^4 x^7 + \alpha^4 x^6 + \alpha^8 x^5 + \alpha^5 x^4 \\
&\quad + \alpha^{14} x^3 + \alpha^{10} x^2 + \alpha^{10} x + \alpha^3 \\
r(x) &= x\alpha^3 + \alpha^6
\end{aligned}
$$

respectively. The codeword polynomial is therefore

$$x^2 i(x) + r(x) = \alpha x^{12} + x^9 + \alpha^7 x^8 + \alpha^2 x^7 + x^4 + \alpha x^3 + \alpha^2 x^2 + \alpha^3 x + \alpha^6$$

which gives the codeword

$$\boldsymbol{c} = (0\ 0\ \alpha\ 0\ 0\ 1\ \alpha^7\ \alpha^2\ 0\ 0\ 1\ \alpha\ \alpha^2\ \alpha^3\ \alpha^6).$$

□

The generator polynomial $g(x)$ of a t-error-correcting Reed–Solomon code has $2t$ linear factors, one for each root $\beta, \beta^2, \ldots, \beta^{2t}$ and the degree of $g(x)$ is therefore $n-k=2t$. Hence the number of parity-check symbols is $2t$ and as such a t-error correcting Reed–Solomon code with blocklength n can be referred to as a $(n, n-2t)$ Reed–Solomon code. Recall that a t-error-correcting code requires a minimum distance of

$$d_{\min} = 2t + 1$$

and so a t-error-correcting Reed–Solomon code has

$$d_{\min} = n - k + 1$$

and therefore the Reed–Solomon codes are maximum-distance codes. Note also that the designed distance, d_0, and minimum distance, $d_{\min}$, of a Reed–Solomon code are the same.

The number of codewords in a non-binary code can be surprisingly large. An (n, k) binary code has 2^k codewords as there are 2^k distinct information words. For example the $(7, 4)$ binary code has 16 codewords. In a t-error-correcting $(n, n-2t)$ Reed–Solomon code over $GF(q)$ each information symbol has q distinct values and there are therefore q^{n-2t} codewords. For example each symbol of the single-error-correcting $(7, 5)$ Reed–Solomon code has 8 distinct values and the code has 32 768 codewords. Clearly decoding algorithms that avoid the use of look-up tables are necessary with codes with such large numbers of codewords.

Decoding Reed–Solomon codes is achieved by first determining the error positions and then the error magnitudes. The methods used for locating errors in binary codes can also be used in Reed–Solomon codes, the only additional theory required is for finding error magnitudes. In a binary code an error pattern of μ errors can be represented by the error polynomial

$$e(x) = x^{p_1} + x^{p_2} + \cdots + x^{p_\mu}$$

where $p_1, p_2, \ldots, p_\mu$ are the error positions. Taking error magnitudes into account, the error polynomial becomes

$$e(x) = y_{p_1}x^{p_1} + y_{p_2}x^{p_2} + \cdots + y_{p_\mu}x^{p_\mu} \tag{7.38}$$

where y_{p_i} is the error magnitude at the position p_i. The decoder input is $v(x) = c(x) + e(x)$ where $c(x)$ is the codeword polynomial incurring the errors. For a t-error correcting code the error syndromes calculated by the decoder are

$$S_i = v(\alpha^i) = c(\alpha^i) + e(\alpha^i) = e(\alpha^i)$$

so giving

$$\begin{aligned} S_1 &= y_{p_1}\alpha^{p_1} + y_{p_2}\alpha^{p_2} + \cdots + y_{p_\mu}\alpha^{p_\mu} \\ S_2 &= y_{p_1}\alpha^{2p_1} + y_{p_2}\alpha^{2p_2} + \cdots + y_{p_\mu}\alpha^{2p_\mu} \\ S_3 &= y_{p_1}\alpha^{3p_1} + y_{p_2}\alpha^{3p_2} + \cdots + y_{p_\mu}\alpha^{3p_\mu} \\ &\vdots \\ S_{2t} &= y_{p_1}\alpha^{2tp_1} + y_{p_2}\alpha^{2tp_2} + \cdots + y_{p_\mu}\alpha^{2tp_\mu}. \end{aligned} \tag{7.39}$$

Recall that for binary codes we defined the error-location number $X_i = \alpha^{p_i}$. Here we define an additional term $Y_i = y_{p_i}$ known as the *error magnitude* of the *i*th error-location number, and in doing so we can express the syndrome equations in the more convenient form

$$\begin{aligned} S_1 &= Y_1X_1 + Y_2X_2 + \cdots + Y_\mu X_\mu \\ S_2 &= Y_1X_1^2 + Y_2X_2^2 + \cdots + Y_\mu X_\mu^2 \\ S_3 &= Y_1X_1^3 + Y_2X_2^3 + \cdots + Y_\mu X_\mu^3 \\ &\vdots \\ S_{2t} &= Y_1X_1^{2t} + Y_2X_2^{2t} + \cdots + Y_\mu X_\mu^{2t}. \end{aligned} \tag{7.40}$$

We have already seen that for a binary code $S_2 = S_1^2$, this though does not apply to non-binary codes. Taking S_1^2 gives

$$\begin{aligned} S_1^2 &= (Y_1X_1 + Y_2X_2 + \cdots + Y_\mu X_\mu)^2 \\ &= Y_1^2X_1^2 + Y_2^2X_2^2 + \cdots + Y_\mu^2 X_\mu^2 \neq S_2. \end{aligned}$$

Likewise we can show that $S_4 \neq S_2^2$ and clearly for a non-binary code

$$S_{2i} \neq S_i^2$$

and so the $2t$ error syndromes need to be individually evaluated.

Equations 7.40 consist of $2t$ equations with μ unknown error magnitudes, along with known error syndromes and error-location numbers, which can be solved to give the error magnitudes. Before addressing eqns 7.40 for any value of t we first consider a single-error-correcting code. The decoder of a single-error correcting code determines two syndromes S_1 and S_2. From eqn 7.40 setting $t = 1$ and $\mu = 1$ (as the maximum number of correctable errors is 1) gives

$$\begin{aligned} S_1 &= Y_1X_1 \\ S_2 &= Y_1X_1^2 \end{aligned} \tag{7.41}$$

and dividing S_2 by S_1 gives

$$\frac{S_2}{S_1} = \frac{Y_1X_1^2}{Y_1X_1} = X_1.$$

Substituting $X_1 = S_2/S_1$ into the first expression in eqns 7.41 gives $S_1 = Y_1(S_2/S_1)$ and so $Y_1 = S_1^2/S_2$. Therefore the error-location number X_1 and error magnitude Y_1 of a single-error correcting Reed–Solomon code are given by

$$\begin{aligned} X_1 &= S_2/S_1 \\ Y_1 &= S_1^2/S_2. \end{aligned} \tag{7.42}$$

Note that if we let $S_2 = S_1^2$, then eqns 7.42 give

$$\begin{aligned} X_1 &= S_1^2/S_1 = S_1 \\ Y_1 &= S_1^2/S_1^2 = 1 \end{aligned}$$

which are the correct error-location number and error magnitude for a single-error-correcting binary code.

Example 7.9
Consider the (7, 5) single-error-correcting Reed–Solomon code. Given that $\boldsymbol{v} = (0\ 1\ \alpha^5\ \alpha^2\ 1\ \alpha^6\ \alpha^3)$, where α is an element of $GF(2^3)$, corresponds to a codeword $\boldsymbol{c}$ with a single error, determine the position and magnitude of the error and the codeword $\boldsymbol{c}$.

The polynomial corresponding to $\boldsymbol{c}$ is

$$v(x) = x^5 + \alpha^5 x^4 + \alpha^2 x^3 + x^2 + \alpha^6 x + \alpha^3$$

and in $GF(2^3)$ the error syndromes are

$$S_1 = v(\alpha) = \alpha$$
$$S_2 = v(\alpha^2) = \alpha^3.$$

Using eqns 7.42 gives

$$X_1 = S_2/S_1 = \alpha^3/\alpha = \alpha^2$$
$$Y_1 = S_1^2/S_2 = \alpha^2/\alpha^3 = \alpha^6$$

giving an error location of x^2 and error magnitude α^6. The error incurred by $\boldsymbol{c}$ is therefore $\alpha^6 x^2$ and so the codeword polynomial is

$$\begin{aligned} c(x) &= v(x) + \alpha^6 x^2 \\ &= x^5 + \alpha^5 x^4 + \alpha^2 x^3 + (1 + \alpha^6)x^2 + \alpha^6 x + \alpha^3 \\ &= x^5 + \alpha^5 x^4 + \alpha^2 x^3 + \alpha^2 x^2 + \alpha^6 x + \alpha^3 \end{aligned}$$

giving $\boldsymbol{c} = (0\ 1\ \alpha^5\ \alpha^2\ \alpha^2\ \alpha^6\ \alpha^3)$. □

A decoder for a t-error-correcting Reed–Solomon code determines the number of errors μ and the error-location numbers $X_1, X_2, \ldots, X_\mu$ using any technique that can be used for a binary BCH code. Once the error-location numbers have been found, the μ error magnitudes can be obtained by solving the first μ equations in eqns 7.40 for $Y_1, Y_2, \ldots, Y_\mu$. Note that the error syndromes, given by eqns 7.40, are nonlinear functions of the error-location numbers, but linear functions of the error magnitudes. Hence the error magnitudes can be obtained from the syndrome equations by using a standard matrix inversion method. Defining the column vectors $\boldsymbol{S}$ and $\boldsymbol{Y}$ as

$$\boldsymbol{S} = \begin{bmatrix} S_1 \\ S_2 \\ S_3 \\ \vdots \\ S_\mu \end{bmatrix} \tag{7.43}$$

$$\boldsymbol{Y} = \begin{bmatrix} Y_1 \\ Y_2 \\ Y_3 \\ \vdots \\ Y_\mu \end{bmatrix} \tag{7.44}$$

and the matrix $\boldsymbol{X}$ as

$$\boldsymbol{X} = \begin{bmatrix} X_1 & X_2 & X_3 & \ldots & X_\mu \\ X_1^2 & X_2^2 & x_3^2 & \ldots & X_\mu^2 \\ X_1^3 & X_2^3 & X_3^3 & \ldots & X_\mu^3 \\ \vdots & & & & \vdots \\ X_1^\mu & X_2^\mu & X_3^\mu & \ldots & X_\mu^\mu \end{bmatrix} \tag{7.45}$$

then eqns 7.40 can be written as

$$\boldsymbol{S} = \boldsymbol{XY}.$$

Note that the column vector $\boldsymbol{S}$ given by eqn 7.43 is not the same as $\boldsymbol{S}$ (eqn 7.32) defined for use in the Peterson–Gorenstein–Zierler decoder, for convenience the same notation is used, this should not cause confusion. The matrix $\boldsymbol{X}$ and column vector $\boldsymbol{S}$ are known terms, and so $\boldsymbol{Y}$ can be found by inverting $\boldsymbol{S}=\boldsymbol{XY}$ to give $\boldsymbol{Y}=\boldsymbol{X}^{-1}\boldsymbol{S}$. Therefore the error magnitudes are given by

$$\boldsymbol{Y} = \boldsymbol{X}^{-1}\boldsymbol{S}. \tag{7.46}$$

Note that $\boldsymbol{X}$ cannot be singular because μ nonzero and distinct errors are already known to exist.

Finding error magnitudes is simpler than we may have at first expected. The nonlinear problem faced when determining the error locations does not arise. Instead the error magnitudes are linearly related to the error syndromes and the known error-location numbers. However, as before, we face the computationally inefficient process of matrix inversion. Later we shall see how this matrix inversion can be circumvented (see Section 7.9). Decoding Reed–Solomon codes can be summarized as follows:

1. Find the number of errors μ and error-location numbers $X_1, X_2, \ldots, X_\mu$ by using any technique suitable to binary BCH codes.
2. From the error-location numbers construct the matrix $\boldsymbol{X}$ and determine its inverse $\boldsymbol{X}^{-1}$.
3. The error magnitudes $Y_1, Y_2, \ldots, Y_\mu$ are then given by $\boldsymbol{Y}=\boldsymbol{X}^{-1}\boldsymbol{S}$, where $\boldsymbol{S}$ is the column vector constructed from the error syndromes $S_1, S_2, S_3, \ldots, S_\mu$.

The example that follows considers decoding a (15, 9) triple-error-correcting Reed–Solomon code. The approach used is based on the Peterson–Gorenstein–Zierler decoder with two matrix inversions, one for the error-location numbers and the other for the error magnitudes. Consider a codeword polynomial $c(x)$, belonging to the triple-error-correcting Reed–Solomon (15, 9) code, that has incurred 3 errors so giving

$$v(x) = \alpha^3 x^{12} + x^8 + \alpha^{10} x^7 + \alpha^2 x^5 + \alpha^8 x^4 + \alpha^{14} x^3 + \alpha^6.$$

Over $GF(2^4)$ the error syndromes are

$$S_1 = v(\alpha) = \alpha^{15} + \alpha^8 + \alpha^{17} + \alpha^7 + \alpha^{12} + \alpha^{17} + \alpha^6 = \alpha^6$$

$$S_2 = v(\alpha^2) = \alpha^{27} + \alpha^{16} + \alpha^{24} + \alpha^{12} + \alpha^{16} + \alpha^{20} + \alpha^6 = 0$$

$$S_3 = v(\alpha^3) = \alpha^{39} + \alpha^{24} + \alpha^{31} + \alpha^{17} + \alpha^{20} + \alpha^{23} + \alpha^6 = \alpha^{14}$$

$$S_4 = v(\alpha^4) = \alpha^{51} + \alpha^{32} + \alpha^{38} + \alpha^{22} + \alpha^{24} + \alpha^{26} + \alpha^6 = \alpha^{11}$$

$$S_5 = v(\alpha^5) = \alpha^{63} + \alpha^{40} + \alpha^{45} + \alpha^{27} + \alpha^{28} + \alpha^{29} + \alpha^6 = \alpha^{14}$$

$$S_6 = v(\alpha^6) = \alpha^{75} + \alpha^{48} + \alpha^{52} + \alpha^{32} + \alpha^{32} + \alpha^{32} + \alpha^6 = \alpha^9.$$

The decoder first assumes that the maximum number of correctable errors, 3, have occurred and constructs the matrix

$$\boldsymbol{M} = \begin{bmatrix} S_1 & S_2 & S_3 \\ S_2 & S_3 & S_4 \\ S_3 & S_4 & S_5 \end{bmatrix} = \begin{bmatrix} \alpha^6 & 0 & \alpha^{14} \\ 0 & \alpha^{14} & \alpha^{11} \\ \alpha^{14} & \alpha^{11} & \alpha^{14} \end{bmatrix}.$$

Evaluating the determinate of $\boldsymbol{M}$ gives

$$\det(\boldsymbol{M}) = \alpha^6 \begin{vmatrix} \alpha^{14} & \alpha^{11} \\ \alpha^{11} & \alpha^{14} \end{vmatrix} + 0 \begin{vmatrix} 0 & \alpha^{11} \\ \alpha^{14} & \alpha^{14} \end{vmatrix} + \alpha^{14} \begin{vmatrix} 0 & \alpha^{14} \\ \alpha^{14} & \alpha^{11} \end{vmatrix}$$

$$= \alpha^6(\alpha^{13} + \alpha^7) + \alpha^{14}(\alpha^{14}\alpha^{14}) = \alpha^{11} + \alpha^{12} = 1$$

Hence $\det(\boldsymbol{M}) \neq 0$ and so the decoder assumes that 3 errors have occurred (which we know is correct). The inverse of $\boldsymbol{M}$ is

$$\boldsymbol{M}^{-1} = \frac{\operatorname{adj}(\boldsymbol{M})}{\det(\boldsymbol{M})} = \begin{bmatrix} \alpha^5 & \alpha^{10} & \alpha^{13} \\ \alpha^{10} & \alpha^7 & \alpha^2 \\ \alpha^{13} & \alpha^2 & \alpha^5 \end{bmatrix}.$$

The coefficients of the error-location polynomial are given by

$$\begin{bmatrix} \sigma_3 \\ \sigma_2 \\ \sigma_1 \end{bmatrix} = \begin{bmatrix} \alpha^5 & \alpha^{10} & \alpha^{13} \\ \alpha^{10} & \alpha^7 & \alpha^2 \\ \alpha^{13} & \alpha^2 & \alpha^5 \end{bmatrix} \begin{bmatrix} \alpha^{11} \\ \alpha^{14} \\ \alpha^9 \end{bmatrix} = \begin{bmatrix} \alpha^{16} + \alpha^{24} + \alpha^{22} \\ \alpha^{21} + \alpha^{21} + \alpha^{11} \\ \alpha^{24} + \alpha^{16} + \alpha^{14} \end{bmatrix} = \begin{bmatrix} \alpha^4 \\ \alpha^{11} \\ 1 \end{bmatrix}$$

and so $\sigma_1 = 1$, $\sigma_2 = \alpha^{11}$ and $\sigma_3 = \alpha^4$ giving the error-location polynomial

$$\sigma(x) = 1 + \sigma_1 x + \sigma_2 x^2 + \sigma_3 x^3 = 1 + x + \alpha^{11} x^2 + \alpha^4 x^3.$$

Searching $GF(2^4)$ for the roots of $\sigma(x)$ shows that $x = \alpha^3$, α^9, and α^{14} are roots, and taking the reciprocals of the roots gives the error-location numbers $X_1 = \alpha^{12}$,

$X_2 = \alpha^6$, and $X_3 = \alpha$ respectively. To find the error magnitudes we construct

$$\boldsymbol{X} = \begin{bmatrix} X_1 & X_2 & X_3 \\ X_1^2 & X_2^2 & X_3^2 \\ X_1^3 & X_2^3 & X_3^3 \end{bmatrix} = \begin{bmatrix} \alpha^{12} & \alpha^6 & \alpha \\ \alpha^9 & \alpha^{12} & \alpha^2 \\ \alpha^6 & \alpha^3 & \alpha^3 \end{bmatrix}.$$

The determinant of $\boldsymbol{X}$ is $\det(\boldsymbol{X}) = \alpha^2$ and its inverse is

$$\boldsymbol{X}^{-1} = \begin{bmatrix} \alpha^8 & \alpha^{12} & \alpha \\ \alpha^7 & \alpha^7 & \alpha^9 \\ \alpha^8 & \alpha^9 & \alpha^5 \end{bmatrix}.$$

Using $\boldsymbol{Y} = \boldsymbol{X}^{-1}\boldsymbol{S}$ the error magnitudes Y_1, Y_2, and Y_3 are given by

$$\begin{bmatrix} Y_1 \\ Y_2 \\ Y_3 \end{bmatrix} = \begin{bmatrix} \alpha^8 & \alpha^{12} & \alpha \\ \alpha^7 & \alpha^7 & \alpha^9 \\ \alpha^8 & \alpha^9 & \alpha^5 \end{bmatrix} \begin{bmatrix} \alpha^6 \\ 0 \\ \alpha^{14} \end{bmatrix} = \begin{bmatrix} \alpha^{14} + \alpha^{15} \\ \alpha^{13} + \alpha^{23} \\ \alpha^{14} + \alpha^{19} \end{bmatrix} = \begin{bmatrix} \alpha^3 \\ \alpha^3 \\ \alpha^9 \end{bmatrix}$$

and so $Y_1 = \alpha^3$, $Y_2 = \alpha^3$, and $Y_3 = \alpha^9$. The error-location numbers $X_1 = \alpha^{12}$, $X_2 = \alpha^6$, and $X_3 = \alpha$ correspond to errors in positions x^{12}, x^6, and x respectively, the error pattern is therefore

$$e(x) = \alpha^3 x^{12} + \alpha^3 x^6 + \alpha^9 x$$

and adding this to $v(x)$ gives the codeword polynomial

$$c(x) = x^8 + \alpha^{10}x^7 + \alpha^3 x^6 + \alpha^2 x^5 + \alpha^8 x^4 + \alpha^{14}x^3 + \alpha^9 x + \alpha^6.$$

Example 7.10

A codeword $c(x)$ belonging to the triple-error-correcting (15, 9) Reed–Solomon code incurs errors so giving

$$v(x) = x^{10} + \alpha^3 x^8 + \alpha^{11}x^7 + \alpha^8 x^6 + \alpha^6 x^5 + \alpha^4 x^4 + \alpha^5 x^2 + \alpha^9 x + \alpha^6$$

determine $c(x)$.

The error syndromes corresponding to $v(x)$ are

$$S_1 = \alpha^4, S_2 = 1, S_3 = \alpha^{10}, S_4 = \alpha^7, S_5 = 0, S_6 = \alpha^{14}$$

over $GF(2^4)$. Taking $\mu = 3$ and substituting the error syndromes into eqn 7.31 gives

$$\boldsymbol{M} = \begin{bmatrix} S_1 & S_2 & S_3 \\ S_2 & S_3 & S_4 \\ S_3 & S_4 & S_5 \end{bmatrix} = \begin{bmatrix} \alpha^4 & 1 & \alpha^{10} \\ 1 & \alpha^{10} & \alpha^7 \\ \alpha^{10} & \alpha^7 & 0 \end{bmatrix}$$

the inverse of which is found to be

$$\boldsymbol{M}^{-1} = \begin{bmatrix} 1 & \alpha^3 & \alpha^{14} \\ \alpha^3 & \alpha^6 & 1 \\ \alpha^{14} & 1 & \alpha^4 \end{bmatrix}.$$

Furthermore

$$\boldsymbol{S}=\begin{bmatrix}S_4\\S_5\\S_6\end{bmatrix}=\begin{bmatrix}\alpha^7\\0\\\alpha^{14}\end{bmatrix}$$

and substituting $\boldsymbol{M}^{-1}$ and $\boldsymbol{S}$ into eqn 7.34 gives the coefficients of the error-location polynomial

$$\begin{bmatrix}\sigma_3\\\sigma_2\\\sigma_1\end{bmatrix}=\boldsymbol{M}^{-1}\boldsymbol{S}=\begin{bmatrix}1&\alpha^3&\alpha^{14}\\\alpha^3&\alpha^6&1\\\alpha^{14}&1&\alpha^4\end{bmatrix}\begin{bmatrix}\alpha^7\\0\\\alpha^{14}\end{bmatrix}=\begin{bmatrix}\alpha^5\\\alpha^{11}\\\alpha^2\end{bmatrix}.$$

Therefore the error-location polynomial is

$$\sigma(x)=1+\alpha^2x+\alpha^{11}x^2+\alpha^5x^3$$

the roots of which are α^5, α^8, α^{12} which correspond to error-location numbers $X_1=\alpha^{10}$, $X_2=\alpha^7$, and $x_3=\alpha^3$ respectively. To determine the error magnitudes we construct

$$\boldsymbol{X}=\begin{bmatrix}X_1&X_2&X_3\\X_1^2&X_2^2&X_3^2\\X_1^3&X_2^3&X_3^3\end{bmatrix}=\begin{bmatrix}\alpha^{10}&\alpha^7&\alpha^3\\\alpha^5&\alpha^{14}&\alpha^6\\1&\alpha^6&\alpha^9\end{bmatrix}$$

and its inverse

$$\boldsymbol{X}^{-1}=\begin{bmatrix}\alpha^{12}&\alpha^6&\alpha^2\\\alpha^{11}&\alpha^{10}&\alpha^{13}\\\alpha^{13}&\alpha^2&\alpha^{11}\end{bmatrix}$$

along with

$$\boldsymbol{S}=\begin{bmatrix}S_1\\S_2\\S_3\end{bmatrix}=\begin{bmatrix}\alpha^4\\1\\\alpha^{10}\end{bmatrix}$$

Using eqn 7.46 the error magnitudes are given by

$$\begin{bmatrix}Y_1\\Y_2\\Y_3\end{bmatrix}=\begin{bmatrix}\alpha^{12}&\alpha^6&\alpha^2\\\alpha^{11}&\alpha^{10}&\alpha^{13}\\\alpha^{13}&\alpha^2&\alpha^{11}\end{bmatrix}\begin{bmatrix}\alpha^4\\1\\\alpha^{10}\end{bmatrix}=\begin{bmatrix}1\\\alpha^4\\\alpha^6\end{bmatrix}$$

and so $Y_1=1$, $Y^2=\alpha^4$, and $Y_3=\alpha^6$. The error-location numbers $X_1=\alpha^{10}$, $X_2=\alpha^7$, and $X_3=\alpha^3$ correspond to errors in positions x^{10}, x^7, and x^3 respectively, the error pattern is therefore

$$e(x)=x^{10}+\alpha^4x^7+\alpha^6x^3$$

and adding this to $v(x)$ gives

$$c(x) = \alpha^3 x^8 + \alpha^{13} x^7 + \alpha^8 x^6 + \alpha^6 x^5 + \alpha^4 x^4 + \alpha^6 x^3 + \alpha^5 x^2 + \alpha^9 x + \alpha^6$$

as the required codeword polynomial. □

7.8 The Berlekamp algorithm

The Peterson–Gorenstein–Zierler decoder is fundamental to decoding BCH codes, it overcomes the problem of solving the nonlinear syndrome equations by the use of an error-location polynomial whose coefficients can be obtained by standard linear matrix-inversion methods. This though is the weak link in the decoder, for matrix inversion requires excessive computation, especially for large matrices. In particular for non-binary codes for which a second matrix inversion is required to obtain the error magnitudes. As such the Peterson–Gorenstein–Zierler decoder is quite inefficient and whilst it is of prime importance in illustrating the principles of decoding BCH codes it is, nevertheless, of limited practical use.

The Berlekamp algorithm is a fast and efficient algorithm for decoding BCH codes. Like the Peterson–Gorenstein–Zierler decoder it uses the error-location polynomial, but it avoids the need for matrix inversion when determining the polynomial coefficients. The algorithm is generally considered to be significantly more complex than the Peterson–Gorenstein–Zierler decoder. However its complexity lies mainly in the proof of the algorithm, which we omit.

The algorithm uses an iterative technique to find an error-location polynomial $\sigma(x)$ whose coefficients satisfy Newton's identities, as given by eqns 7.23 and 7.24. The algorithm starts by finding a polynomial $\sigma^{(1)}(x)$, whose coefficients satisfy the first of Newton's identities. A suitable set of initial conditions is required to achieve this, otherwise the algorithm may fail to carry out the required number of iterations. The polynomial $\sigma^{(1)}(x)$ must not only satisfy the first identity, but it must be the polynomial of least degree that meets the requirement. Note that the superscript 1 in $\sigma^{(1)}(x)$ is enclosed in parenthesis to avoid any possible ambiguity with powers of $\sigma(x)$. Next a polynomial $\sigma^{(2)}(x)$ is found that satisfies the first and second identities, again the polynomial must be the polynomial of least degree that meets the requirement. To find $\sigma^{(2)}(x)$ we first check if $\sigma^{(1)}(x)$ meets the requirement, if it does then we let $\sigma^{(2)}(x) = \sigma^{(1)}(x)$. Otherwise $\sigma^{(1)}(x)$ is modified by adding a suitable correction term such that the resulting polynomial has coefficients that satisfy the first two equations of Newton's identities. The modification to $\sigma^{(1)}(x)$ must ensure that $\sigma^{(2)}(x)$ is the polynomial of least degree that meets the requirements. The process continues iteratively. Check to see if $\sigma^{(2)}(x)$ satisfies the first three identities, if it does then $\sigma^{(3)}(x) = \sigma^{(2)}(x)$, otherwise modify $\sigma^{(2)}(x)$ to obtain $\sigma^{(3)}(x)$. Then generate $\sigma^{(4)}(x)$, $\sigma^{(5)}(x), \ldots$ and so forth, until a polynomial $\sigma^{(2t)}(x)$, with coefficients satisfying all the Newton's identities, is obtained. The error-location polynomial is then

$$\sigma(x) = \sigma^{(2t)}(x)$$

and the error-location numbers are found in the usual manner of finding the inverse roots of $\sigma(x)$.

Recall that Newton's identities relate the error syndromes to the coefficients of the error-location polynomial. At each iteration in the algorithm the polynomial coefficients are used to estimate the error syndrome of following iteration. Let $\sigma^{(i)}(x)$ be the polynomial of least degree whose coefficients satisfy the first i Newton's identities, we can express $\sigma^{(i)}(x)$ as

$$\sigma^{(i)}(x) = 1 + \sigma_1^{(i)}(x) + \sigma_2^{(i)}x^2 + \cdots + \sigma_{r_i}^{(i)}x^{r_i}$$

where r_i is the degree of $\sigma^{(i)}(x)$. To test whether the coefficients of $\sigma^{(i)}(x)$ satisfy the $(i+1)$th identity we compute

$$\hat{S}_{i+1} = \sigma_1^{(i)}S_i + \sigma_2^{(i)}S_{i-1} + \cdots + \sigma_{r_i}^{(i)}S_{i+1-r_i} \tag{7.47}$$

which can be thought of as the $(i+1)$th syndrome as estimated or predicted by $\sigma^{(i)}(x)$. Adding $\hat{S}_{i+1}$ to the error syndrome S_{i+1} gives

$$d_i = S_{i+1} + \hat{S}_{i+1} \tag{7.48}$$

where d_i is known as the *ith discrepancy*. Both S_{i+1} and $\hat{S}_{i+1}$ are elements in the same field $GF(2^m)$ and the discrepancy is therefore a field element that gives a measure of the difference between the two error syndromes (as its name obviously implies). If $d_i = 0$ then the coefficients of $\sigma^{(i)}(x)$ satisfy the first $(i+1)$ identities and $\sigma^{(i)}(x)$ is taken as the next polynomial $\sigma^{(i+1)}(x)$, and therefore

$$\sigma^{(i+1)}(x) = \sigma^{(i)}(x). \tag{7.49}$$

The degree r_{i+1} of $\sigma^{(i+1)}(x)$ is the same as that of $\sigma^{(i)}(x)$, and so

$$r_{i+1} = r_i.$$

If $d_i \neq 0$ then the coefficients of $\sigma^{(i)}(x)$ fail to satisfy the $(i+1)$th identity and $\sigma^{(i)}(x)$ has to be modified by adding a suitable correction term. The correction term is a polynomial that depends upon one of the previous polynomials $\sigma^{(k)}(x)$, such that:

(1) the discrepancy $d_k \neq 0$;

(2) n_k has the largest value

where $n_k = k - r_k$ and r_k is the degree of $\sigma^{(k)}(x)$. The polynomial required is then given by

$$\sigma^{(i+1)}(x) = \sigma^{(i)}(x) + \left(\frac{x^i d_i}{x^k d_k}\right)\sigma^k(x) \tag{7.50}$$

and is the polynomial of least degree whose coefficients satisfy Newton's identities up to the $(i+1)$th. The correction term in eqn 7.50 has degree $i + r_k - k = i - n_k$ and so r_{i+1}, the degree of $\sigma^{(i+1)}(x)$, is r_i or $i - n_k$ depending upon which has the largest value, and we can express this as

$$r_{i+1} = \max(r_i, i - n_k). \tag{7.51}$$

Once $\sigma^{(i+1)}(x)$ and r_{i+1} are found, the coefficients of $\sigma^{(i+1)}(x)$ are used to estimate the next error syndrome

$$\hat{S}_{i+2} = \sigma_1^{(i+1)} S_{i+1} + \sigma_2^{(i+1)} S_i + \cdots + \sigma_{r_{i+1}}^{(i+1)} S_{i+2-r_{i+1}} \tag{7.52}$$

along with the discrepancy

$$d_{i+1} = S_{i+2} + \hat{S}_{i+2}.$$

A note of d_{i+1} is made in a table, along with the value of

$$n_{i+1} = i + 1 - r_{i+1}.$$

Note that the degree of the correction term in eqn 7.50 decreases with increasing n_k and selecting the polynomial with the largest n_k value ensures that the resulting polynomial is the polynomial of least degree that meets the required identities.

For the algorithm to work correctly a carefully selected set of initial conditions are required. Such a set is one that defines two iterations, labelled $i = -1$ and $i = 0$, as shown in Table 7.2.

A step-by-step description should help to clarify the algorithm. Given that the ith has been completed so that $\sigma^{(i)}(x)$ and d_i are known, then at the $(i+1)$th iteration the polynomial $\sigma^{(i+1)}(x)$ is obtained as follows:

Step 1 If $d_i = 0$:
Take $\sigma^{(i)}(x)$ as the next polynomial, and so

$$\sigma^{(i+1)}(x) = \sigma^{(i)}(x)$$
$$r_{i+1} = r_i.$$

Go to Step 3.

Step 2 If $d_i \neq 0$:
Find a previous polynomial $\sigma^{(k)}(x)$ such that n_k has the largest value and $d_k \neq 0$. Then

$$\sigma^{(i+1)}(x) = \sigma^{(i)}(x) + \left(\frac{x^i d_i}{x^k d_k}\right)\sigma^k(x)$$
$$r_{i+1} = \max(r_i, i - n_k).$$

Table 7.2
Initial conditions for the Berlekamp algorithm

i	$\sigma^i(x)$	r_i	n_i	d_i
−1	1	0	−1	1
0	1	0	0	S_1

Step 3 The polynomial $\sigma^{(i+1)}(x)$ is checked to see if its coefficients are consistent with the next Newton's identity, so determine

$$\hat{S}_{i+2} = \sigma_1^{(i+1)} S_{i+1} + \sigma_2^{(i+1)} S_i + \cdots + \sigma_{r_{i+1}}^{(i+1)} S_{i+2-r_{i+1}}$$

and the discrepancy

$$d_{i+1} = S_{i+2} + \hat{S}_{i+2}$$

along with

$$n_{i+1} = i + 1 - r_{i+1}.$$

Make a note of

$$\sigma^{(i+1)}(x)$$
$$r_{i+1}$$
$$n_{i+1}$$
$$d_{i+1}$$

in the table containing the initial conditions.

Repeat iterations until $\sigma^{(2t)}(x)$ is obtained.

Note that at each iteration the index i denotes the results of the previous iteration and the algorithm determines the polynomial for the current iteration, i.e. the $(i+1)$th, and then checks how well the polynomial satisfies the conditions for the next iteration, i.e. the $(i+2)$th. For a t-error-correcting code a maximum of $2t$ iterations are required. Furthermore note that if the degree of $\sigma^{(2t)}(x)$ is greater than t then an uncorrectable number of errors has occurred.

In Example 7.6 the Peterson–Gorenstein–Zierler decoder was used to decode a (15, 5) triple-error-correcting code in which the error syndromes were $S_1 = \alpha^{12}$, $S_2 = \alpha^9$, $S_3 = \alpha^{10}$, $S_4 = \alpha^3$, $S_5 = \alpha^5$, and $S_6 = \alpha^5$ where α is an element of $GF(2^4)$. Let's now repeat the example but this time using the Berlekamp algorithm to determine the error-location polynomial. The initial conditions required are shown in the first two rows of Table 7.3.

Table 7.3
The Berlekamp algorithm applied to a triple-error correcting code

i	$\sigma^{(i)}(x)$	r_i	n_i	d_i
−1	1	0	−1	1
0	1	0	0	α^{12}
1	$1+\alpha^{12}x$	1	0	0
2	$1+\alpha^{12}x$	1	1	α^7
3	$1+\alpha^{12}x+\alpha^{10}x^2$	2	1	0
4	$1+\alpha^{12}x+\alpha^{10}x^2$	2	2	1
5	$1+\alpha^{12}x+\alpha x^2+\alpha^5x^3$	3	2	0
6	$1+\alpha^{12}x+\alpha x^2+\alpha^5x^3$	—	—	—

First iteration

The $i=0$ row in Table 7.3 defines the previous iteration and so we start with $i=0$ and determine $\sigma^{(i+1)}(x)=\sigma^{(1)}(x)$. From Table 7.3 we see that the discrepancy $d_0=\alpha^{12}\neq 0$ and so we skip Step 1 and go to Step 2. We need to find a polynomial prior to $\sigma^{(0)}(x)$ that has $d_k\neq 0$ and the largest value of n_k. There is only one polynomial before $\sigma^{(0)}(x)$, namely $\sigma^{(-1)}(x)$ and so this must be used and therefore $k=-1$. The discrepancy $d_{-1}=1\neq 0$ and the required polynomial is

$$\begin{aligned}\sigma^{(1)}(x) &= \sigma^{(0)}(x)+\left(\frac{x^0 d_0}{x^{-1}d_{-1}}\right)\sigma^{-1}(x)\\ &= 1+\left(\frac{\alpha^{12}}{x^{-1}}\right)\\ &= 1+\alpha^{12}x.\end{aligned}$$

We can express $\sigma^{(1)}(x)$ as

$$\sigma^{(1)}(x)=1+\sigma_1^{(1)}x$$

where $\sigma_1^{(1)}=\alpha^{12}$. The degree of $\sigma^{(1)}(x)$ is

$$r_1=\max(r_0, 0-n_{-1})=\max(0,1)=1.$$

Proceeding now to Step 3 we need to find d_1 by first using the coefficient of $\sigma^{(1)}(x)$ to estimate the next error syndrome

$$\hat{S}_2=\sigma_1^{(1)}S_1=\alpha^{12}\alpha^{12}=\alpha^9$$

and so

$$d_1=S_2+\hat{S}_2=\alpha^9+\alpha^9=0.$$

Finally we evaluate

$$n_1=0+1-r_1=0$$

and make a note of $\sigma^{(1)}(x)$, r_1, n_1, and d_1 in Table 7.3 (row $i=1$).

Second iteration

Here $i=1$ and we start by looking at the previous discrepancy d_1. Since $d_1=0$ we carry out Step 1

$$\begin{aligned}\alpha^{(2)}(x) &= \sigma^{(1)}(x)=1+\alpha^{12}x\\ r_2 &= r_1=1.\end{aligned}$$

Omitting Step 2 and carrying out Step 3 gives

$$\hat{S}_3 = \sigma_1^{(2)} S_2 + \sigma_2^{(2)} S_1 = \alpha^{12}\alpha^9 = \alpha^6$$

as $\sigma_1^{(2)} = \alpha^{12}$ and $\sigma_2^{(2)} = 0$. The discrepancy d_2 is

$$d_2 = S_3 + \hat{S}_3 = \alpha^{10} + \alpha^6 = \alpha^7$$

and $n_2 = 1$. The results are noted in Table 7.3 (row $i = 2$).

Third, fourth and fifth iterations

Taking $i = 2$, 3, 4 and proceeding as illustrated in the previous two iterations gives the next three rows shown in Table 7.3.

Sixth iteration

Take $i = 5$, the previous discrepancy is $d_5 = 0$, and so Step 1 gives

$$\sigma^{(6)}(x) = \sigma^{(5)}(x) = 1 + \alpha^{12}x + \alpha x^2 + \alpha^5 x^3.$$

There is no need to continue further, for a triple-error-correcting code has $t = 3$ and so $\sigma^{(2t)}(x) = \sigma^{(6)}(x)$ is the last polynomial to be determined. The required error-location polynomial is therefore

$$\sigma(x) = 1 + \alpha^{12}x + \alpha x^2 + \alpha^5 x^3$$

which is the same as that obtained in Example 7.6.

If the number of errors μ is less than t, then not all $2t$ iterations are required. If the discrepancy d_i and the following $m_i = t - r_i - 1$ discrepancies are zero, then $\sigma^i(x)$ is the required error-location polynomial. Adding this to the algorithm is straightforward, each time $d_i = 0$ we start counting successive zero discrepancies. If the count reaches the required number m_i then the algorithm stops. Otherwise the count stops, is initialized to zero and counting recommences at the next zero discrepancy. Consider, for example, a triple-error-correcting code. Assuming the occurrence of at least one error, the first iteration gives

$$\sigma^{(1)}(x) = 1 + S_1 x$$
$$r_1 = 1$$
$$n_1 = 0$$
$$d_1 = 0$$

and $m_1 = t - r_1 - 1 = 1$ as $t = 3$. Therefore if the next discrepancy is zero there is no need to continue with the remaining iterations. In the event of there being just a single error the next discrepancy will be zero, no more iterations are required and the error-location polynomial will be $\sigma(x) = \sigma^{(1)}(x)$. The next example illustrates this.

Example 7.11
Consider the codeword $c(x) = x^{12} + x^{11} + x^9 + x^8 + x^7 + x^2 + 1$ belonging to the (15, 5) triple-error-correcting binary BCH code. Introducing the single error $e(x) = x^{11}$ gives $v(x) = x^{12} + x^9 + x^8 + x^7 + x^2 + 1$. Show that the Berlekamp algorithm requires fewer than $2t$ interactions.

Over $GF(2^4)$ the error syndromes are

$$S_1 = \alpha^{11},\ S_2 = \alpha^7,\ S_3 = \alpha^3,\ S_4 = \alpha^{14},\ S_5 = \alpha^{10},\ \text{and}\ S_6 = \alpha^6.$$

The initial conditions are shown in the first two rows of Table 7.4.

At the first iteration we take the previous discrepancy $d_0 = \alpha^{11} \neq 0$. The only iteration prior to this is $i = -1$, so taking $k = -1$ gives

$$\sigma^{(1)}(x) = \sigma^{(0)}(x) + \left(\frac{x^0 d_0}{x^{-1} d_{-1}}\right)\sigma^{(-1)}(x) = 1 + \alpha^{11}x.$$

The values of r_1 and n_1 are

$$r_1 = \max(r_0, 0 - n_{-1}) = 1$$
$$n_1 = 1 - r_1 = 0$$

as $r_0 = 0$ and $n_{-1} = -1$. To find d_1 we first determine

$$\hat{S}_2 = \sigma_1^{(1)} S_1 = \alpha^{11}\alpha^{11} = \alpha^7$$

and so

$$d_1 = S_2 + \hat{S}_2 = \alpha^7 + \alpha^7 = 0.$$

Now $m_1 = t - r_1 - 1 = 1$ as the error-correction limit $t = 3$ and therefore if the next discrepancy is zero we can stop. As $d_1 = 0$ the next iteration gives

$$\sigma^{(2)}(x) = \sigma^{(1)}(x) = 1 + \alpha^{11}x$$

and

$$\hat{S}_3 = \sigma_1^{(2)} S_2 + \sigma_2^{(2)} S_1 = \alpha^{11}\alpha^7 = \alpha^3$$

Table 7.4
The Berlekamp algorithm applied to a triple-error correcting code with one error

i	$\sigma^{(i)}(x)$	r_i	n_i	d_i
−1	1	0	−1	1
0	1	0	0	α^{11}
1	$1+\alpha^{11}x$	1	0	0
2	$1+\alpha^{11}x$	1	1	0

as $\sigma_1^{(2)} = \alpha^{11}$ and $\sigma_2^{(2)} = 0$. The resulting discrepancy is

$$d_2 = S_3 + \hat{S}_3 = \alpha^3 + \alpha^3 = 0$$

so therefore no further iterations are needed and

$$\sigma(x) = \sigma^{(2)}(x) = 1 + \alpha^{11}x$$

is the required error-location polynomial. This is clearly correct because its root is $1/\alpha^{11}$, the reciprocal of which gives the error-location number α^{11} and error pattern $e(x) = x^{11}$. All the remaining discrepancies must equal zero, we can check this by calculating $\hat{S}_4$, $\hat{S}_5$, and $\hat{S}_6$. Note that $\sigma(x) = 1 + \sigma^{11}x$ gives $\sigma_1 = \alpha^{11}$, $\sigma_2 = \sigma_3 = 0$ and so

$$\begin{aligned}
\hat{S}_4 &= \sigma_1 S_3 + \sigma_2 S_2 + \sigma_3 S_1 = \alpha^{11}\alpha^3 = \alpha^{14} \\
\hat{S}_5 &= \sigma_1 S_4 + \sigma_2 S_3 + \sigma_3 S_2 = \alpha^{11}\alpha^{14} = \alpha^{10} \\
\hat{S}_6 &= \sigma_1 S_5 + \sigma_2 S_4 + \sigma_3 S_3 = \alpha^{11}\alpha^{10} = \alpha^6
\end{aligned}$$

which give the discrepancies

$$\begin{aligned}
d_3 &= S_4 + \hat{S}_4 = \alpha^{14} + \alpha^{14} = 0 \\
d_4 &= S_5 + \hat{S}_5 = \alpha^{10} + \alpha^{10} = 0 \\
d_5 &= S_6 + \hat{S}_6 = \alpha^6 + \alpha^6 = 0.
\end{aligned}$$

Hence the coefficients of $\sigma(x)$ satisfy the required Newton's identities. □

In practice it is quite likely that the number of errors is less than the error-correction limit t, otherwise the code would not be suitable for the given application. Hence the number of iterations required by the Berlekamp algorithm is normally less than $2t$. The algorithm described here applies to non-binary as well as binary codes. For binary codes the algorithm can be simplified so that a maximum of t, and not $2t$, iterations are required; this is not covered here. Hence offering a further reduction in processing time. The processing time of the Peterson–Gorenstein–Zierler decoder increases significantly as t increases, because the computational requirement for evaluating the inverse of a t by t matrix is dependent on t^3. However, in the Berlekamp algorithm, the maximum number of iterations increases linearly with t and there is therefore no significant degradation in performance when codes with large error-correction limits are used.

7.9 The error-evaluator polynomial

Here we consider how the error magnitudes can be determined without the need for matrix inversion. Consider a t-error-correcting Reed–Solomon code and let's assume that the error syndromes $S_1, S_2, \ldots, S_{2t}$ have been determined along with the number of errors μ and the error-location polynomial

$$\sigma(x) = 1 + \sigma_1 x + \sigma_2 x^2 + \cdots + \sigma_\mu x^\mu$$

To find the error magnitudes we first define the *error-evaluator polynomial*

$$\omega(x) = 1 + \omega_1 x + \omega_2 x^2 + \cdots + \omega_\mu x^\mu. \tag{7.53}$$

where

$$\begin{aligned}
\omega_1 &= S_1 + \sigma_1 \\
\omega_2 &= S_2 + \sigma_1 S_1 + \sigma_2 \\
&\vdots \\
\omega_\mu &= S_\mu + \sigma_1 S_{\mu-1} + \sigma_2 S_{\mu-2} + \cdots + \sigma_\mu.
\end{aligned}$$

Once the error-evaluator polynomial is determined, the magnitude Y_j corresponding to the error-location number X_j is given by

$$Y_j = \frac{\omega(X_j^{-1})}{(1 + X_j^{-1}X_1)(1 + X_j^{-1}X_2)\cdots(1 + X_j^{-1}X_\mu)} \tag{7.54}$$

where the product on the denominator excludes the term $(1 + X_j^{-1}X_j)$. To illustrate the algorithm we reconsider the example covered in Section 7.7 in which the Peterson–Gorenstein–Zierler decoder was used to decode a triple-error-correcting (15, 9) Reed–Solomon code. Here

$$v(x) = \alpha^3 x^{12} + x^8 + \alpha^{10} x^7 + \alpha^2 x^5 + \alpha^8 x^4 + \alpha^{14} x^3 + \alpha^6$$

and the error syndromes were found to be

$$\begin{aligned}
&S_1 = \alpha^6, \quad S_2 = 0, \quad S_3 = \alpha^{14} \\
&S_4 = \alpha^{11}, \quad S_5 = \alpha^{14}, \quad S_6 = \alpha^9.
\end{aligned}$$

The error-location polynomial was found to be

$$\sigma(x) = 1 + x + \alpha^{11} x^2 + \alpha^4 x^3$$

which gave $\mu = 3$ errors with error-location numbers $X_1 = \alpha^{12}$, $X_2 = \alpha^6$, and $X_3 = \alpha$. For $\mu = 3$ the error-evaluator polynomial, given by eqn 7.53, reduces to

$$\omega(x) = 1 + \omega_1 x + \omega_2 x^2 + \omega_3 x^3 \tag{7.55}$$

where

$$\begin{aligned}
\omega_1 &= S_1 + \sigma_1 \\
\omega_2 &= S_2 + \sigma_1 S_1 + \sigma_2 \\
\omega_3 &= S_3 + \sigma_1 S_2 + \sigma_2 S_1 + \sigma_3.
\end{aligned} \tag{7.56}$$

The coefficients of $\sigma(x)$ are $\sigma_1 = 1$, $\sigma_2 = \alpha^{11}$, $\sigma_3 = \alpha^4$ and substituting these, along with the error syndromes S_1, S_2, S_3 into eqns 7.56 gives $\omega_1 = \alpha^{13}$, $\omega_2 = \alpha$, $\omega_3 = \alpha^{11}$ and therefore

$$\omega(x) = 1 + \alpha^{13}x + \alpha x^2 + \alpha^{11}x^3$$

is the required error-evaluator polynomial. To determine the error magnitudes we first evaluate $\omega(x)$ at the inverse error-location numbers, this gives

$$\begin{aligned}
\omega(X_1^{-1}) &= \omega(\alpha^{-12}) = 1 + \alpha^{13}\alpha^{-12} + \alpha\alpha^{-24} + \alpha^{11}\alpha^{-36} = \alpha^{11} \\
\omega(X_2^{-1}) &= \omega(\alpha^{-6}) = 1 + \alpha^{13}\alpha^{-6} + \alpha\alpha^{-12} + \alpha^{11}\alpha^{-18} = \alpha^{6} \\
\omega(X_3^{-1}) &= \omega(\alpha^{-1}) = 1 + \alpha^{13}\alpha^{-1} + \alpha\alpha^{-2} + \alpha^{11}\alpha^{-3} = \alpha.
\end{aligned}$$

Taking $j = 1, 2, 3$ and substituting $\omega(X_1^{-1})$, $\omega(X_2^{-1})$, $\omega(X_3^{-1})$ and X_1, X_2, X_3 into eqn 7.54 gives the error magnitudes

$$\begin{aligned}
Y_1 &= \frac{\omega(X_1^{-1})}{(1 + X_1^{-1}X_2)(1 + X_1^{-1}X_3)} = \frac{\alpha^{11}}{(1 + \alpha^{-12}\alpha^{6})(1 + \alpha^{-12}\alpha)} = \alpha^3 \\
Y_2 &= \frac{\omega(X_2^{-1})}{(1 + X_2^{-1}X_1)(1 + X_2^{-1}X_3)} = \frac{\alpha^{6}}{(1 + \alpha^{-6}\alpha^{12})(1 + \alpha^{-6}\alpha)} = \alpha^3 \\
Y_3 &= \frac{\omega(X_3^{-1})}{(1 + X_3^{-1}X_1)(1 + X_3^{-1}X_2)} = \frac{\alpha}{(1 + \alpha^{-1}\alpha^{12})(1 + \alpha^{-1}\alpha^{6})} = \alpha^9
\end{aligned}$$

as obtained previously.

Example 7.12
Repeat Example 7.10 but now using the error-evaluator polynomial to determine the error magnitudes.

From Example 7.10 we found that the error syndromes, error-location numbers and coefficients of the error-location polynomial were

$$\begin{aligned}
&S_1 = \alpha^4,\ S_2 = 1,\ S_3 = \alpha^{10},\ S_4 = \alpha^7,\ S_5 = 0,\ S_1 = \alpha^{14} \\
&X_1 = \alpha^{10},\ X_2 = \alpha^7,\ X_3 = \alpha^3 \\
&\sigma_1 = \alpha^2,\ \sigma_2 = \alpha^{11},\ \sigma_3 = \alpha^5.
\end{aligned}$$

The error-evaluator polynomial for 3 errors is

$$\omega(x) = 1 + \omega_1 x + \omega_2 x^2 + \omega_3 x^3$$

where

$$\begin{aligned}
\omega_1 &= S_1 + \sigma_1 = \alpha^{10} \\
\omega_2 &= S_2 + \sigma_1 S_1 + \sigma_2 = \alpha^4 \\
\omega_3 &= S_3 + \sigma_1 S_2 + \sigma_2 S_1 + \sigma_3 = \alpha^2
\end{aligned}$$

and so

$$\omega(x) = 1 + \alpha^{10}x + \alpha^4 x^2 + \alpha^2 x^3.$$

Evaluating $\omega(x)$ at the inverse error-location numbers gives

$$\omega(X_1^{-1}) = \omega(\alpha^{-10}) = \alpha^{13}$$
$$\omega(X_2^{-1}) = \omega(\alpha^{-7}) = 1$$
$$\omega(X_3^{-1}) = \omega(\alpha^{-3}) = \alpha$$

and the error magnitudes are given by

$$Y_1 = \frac{\omega(X_1^{-1})}{(1 + X_1^{-1}X_2)(1 + X_1^{-1}X_3)} = \frac{\alpha^{13}}{(1 + \alpha^{-10}\alpha^7)(1 + \alpha^{-10}\alpha^3)} = 1$$
$$Y_2 = \frac{\omega(X_2^{-1})}{(1 + X_2^{-1}X_1)(1 + X_2^{-1}X_3)} = \frac{1}{(1 + \alpha^{-7}\alpha^{10})(1 + \alpha^{-7}\alpha^3)} = \alpha^4$$
$$Y_3 = \frac{\omega(X_3^{-1})}{(1 + X_3^{-1}X_1)(1 + X_3^{-1}X_2)} = \frac{\alpha}{(1 + \alpha^{-3}\alpha^{10})(1 + \alpha^{-3}\alpha^7)} = \alpha^6$$

as obtained in Example 7.10. □

Problems

7.1 Show that α, α^2, α^3, and α^4 are roots of $g(x) = x^8 + x^7 + x^6 + x^4 + 1$, the generator polynomial of the (15, 7) double-error-correcting binary BCH code, where α is a primitive field element of $GF(2^4)$.

7.2 The (15, 11) single-error-correcting binary BCH code has $g(x) = x^4 + x + 1$. Show that α, α^2 and α^4 are roots of $g(x)$ but that α^3 is not a root, where α is a primitive field element of $GF(2^4)$. Find the fourth root of $g(x)$.

7.3 Construct a single-error-correcting binary BCH code over $GF(2^3)$.

7.4 Show that the triple-error-correcting binary BCH code constructed over $GF(2^4)$ has generator polynomial $g(x) = x^{10} + x^8 + x^5 + x^4 + x^2 + x + 1$. Determine the code's blocklength and information length.

7.5 Given that the minimal polynomials of α and α^3 in $GF(2^5)$ are

$$m_1(x) = x^5 + x^2 + 1$$
$$m_3(x) = x^5 + x^4 + x^3 + x^2 + 1$$

respectively, construct a binary double-error-correcting code over $GF(2^5)$.

7.6 Codewords $\boldsymbol{c}_1$ and $\boldsymbol{c}_2$, belonging to the (7, 4) single-error-correcting binary BCH code, incur single errors to give $\boldsymbol{v}_1 = (0\ 0\ 1\ 0\ 1\ 0\ 1)$ and $\boldsymbol{v}_2 = (1\ 0\ 1\ 1\ 0\ 1\ 0)$ respectively. Using eqn 7.13 find $\boldsymbol{c}_1$ and $\boldsymbol{c}_2$.

7.7 Given the (15, 7) double-error-correcting binary BCH code and decoder inputs

$$\boldsymbol{v}_1 = (1\ 0\ 0\ 0\ 1\ 1\ 0\ 1\ 1\ 0\ 0\ 0\ 1\ 0\ 1)$$
$$\boldsymbol{v}_2 = (0\ 0\ 1\ 0\ 1\ 0\ 1\ 1\ 1\ 0\ 1\ 0\ 0\ 0\ 1)$$

use eqns 7.18 to determine the resulting error-location polynomials. Hence show that $\boldsymbol{v}_1$ contains an uncorrectable error pattern and $\boldsymbol{v}_2$ contains 2 errors. Find the codeword that $\boldsymbol{v}_2$ is most likely to represent.

7.8 The Peterson–Gorenstein–Zierler decoder is used for decoding the triple-error-correcting (15, 5) binary code. Given the decoder input
$v(x) = x^{12} + x^{10} + x^7 + x^6 + x^4 + x^2 + 1$
find the resulting error-location polynomial. Hence find the codeword polynomial that $v(x)$ corresponds to.

7.9 Show that the single-error-correcting (7, 5) Reed–Solomon code has generator polynomial $g(x) = x^2 + \alpha^4 x + \alpha^3$ where α is a field element of $GF(2^3)$. Determine the systematic codeword corresponding to $\boldsymbol{i}_1 = (0\ \alpha\ 1\ 0\ \alpha^2)$ and the nonsystematic codeword for $\boldsymbol{i}_2 = (\alpha^2\ 1\ 0\ \alpha^6\ \alpha^3)$. Given that α is a root of $g(x)$ determine whether $\boldsymbol{v}_1 = (1\ 0\ \alpha\ \alpha^5\ \alpha^2\ 1\ 0)$ and $\boldsymbol{v}_2 = (0\ \alpha^2\ \alpha^4\ \alpha^3\ 0\ 0\ 1)$ are codewords.

7.10 Determine the generator polynomials of the double-error-correcting (15, 11) and triple-error-correcting (15, 9) Reed–Solomon codes.

7.11 A decoder for the double-error-correcting (15, 11) Reed–Solomon code uses the Peterson–Gorenstein–Zierler decoder to determine the position and magnitude of errors. Given that the input to the decoder is $\boldsymbol{v} = (1\ \alpha^{13}\ \alpha^{13}\ \alpha^8\ \alpha^7\ \alpha^3\ \alpha^{10}\ 0\ 0\ 0\ 0\ \alpha\ 0\ 0\ 0)$ determine the decoding decision.

7.12 Repeat Problem 7.11 for a decoder that uses the Berlekamp algorithm for determining the error locations and the error-evaluator polynomial to determine the error magnitudes.

Convolutional codes

8

We have seen that with an (n, k) block code, codewords are constructed independently of each other. For a binary code the n bit output of an encoder depends solely on the k bits entering the encoder. The $n-k$ parity bits of a codeword depend on the k information bits that enter the encoder, each information bit affects only one codeword. Convolutional codes differ from block codes in that the encoder output is constructed not from a single input but also using some of the previous encoder inputs. Clearly memory is required to achieve this, to store inputs for further use. A convolutional code that at a given time generates n outputs from k inputs and m previous inputs is referred to as an (n, k, m) *convolutional code.*

8.1 Convolution

As their name implies, convolutional codes are based on a convolution operation and it is useful to take a look at this before considering convolutional codes. The mathematical operation of convolution can be applied to analogue functions and to discrete functions, here we address only discrete convolution. Figure 8.1 shows a shift register consisting of 4 *stages* which feed into a modulo-2 adder via the links g_1, g_2, g_3, and g_4, each stage is a 1-bit storage device. The input to the first stage also feeds into the modulo-2 adder, via the link g_0. The terms g_0, g_1, g_2, g_3, and g_4 have values 0 or 1 if the link is absent or present respectively. If all the links are present, then $g_0 = g_1 = g_2 = g_3 = g_4 = 1$. The shift register shown in Fig. 8.1 is an example of a *linear feed-forward shift register* as bits are fed towards the output and not back into the shift register.

Consider the sequence of bits

$$\boldsymbol{u} = (u_0\ u_1\ u_2 \ldots)$$

entering the shift register shown in Fig. 8.1, and let

$$\boldsymbol{v} = (v_0\ v_1\ v_2 \ldots)$$

be the sequence of bits leaving the modulo-2 adder as $\boldsymbol{u}$ enters. The sequences $\boldsymbol{u}$ and $\boldsymbol{v}$ are referred to as the *input sequence* and the *output sequence* respectively. If we now

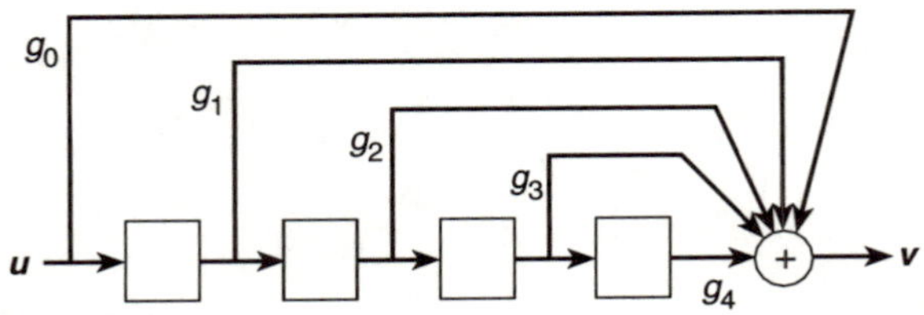

Fig. 8.1 Linear-feedforward shift register with 4 stages.

assume that not all the links in Fig. 8.1 are necessarily present, so that g_0, g_1, g_2, g_3, and g_4 are 0 or 1, then once u_0 enters the register, the output from the modulo-2 adder will be $v_0 = u_0 g_0$ since all the other inputs to the adder are zero. If $u_0 = 0$, or if the link g_0 is absent, so that $g_0 = 0$, then $v_0 = 0$, otherwise $v_0 = 1$. When u_1 enters the register, the bit u_0 moves into the second stage. If $g_1 = 1$, then u_0 is fed into the adder and likewise, if $g_0 = 1$, then u_0 is also fed into the adder. The output from the adder is therefore $v_1 = u_1 g_0 + u_0 g_1$. At the next input the adder output is $v_2 = u_2 g_0 + u_1 g_1 + u_0 g_2$ followed by $v_3 = u_3 g_0 + u_2 g_1 + u_1 g_2 + u_0 g_3$ and $v_4 = u_4 g_0 + u_3 g_1 + u_2 g_2 + u_1 g_3 + u_0 g_4$. In summary the register output is:

$$
\begin{aligned}
v_0 &= u_0 g_0 \\
v_1 &= u_1 g_0 + u_0 g_1 \\
v_2 &= u_2 g_0 + u_1 g_1 + u_0 g_2 \\
v_3 &= u_3 g_0 + u_2 g_1 + u_1 g_2 + u_0 g_3 \\
v_4 &= u_4 g_0 + u_3 g_1 + u_2 g_2 + u_1 g_3 + u_0 g_4.
\end{aligned} \tag{8.1}
$$

Note that the first bit u_0 to enter the register has affected all 5 outputs. As the next bit u_5 enters the shift register u_0 will cease to contribute to the output sequence. In a register with m stages each input contributes towards $m + 1$ outputs.

Example 8.1

Consider the shift register shown in Fig. 8.1 and let $g_0 = 1$, $g_1 = 1$, $g_2 = 0$, $g_3 = 0$ and $g_4 = 1$. Determine the output sequence for the input sequence $\boldsymbol{u} = (1\ 1\ 0\ 1\ 0\ 1\ \ldots)$.

The first 5 inputs are $u_0 = 1$, $u_1 = 1$, $u_2 = 0$, $u_3 = 1$, and $u_4 = 0$. Using eqns 8.1 gives

$$
\begin{aligned}
v_0 &= 1 \cdot 1 = 1 \\
v_1 &= 1 \cdot 1 + 1 \cdot 1 = 0 \\
v_2 &= 0 \cdot 1 + 1 \cdot 1 + 1 \cdot 0 = 1 \\
v_3 &= 1 \cdot 1 + 0 \cdot 1 + 1 \cdot 0 + 1 \cdot 0 = 1 \\
v_4 &= 0 \cdot 1 + 1 \cdot 1 + 0 \cdot 0 + 1 \cdot 0 + 1 \cdot 1 = 0
\end{aligned}
$$

as the first 5 bits of the output sequence. The 6th bit is given by

$$
\begin{aligned}
v_5 &= u_5 g_0 + u_4 g_1 + u_3 g_2 + u_2 g_3 + u_1 g_4 \\
&= 1 \cdot 1 + 0 \cdot 1 + 1 \cdot 0 + 0 \cdot 0 + 1 \cdot 1 \\
&= 0.
\end{aligned}
$$

The output sequence is therefore $(1\ 0\ 1\ 1\ 0\ 0 \ldots)$. □

For a finite input sequence, we need to ensure that the last input feeds through the register and appears at the output. To achieve this an additional m zero inputs are required. Consider Fig. 8.1 again, with the input sequence $\boldsymbol{u} = (u_0\, u_1\, u_2\, u_3\, u_4\, u_5\, u_6\, u_7)$. When u_7 enters the register the output is

$$v_7 = u_7 g_0 + u_6 g_1 + u_5 g_2 + u_4 g_3 + u_3 g_4.$$

Assuming now an additional 4 inputs u_8, u_9, u_{10}, and u_{11} then there are a further 4 outputs

$$\begin{aligned}
v_8 &= u_8g_0 + u_7g_1 + u_6g_2 + u_5g_3 + u_4g_4 \\
v_9 &= u_9g_0 + u_8g_1 + u_7g_2 + u_6g_3 + u_5g_4 \\
v_{10} &= u_{10}g_0 + u_9g_1 + u_8g_2 + u_7g_3 + u_6g_4 \\
v_{11} &= u_{11}g_0 + u_{10}g_1 + u_9g_2 + u_8g_3 + u_7g_4
\end{aligned}$$

and setting $u_8 = u_9 = u_{10} = u_{11} = 0$ gives

$$\begin{aligned}
v_8 &= u_7g_1 + u_6g_2 + u_5g_3 + u_4g_4 \\
v_9 &= u_7g_2 + u_6g_3 + u_5g_4 \\
v_{10} &= u_7g_3 + u_6g_4 \\
v_{11} &= u_7g_4.
\end{aligned}$$

All the 7 inputs have now left the shift register and the output sequence is completed. Any further zero inputs will give a zero output. By feeding in the additional m 0 bits the register is said to be cleared or returned to its *zero state*. The m 0 bits can be appended to the input sequence $\boldsymbol{u}$ but should not be confused with the information bits.

Example 8.2

Figure 8.2 shows a shift register with 3 stages and $g_0 = 1$, $g_1 = 0$, $g_2 = 1$, and $g_3 = 1$. Determine the output sequence given the input sequence $\boldsymbol{u} = (1\ 0\ 0\ 1)$.

As $\boldsymbol{u} = (u_0\, u_1\, u_2\, u_3)$ enters the shift register, the output from the adder is

$$\begin{aligned}
v_0 &= u_0g_0 \\
v_1 &= u_1g_0 + u_0g_1 \\
v_2 &= u_2g_0 + u_1g_1 + u_0g_2 \\
v_3 &= u_3g_0 + u_2g_1 + u_1g_2 + u_0g_3.
\end{aligned}$$

Substituting the values of g_0, g_1, g_2, and g_3 gives

$$\begin{aligned}
v_0 &= u_0 \\
v_1 &= u_1 \\
v_2 &= u_2 + u_0 \\
v_3 &= u_3 + u_1 + u_0
\end{aligned}$$

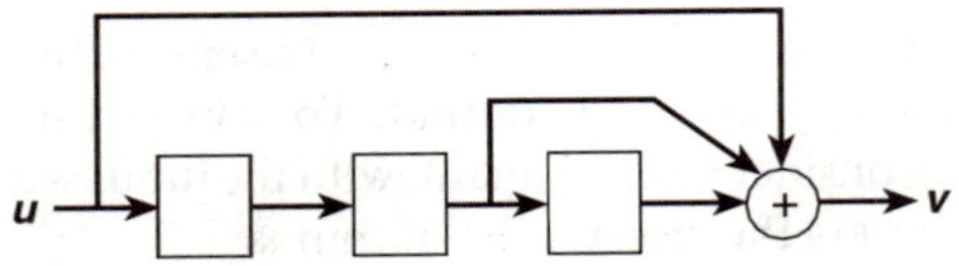

Fig. 8.2 Register with 3 stages.

and for $u_0 = 1$, $u_1 = 0$, $u_2 = 0$, and $u_3 = 1$, we get $v_0 = 1$, $v_1 = 0$, $v_2 = 1$, and $v_3 = 0$. To return the register to its zero state we need another 3 inputs, $u_4 = u_5 = u_6 = 0$. This gives

$$\begin{aligned} v_4 &= u_4 g_0 + u_3 g_1 + u_2 g_2 + u_1 g_3 = 0 \\ v_5 &= u_5 g_0 + u_4 g_1 + u_3 g_2 + u_2 g_3 = 1 \\ v_6 &= u_6 g_0 + u_5 g_1 + u_4 g_2 + u_3 g_3 = 1. \end{aligned}$$

The output sequence is therefore $\boldsymbol{v} = (v_0\, v_1\, v_2\, v_3\, v_4\, v_5\, v_6) = (1\,0\,1\,0\,0\,1\,1)$. □

Returning now to eqns 8.1, for the shift register shown in Fig. 8.1, given the input u_j the output from the modulo-2 adder is

$$v_j = u_j g_0 + u_{j-1} g_1 + u_{j-2} g_2 + u_{j-3} g_3 + u_{j-4} g_4$$

or more concisely

$$v_j = \sum_{r=0}^{4} u_{j-r} g_r$$

where $u_{j-r} = 0$ if $j < r$ (this defines $u_{-1} = u_{-2} = u_{-3} = u_{-4} = 0$). For a shift register with m stages and links $g_0, g_1, g_2, \ldots, g_m$ feeding into a modulo-2 adder, the output from the adder is given by

$$v_j = \sum_{r=0}^{m} u_{j-r} g_r \tag{8.2}$$

where again $u_{j-r} = 0$ if $j < r$. Equation 8.2 shows that each output v_j is a *convolution* of $(m+1)$ inputs with $g_0, g_1, \ldots, g_m$. We can think of convolution as a process in which a *window* of $(m+1)$ bits are multiplied by $g_0, g_1, \ldots, g_m$ and added together. The window is then slid along the input sequence by 1 bit and multiplication and summation repeated. The process continues until the window has covered the entire input sequence. Equation 8.2 can be expressed in terms of the input and output sequences as

$$\boldsymbol{v} = \boldsymbol{u} * \boldsymbol{g} \tag{8.3}$$

where the operation $*$ denotes convolution and $\boldsymbol{g} = (g_0\, g_1 \ldots g_m)$ is referred to as a *generator sequence*. The generator sequence can be obtained by inspection of the shift register or alternatively by considering the *impulse response* when the input sequence is a 1 followed by m 0s, i.e. $\boldsymbol{u} = (1\,0\,0\,0 \ldots 0)$. For example, consider again Fig. 8.1 this time with input sequence $\boldsymbol{u} = (u_0\, u_1\, u_2\, u_3\, u_4)$ where $u_0 = 1$ and $u_1 = u_2 = u_3 = u_4 = 0$. Substituting u_0, u_1, u_2, u_3, and u_4 into eqns 8.1 gives

$$\begin{aligned} v_0 &= 1g_0 = g_0 \\ v_1 &= 0g_0 + 1g_1 = g_1 \\ v_2 &= 0g_0 + 0g_1 + 1g_2 = g_2 \\ v_3 &= 0g_0 + 0g_1 + 0g_2 + 1g_3 = g_3 \\ v_4 &= 0g_0 + 0g_1 + 0g_2 + 0g_3 + 1g_4 = g_4. \end{aligned}$$

Hence the output sequence is

$$\boldsymbol{v} = (v_0\ v_1\ v_2\ v_3\ v_4) = (g_0\ g_1\ g_2\ g_3\ g_4) = \boldsymbol{g}$$

and therefore the impulse response gives the generator sequence $\boldsymbol{g}$. The generator sequence $\boldsymbol{g}$ should not be interpreted as the generator sequence of the shift register but rather of the output from the modulo-2 adder. A line of stages feeding into more than one modulo-2 adder will have a generator sequence for each adder. Example 8.3 below illustrates this.

Example 8.3

Figure 8.3 shows a shift register with 3 stages feeding into two modulo-2 adders. The output $\boldsymbol{v}_1$ does not have an input directly from $\boldsymbol{u}$ nor from the first stage, but has inputs from the second and third stages. Hence the generator sequence of $\boldsymbol{v}_1$ is $\boldsymbol{g}_1 = (0\ 0\ 1\ 1)$. Likewise we can see that the generator sequence for $\boldsymbol{v}_2$ is $\boldsymbol{g}_2 = (1\ 0\ 1\ 1)$. Using eqn 8.2 gives

$$\begin{aligned}
v_0 &= u_0g_0 + u_{-1}g_1 + u_{-2}g_2 + u_{-3}g_3 \\
v_1 &= u_1g_0 + u_0g_1 + u_{-1}g_2 + u_{-2}g_3 \\
v_2 &= u_2g_0 + u_1g_1 + u_0g_2 + u_{-1}g_3 \\
v_3 &= u_3g_0 + u_2g_1 + u_1g_2 + u_0g_3 \\
v_4 &= u_4g_0 + u_3g_1 + u_2g_2 + u_1g_3
\end{aligned}$$

for an input $\boldsymbol{u} = (u_0\ u_1\ u_2\ u_3\ u_4)$ and substituting the values of $\boldsymbol{g}_1$ and $\boldsymbol{g}_2$ we get

$$\begin{aligned}
v_0 &= 0 \\
v_1 &= 0 \\
v_2 &= u_0 \\
v_3 &= u_1 + u_0 \\
v_4 &= u_2 + u_1
\end{aligned}$$

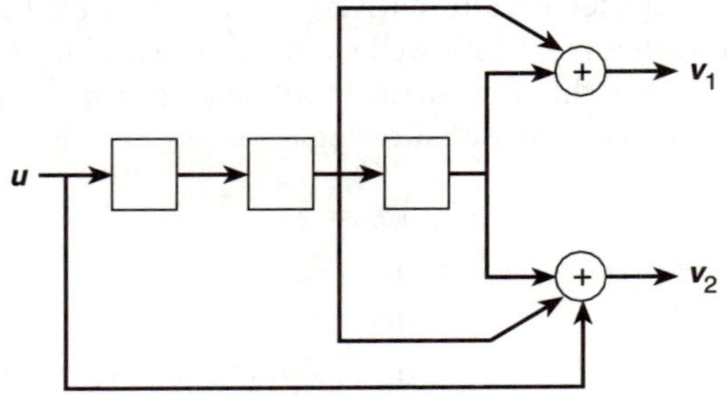

Fig. 8.3 Register with 2 outputs.

and

$$v_0 = u_0$$
$$v_1 = u_1$$
$$v_2 = u_2 + u_0$$
$$v_3 = u_3 + u_1 + u_0$$
$$v_4 = u_4 + u_2 + u_1$$

as the components of $\boldsymbol{v}_1$ and $\boldsymbol{v}_2$ respectively. □

8.2 Encoding convolutional codes

With the notion of convolution now established, we move on to looking at encoding convolutional codes. Figure 8.4 shows an encoder for an arbitrary (n, k, m) convolutional code. The encoder consists of a bank of k linear-feedforward shift registers, which do not necessarily have the same number of stages. For an (n, k, m) convolutional code, the maximum number of stages m in any shift register is known as the *memory order* of the code. The encoder output is taken from n modulo-2

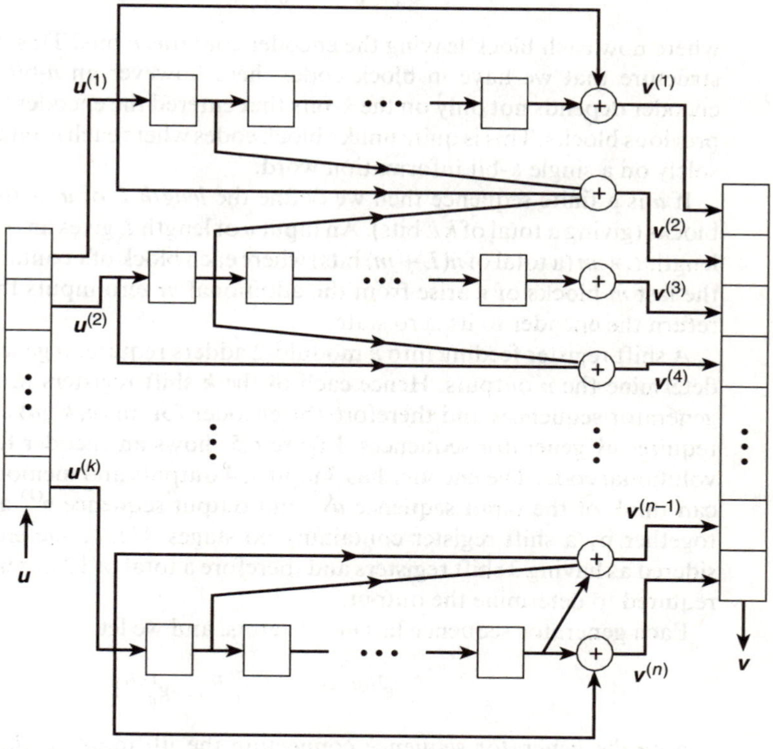

Fig. 8.4 Encoder for an (n, k, m) convolutional code.

adders whose inputs are taken from various stages (depending upon the particular code). Each bit in the encoder can affect any of the n outputs, and given that a bit can stay in the encoder for up to $m+1$ inputs then each bit can affect up to $n(m+1)$ output bits. This measure of the extent to which an input bit affects the output is referred to as the *constraint length.*

The input sequence to the ith shift register is denoted by

$$\boldsymbol{u}^{(i)} = (u_0^{(i)} u_1^{(i)} u_2^{(i)} \ldots) \tag{8.4}$$

and bits are fed into their respective shift register one bit at a time. Viewing the encoder on the whole, bits enter in blocks of k bits at a time, and we can express the input sequence to the encoder as

$$\boldsymbol{u} = (u_0^{(1)} u_0^{(2)} \ldots u_0^{(k)}, u_1^{(1)} u_1^{(2)} \ldots u_1^{(k)}, \ldots). \tag{8.5}$$

Here the first input to the encoder is the block of k bits $u_0^{(1)} u_0^{(2)} \ldots u_0^{(k)}$ followed by the block $u_1^{(1)} u_1^{(2)} \ldots u_1^{(k)}$ and so forth. The output sequence of the jth modulo-2 adder is

$$\boldsymbol{v}^{(j)} = (v_0^{(j)} v_1^{(j)} v_2^{(j)} \ldots) \tag{8.6}$$

and again viewing the encoder on the whole the output sequence is

$$\boldsymbol{v} = (v_0^{(1)} v_0^{(2)} \ldots v_0^{(n)}, v_1^{(1)} v_1^{(2)} \ldots v_1^{(n)}, \ldots) \tag{8.7}$$

where now each block leaving the encoder contains n bits. This may sound like the structure that we have in block codes, here however an n-bit block leaving the encoder depends not only on the k-bits that entered the encoder but also on up to m previous blocks. This is quite unlike block codes where each n-bit codeword depends solely on a single k-bit information word.

If $\boldsymbol{u}$ is a finite sequence then we define the *length* L of $\boldsymbol{u}$ as the number of k-bit blocks (giving a total of kL bits). An input $\boldsymbol{u}$ of length L gives an output sequence $\boldsymbol{v}$ of length $L+m$ (a total of $n(L+m)$ bits) where each block of $\boldsymbol{v}$ contains n bits and where the last m blocks of $\boldsymbol{v}$ arise from the additional m zero inputs that are required to return the encoder to its zero state.

A shift register feeding into n modulo-2 adders requires n generator sequences to determine the n outputs. Hence each of the k shift registers in Fig. 8.4 requires n generator sequences and therefore the encoder for an (n, k, m) convolutional code requires nk generator sequences. Figure 8.5 shows an encoder for the $(4, 3, 2)$ convolutional code. The encoder has 3 inputs, 4 outputs and memory order $m=2$. We can think of the input sequence $\boldsymbol{u}^{(1)}$ and output sequence $\boldsymbol{v}^{(1)}$ as being connected together by a shift register containing no stages. Hence the encoder can be considered as having 3 shift registers and therefore a total of 12 generator sequences are required to determine the output.

Each generator sequence has $m+1$ terms, and we let

$$\boldsymbol{g}^{(i,j)} = (g_0^{(i,j)} g_1^{(i,j)} \ldots g_m^{(i,j)})$$

denote the generator sequence connecting the ith input to the jth output. Then by considering the impulse response of each shift register or by inspecting the

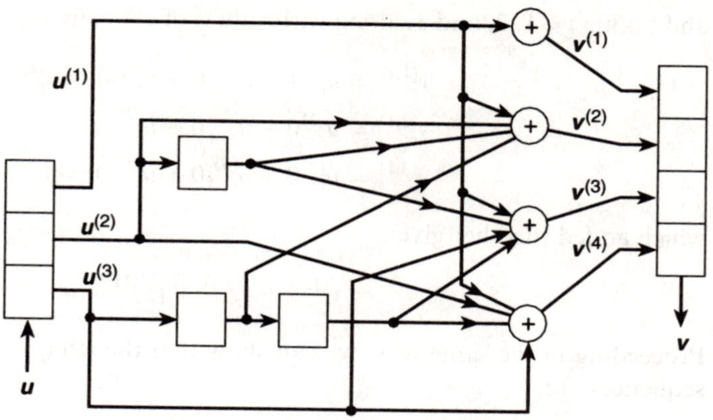

Fig. 8.5 Encoder for the (4, 3, 2) convolutional code.

connections to each adder we find that

$$\begin{aligned} g^{(1,1)} &= (1\ 0\ 0),\ g^{(1,2)} = (1\ 0\ 0),\ g^{(1,3)} = (1\ 0\ 0),\ g^{(1,4)} = (1\ 0\ 0) \\ g^{(2,1)} &= (0\ 0\ 0),\ g^{(2,2)} = (1\ 1\ 0),\ g^{(2,3)} = (0\ 1\ 0),\ g^{(2,4)} = (1\ 0\ 0) \\ g^{(3,1)} &= (0\ 0\ 0),\ g^{(3,2)} = (0\ 1\ 0),\ g^{(3,3)} = (1\ 0\ 1),\ g^{(3,4)} = (1\ 0\ 1). \end{aligned} \tag{8.8}$$

In Section 8.1 we saw that given a generator sequence $\boldsymbol{g}$, the output sequence $\boldsymbol{v}$ for an input sequence $\boldsymbol{u}$ is $\boldsymbol{v} = \boldsymbol{u} * \boldsymbol{g}$ (eqn 8.3). For the encoder shown in Fig. 8.5, each output sequence has contributions from up to 3 inputs and therefore eqn 8.3 has to be modified to take the additional inputs into account. This is achieved by adding the term $\boldsymbol{u}^{(i)} * \boldsymbol{g}^{(i,j)}$ term to $\boldsymbol{v}^{(j)}$ for each input $\boldsymbol{u}^{(i)}$. For convenience we define

$$\boldsymbol{v}^{(i,j)} = \boldsymbol{u}^{(i)} * \boldsymbol{g}^{(i,j)} \tag{8.9}$$

as the jth output arising from the ith input. The rth component of $\boldsymbol{v}^{(i,j)}$ is

$$v_r^{(i,j)} = u_r^{(i)} g_0^{(i,j)} + u_{r-1}^{(i)} g_1^{(i,j)} + u_{r-2}^{(i)} g_2^{(i,j)}. \tag{8.10}$$

The jth output sequence can be expressed as

$$\boldsymbol{v}^{(j)} = \boldsymbol{v}^{(1,j)} + \boldsymbol{v}^{(2,j)} + \boldsymbol{v}^{(3,j)} \tag{8.11}$$

which has

$$v_r^{(j)} = v_r^{(1,j)} + v_r^{(2,j)} + v_r^{(3,j)} \tag{8.12}$$

as its rth component. From eqns 8.10 and 8.12 the components of each output sequence can be determined. Consider first the components of the output sequence $\boldsymbol{v}^{(1)}$. The rth component of $\boldsymbol{v}^{(1)}$ arising from the ith input is

$$v_r^{(i,1)} = u_r^{(i)} g_0^{(i,1)} + u_{r-1}^{(i)} g_1^{(i,1)} + u_{r-2}^{(i)} g_2^{(i,1)}$$

and taking $i = 1, 2$, and 3, along with values of $g^{(i,j)}$ already given, we obtain

$$\begin{aligned} v_r^{(1,1)} &= u_r^{(1)}1 + u_{r-1}^{(1)}0 + u_{r-2}^{(1)}0 = u_r^{(1)} \\ v_r^{(2,1)} &= u_r^{(2)}0 + u_{r-1}^{(2)}0 + u_{r-2}^{(2)}0 = 0 \\ v_r^{(3,1)} &= u_r^{(3)}0 + u_{r-1}^{(3)}0 + u_{r-2}^{(3)}0 = 0 \end{aligned}$$

which added together give

$$v_r^{(1)} = v_r^{(1,1)} + v_r^{(2,1)} + v_r^{(3,1)} = u_r^{(1)}.$$

Proceeding in the same way, we can show that the rth component of all 4 output sequences are

$$\begin{aligned} v_r^{(1)} &= u_r^{(1)} \\ v_r^{(2)} &= u_r^{(1)} + u_r^{(2)} + u_{r-1}^{(2)} + u_{r-1}^{(3)} \\ v_r^{(3)} &= u_r^{(1)} + u_{r-1}^{(2)} + u_r^{(3)} + u_{r-2}^{(3)} \\ v_r^{(4)} &= u_r^{(1)} + u_r^{(2)} + u_r^{(3)} + u_{r-2}^{(3)} \end{aligned} \qquad (8.13)$$

and the output sequence leaving the encoder is

$$\boldsymbol{v} = (v_0^{(1)}v_0^{(2)}v_0^{(3)}v_0^{(4)}, v_1^{(1)}v_1^{(2)}v_1^{(3)}v_1^{(4)}, v_2^{(1)}v_2^{(2)}v_2^{(3)}v_2^{(4)} \ldots).$$

Example 8.4
Determine the output sequence from the (4, 3, 2) convolutional encoder, shown in Fig. 8.5, given the input sequences $\boldsymbol{u}^{(1)} = (1\ 0\ 1)$, $\boldsymbol{u}^{(2)} = (1\ 1\ 0)$, and $\boldsymbol{u}^{(3)} = (0\ 1\ 1)$.

The input sequences are of the form $\boldsymbol{u} = (u_0\, u_1\, u_2)$ with $u_r = 0$ if $r < 0$. We start by taking $r = 0$, then from eqns 8.13

$$\begin{aligned} v_0^{(1)} &= u_0^{(1)} \\ v_0^{(2)} &= u_0^{(1)} + u_0^{(2)} \\ v_0^{(3)} &= u_0^{(1)} + u_0^{(3)} \\ v_0^{(4)} &= u_0^{(1)} + u_0^{(2)} + u_0^{(3)} \end{aligned}$$

and substituting $u_0^{(1)} = 1$, $u_0^{(2)} = 1$, $u_0^{(3)} = 0$ gives $v_0^{(1)} = 1$, $v_0^{(2)} = 0$, $v_0^{(3)} = 1$, and $v_0^{(4)} = 0$.

Next take $r = 1$, this gives

$$\begin{aligned} v_1^{(1)} &= u_1^{(1)} = 0 \\ v_1^{(2)} &= u_1^{(1)} + u_1^{(2)} + u_0^{(2)} + u_0^{(3)} = 0 \\ v_1^{(3)} &= u_1^{(1)} + u_0^{(2)} + u_1^{(3)} = 0 \\ v_1^{(4)} &= u_1^{(1)} + u_1^{(2)} + u_1^{(3)} = 0 \end{aligned}$$

and for $r=2$

$$\begin{aligned}
v_2^{(1)} &= u_2^{(1)} = 1 \\
v_2^{(2)} &= u_2^{(1)} + u_2^{(2)} + u_1^{(2)} + u_1^{(3)} = 1 \\
v_2^{(3)} &= u_2^{(1)} + u_1^{(2)} + u_2^{(3)} + u_0^{(3)} = 1 \\
v_2^{(4)} &= u_2^{(1)} + u_2^{(2)} + u_2^{(3)} + u_0^{(3)} = 0.
\end{aligned}$$

Because the inputs are finite sequences and $m=2$ we need to consider 2 more 0 inputs to return the encoder back to the zero state. This requires

$$\begin{aligned}
u_3^{(1)} &= u_3^{(2)} = u_3^{(3)} = 0 \\
u_4^{(1)} &= u_4^{(2)} = u_4^{(3)} = 0.
\end{aligned}$$

For $r=3$ we get

$$\begin{aligned}
v_3^{(1)} &= u_3^{(1)} = 0 \\
v_3^{(2)} &= u_3^{(1)} + u_3^{(2)} + u_2^{(2)} + u_2^{(3)} = 1 \\
v_3^{(3)} &= u_3^{(1)} + u_2^{(2)} + u_3^{(3)} + u_1^{(3)} = 1 \\
v_3^{(4)} &= u_3^{(1)} + u_3^{(2)} + u_3^{(3)} + u_1^{(3)} = 1
\end{aligned}$$

and $r=4$ gives

$$\begin{aligned}
v_4^{(1)} &= u_4^{(1)} = 0 \\
v_4^{(2)} &= u_4^{(1)} + u_4^{(2)} + u_3^{(2)} + u_3^{(3)} = 0 \\
v_4^{(3)} &= u_4^{(1)} + u_3^{(2)} + u_4^{(3)} + u_2^{(3)} = 1 \\
v_4^{(4)} &= u_4^{(1)} + u_4^{(2)} + u_4^{(3)} + u_2^{(3)} = 1.
\end{aligned}$$

Therefore the output sequence is

$$\boldsymbol{v} = (1\,0\,1\,0,\ 0\,0\,0\,0,\ 1\,1\,1\,0,\ 0\,1\,1\,1,\ 0\,0\,1\,1). \qquad \square$$

The rth component of the output sequence, given by eqns 8.13, can be obtained directly from the encoder shown in Fig. 8.5. Let's assume that the input sequence is at the rth component, so that $u_r^{(1)}$, $u_r^{(2)}$, and $u_r^{(3)}$ are the inputs, then the 4 outputs can be determined by inspecting the encoder. The output $\boldsymbol{v}^{(1)}$ has only one input to its modulo-2 adder, namely that which arrives directly from $\boldsymbol{u}^{(1)}$ and so $v_r^{(1)} = u_r^{(1)}$, as given by eqns 8.13. The output $\boldsymbol{v}^{(2)}$ has:

(1) inputs directly from $\boldsymbol{u}^{(1)}$, so contributing $u_r^{(1)}$ to $\boldsymbol{v}^{(2)}$;
(2) inputs directly from $\boldsymbol{u}^{(2)}$, which contributes $u_r^{(2)}$;
(3) a contribution from $\boldsymbol{u}^{(2)}$ that is delayed by 1 stage and therefore contributes $u_{r-1}^{(2)}$;
(4) a contribution from $\boldsymbol{u}^{(3)}$ that is also delayed by 1 stage and therefore contributes $u_{r-1}^{(3)}$ to $\boldsymbol{v}^{(2)}$.

Adding together the four contributions gives $u_r^{(1)} + u_r^{(2)} + u_{r-1}^{(2)} + u_{r-1}^{(3)}$ as the rth component of $\boldsymbol{v}^{(2)}$, which again is in agreement with that given by eqns 8.13. Likewise we can verify that $v_r^{(3)}$ and $v_r^{(4)}$ are as given by eqns 8.13.

As with blockcodes the error-control properties of convolutional codes depend on the distance characteristics of the resulting encoded sequences. However with convolutional codes there are several minimum distance measures, the most important measure being the minimum *free distance* defined as the minimum distance between any two encoded sequences. If two sequences $\boldsymbol{v}_1$ and $\boldsymbol{v}_2$ are of different length then zeros are added to the shorter corresponding input sequence $\boldsymbol{u}_1$ or $\boldsymbol{u}_2$ so that the sequences are the same length. Hence the definition of free distance includes all sequences and not just sequences with the same length. Convolutional codes are linear codes and therefore the sum of two encoded sequences $\boldsymbol{v}_1$ and $\boldsymbol{v}_2$ gives another encoded sequence $\boldsymbol{v}_3$ where the weight of $\boldsymbol{v}_3$ equals the distance between $\boldsymbol{v}_1$ and $\boldsymbol{v}_2$, and $\boldsymbol{v}_3 \neq 0$ if $\boldsymbol{v}_1 \neq \boldsymbol{v}_2$. Hence the free distance of a convolutional code is given by the minimum-weight sequence of any length produced by a nonzero input sequence.

8.3 Generator matrices for convolutional codes

The use of convolution to derive the output sequences is rather tedious and as might be expected, the use of matrices can help to simplify encoding. We have seen that the output sequence $\boldsymbol{v} = \boldsymbol{u} * \boldsymbol{g}$, where $\boldsymbol{u}$ is an input sequence, $\boldsymbol{g}$ is a generator sequence and $*$ is the convolution operation. This equation can be expressed in matrix form, and the convolution operation replaced by matrix multiplication, by defining a *generator matrix G* such that

$$\boldsymbol{v} = \boldsymbol{uG}. \tag{8.14}$$

The matrix $\boldsymbol{G}$ is constructed from the components of the generator sequences and for an (n, k, m) convolutional code

$$\boldsymbol{G} = \begin{bmatrix} \boldsymbol{G}_0 & \boldsymbol{G}_1 & \boldsymbol{G}_2 & \cdots & \boldsymbol{G}_m & \boldsymbol{0} & \boldsymbol{0} & \cdots \\ \boldsymbol{0} & \boldsymbol{G}_0 & \boldsymbol{G}_1 & \boldsymbol{G}_2 & \cdots & \boldsymbol{G}_m & \boldsymbol{0} & \cdots \\ \boldsymbol{0} & \boldsymbol{0} & \boldsymbol{G}_0 & \boldsymbol{G}_1 & \boldsymbol{G}_2 & \cdots & \boldsymbol{G}_m & \cdots \\ \cdot & & & & & \cdots & & \\ \cdot & & & & & & \cdots & \\ \cdot & & & & & & & \cdots \end{bmatrix} \tag{8.15}$$

where $\boldsymbol{G}_r$ is the k by n matrix

$$\boldsymbol{G}_r = \begin{bmatrix} g_r^{(1,1)} & g_r^{(1,2)} & \cdots & g_r^{(1,n)} \\ g_r^{(2,1)} & g_r^{(2,2)} & \cdots & g_r^{(2,n)} \\ \vdots & & & \vdots \\ g_r^{(k,1)} & g_r^{(k,2)} & \cdots & g_r^{(k,n)} \end{bmatrix} \tag{8.16}$$

and $\mathbf{0}$ is a k by n zero matrix. Each row in $\boldsymbol{G}$ is obtained from the previous row by shifting all the matrices one place to the right. If $\boldsymbol{u}$ has finite length L then $\boldsymbol{G}$ has L rows and $L+m$ columns. On substituting eqn 8.16 into 8.15 the matrix $\boldsymbol{G}$ has kL rows and $n(L+m)$ columns. For the (4, 3, 2) code already considered, we have

$$\boldsymbol{G}_r = \begin{bmatrix} g_r^{(1,1)} & g_r^{(1,2)} & g_r^{(1,3)} & g_r^{(1,4)} \\ g_r^{(2,1)} & g_r^{(2,2)} & g_r^{(2,3)} & g_r^{(2,4)} \\ g_r^{(3,1)} & g_r^{(3,2)} & g_r^{(3,3)} & g_r^{(3,4)} \end{bmatrix}.$$

Using the generator sequences already given for the (4, 3, 2) code (see section 8.2) we get:

$$\boldsymbol{G}_0 = \begin{bmatrix} 1 & 1 & 1 & 1 \\ 0 & 1 & 0 & 1 \\ 0 & 0 & 1 & 1 \end{bmatrix}$$

$$\boldsymbol{G}_1 = \begin{bmatrix} 0 & 0 & 0 & 0 \\ 0 & 1 & 1 & 0 \\ 0 & 1 & 0 & 0 \end{bmatrix}$$

$$\boldsymbol{G}_2 = \begin{bmatrix} 0 & 0 & 0 & 0 \\ 0 & 0 & 0 & 0 \\ 0 & 0 & 1 & 1 \end{bmatrix}.$$

The generator matrix for the (4, 3, 2) convolutional code is therefore

$$\boldsymbol{G} = \left[\begin{array}{ccccc} 1\,1\,1\,1 & 0\,0\,0\,0 & 0\,0\,0\,0 & \mathbf{0} & \mathbf{0}\ \ldots \\ 0\,1\,0\,1 & 0\,1\,1\,0 & 0\,0\,0\,0 & & \\ 0\,0\,1\,1 & 0\,1\,0\,0 & 0\,0\,1\,1 & & \\ \mathbf{0} & 1\,1\,1\,1 & 0\,0\,0\,0 & 0\,0\,0\,0 & \mathbf{0}\ \ldots \\ & 0\,1\,0\,1 & 0\,1\,1\,0 & 0\,0\,0\,0 & \\ & 0\,0\,1\,1 & 0\,1\,0\,0 & 0\,0\,1\,1 & \\ \mathbf{0} & \mathbf{0} & 1\,1\,1\,1 & 0\,0\,0\,0 & 0\,0\,0\,0 \\ & & 0\,1\,0\,1 & 0\,1\,1\,0 & 0\,0\,0\,0 \\ \cdot & \cdot & 0\,0\,1\,1 & 0\,1\,0\,0 & 0\,0\,1\,1 \\ \cdot & \cdot & & \ldots & \\ \cdot & \cdot & & & \ldots \\ & & & & \ldots \end{array}\right] \qquad (8.17)$$

where $\mathbf{0}$ is a 3 by 4 zero matrix. Once the generator matrix is established finding the output sequence for a given input sequence is straightforward. Reconsider Example 8.4 in which the output sequence for the input sequences $\boldsymbol{u}^{(1)} = (1\ 0\ 1)$, $\boldsymbol{u}^{(2)} = (1\ 1\ 0)$ and $\boldsymbol{u}^{(3)} = (0\ 1\ 1)$ was found using convolution. The input sequence to the register is

$$\begin{aligned} \boldsymbol{u} &= (u_0^{(1)} u_0^{(2)} u_0^{(3)}, u_1^{(1)} u_1^{(2)} u_1^{(3)}, u_2^{(1)} u_2^{(2)} u_2^{(3)}) \\ &= (1\ 1\ 0,\ 0\ 1\ 1,\ 1\ 0\ 1). \end{aligned}$$

The generator matrix $\boldsymbol{G}$ is the 9 by 20 matrix

$$\boldsymbol{G} = \begin{bmatrix} 1111 & 0000 & 0000 & 0000 & 0000 \\ 0101 & 0110 & 0000 & 0000 & 0000 \\ 0011 & 0100 & 0011 & 0000 & 0000 \\ 0000 & 1111 & 0000 & 0000 & 0000 \\ 0000 & 0101 & 0110 & 0000 & 0000 \\ 0000 & 0011 & 0100 & 0011 & 0000 \\ 0000 & 0000 & 1111 & 0000 & 0000 \\ 0000 & 0000 & 0101 & 0110 & 0000 \\ 0000 & 0000 & 0011 & 0100 & 0011 \end{bmatrix}$$

and using $\boldsymbol{v} = \boldsymbol{u}\boldsymbol{G}$ gives

$$\boldsymbol{v} = (1\,0\,1\,0,\, 0\,0\,0\,0,\, 1\,1\,1\,0,\, 0\,1\,1\,1,\, 0\,0\,1\,1)$$

which is in agreement with the answer obtained in Example 8.4.

8.4 Generator polynomials for convolutional codes

Whilst the use of generator matrices is an improvement over the use of convolution, it is still nevertheless awkward due to the large size of the generator matrices. However, encoding can be further simplified through the use of *generator polynomials*. Consider the input sequence $\boldsymbol{u} = (u_0\, u_1\, u_2 \ldots)$, it can be represented by the polynomial

$$u(D) = u_0 + u_1 D + u_2 D^2 + \cdots \tag{8.18}$$

where D is the *unit-delay operator* and represents a delay of 1 bit, D^2 represents a 2-bit delay and so forth. In eqn 8.18 u_1 is interpreted as being delay by 1 bit relative to u_0, that is, it arrives 1 bit later than u_0. The bit u_2 is delayed by 2 bits relative to u_0 and 1 bit relative to u_1. Given an output sequence $\boldsymbol{v} = (v_0\, v_1\, v_2 \ldots)$ we can likewise write

$$v(D) = v_0 + v_1 D + v_2 D^2 + \cdots \tag{8.19}$$

The polynomials $u(D)$ and $v(D)$ are referred to as the *input polynomial* and *output polynomial* respectively. If $v(D)$ is the output arising from $u(D)$ then

$$v(D) = u(D)g(D) \tag{8.20}$$

where $g(D)$ is a *generator polynomial*. The product $u(D)g(D)$ is constructed using polynomial multiplication subject to modulo-2 arithmetic.

Example 8.5
Given the arbitrary polynomials $u(D) = 1 + D^2 + D^3$ and $g(D) = 1 + D^3$ then

$$\begin{aligned} v(D) &= (1 + D^2 + D^3)(1 + D^3) \\ &= 1 + D^3 + D^2 + D^5 + D^3 + D^6 \\ &= 1 + D^2 + D^5 + D^6. \end{aligned}$$

□

The generator polynomial can be determined directly from the encoder, in the same way as the generator sequences are determined. Each link to a modulo-2 adder contributes a D^r term to the generator polynomial where r is the number of stages that a bit has to pass through to arrive at the adder. The rth component of the generator sequence $\boldsymbol{g} = (g_0\, g_1\, g_2 \ldots g_m)$ is 0 or 1 depending on whether, after the rth stage there is a link to the adder, and therefore the rth component in the generator polynomial can be written as $g_r D^r$. Hence, for m stages, the generator polynomial can be expressed as

$$g(D) = g_0 + g_1 D + g_2 D^2 + \cdots + g_m D^m. \tag{8.21}$$

Example 8.6
Determine the generator polynomial of the shift register shown in Fig. 8.2. Hence find the output sequence given the input sequence $\boldsymbol{u} = (1\ 0\ 0\ 1)$.

We have already seen that the generator sequence of the encoder shown in Fig. 8.2 is $\boldsymbol{g} = (1\ 0\ 1\ 1)$ and the generator polynomial is therefore

$$g(D) = g_0 + g_1 D + g_2 D^2 + g_3 D^3 = 1 + D^2 + D^3.$$

The input polynomial corresponding to the input sequence $\boldsymbol{u} = (1\ 0\ 0\ 1)$ is

$$u(D) = u_0 + u_1 D + u_2 D^2 + u_3 D^3 = 1 + D^3$$

and the output polynomial is therefore

$$\begin{aligned} v(D) &= u(D)g(D) \\ &= (1 + D^3)(1 + D^2 + D^3) \\ &= 1 + D^2 + D^5 + D^6 \end{aligned}$$

giving the output sequence

$$\boldsymbol{v} = (v_0\ v_1\ v_2\ v_3\ v_4\ v_5\ v_6) = (1\ 0\ 1\ 0\ 0\ 1\ 1).$$

The results agree with that obtained using convolution in Example 8.2. □

An (n, k, m) convolutional code has k generator polynomials for each of the n output sequences and therefore a total of nk generator polynomials. Let

$$g^{(i,j)}(D) = \sum_{r=0}^{m} g_r^{(i,j)} D^r \tag{8.22}$$

be the generator polynomial for the jth output arising from the ith input, where the polynomial coefficients $g_r^{(i,j)}$ are the components of the generator sequence $\boldsymbol{g}^{(i,j)}$. The (4, 3, 2) encoder shown in Fig. 8.5 has 12 generator polynomials of the form

$$g^{(i,j)}(D) = g_0^{(i,j)} + g_1^{(i,j)} D + g_2^{(i,j)} D^2.$$

Taking $i = 1, 2$, and 3, and $j = 1, 2, 3$, and 4, along with the generator sequences given in Section 8.2, gives

$$g^{(1,1)}(D) = 1,\ g^{(1,2)}(D) = 1,\ g^{(1,3)}(D) = 1,\ g^{(1,4)}(D) = 1$$
$$g^{(2,1)}(D) = 0,\ g^{(2,2)}(D) = 1 + D,\ g^{(2,3)}(D) = D,\ g^{(2,4)}(D) = 1$$
$$g^{(3,1)}(D) = 0,\ g^{(3,2)}(D) = D,\ g^{(3,3)}(D) = 1 + D^2,\ g^{(3,4)}(D) = 1 + D^2.$$

The jth output polynomial is a linear combination of the contributions from each of the k inputs and is given by

$$v^{(j)}(D) = \sum_{i=1}^{k} u^{i}(D) g^{(i,j)}(D). \tag{8.23}$$

Example 8.7
Consider again the (4, 3, 2) code with input sequence $\boldsymbol{u}^{(1)} = (1\ 0\ 1)$, $\boldsymbol{u}^{(2)} = (1\ 1\ 0)$, and $\boldsymbol{u}^{(3)} = (0\ 1\ 1)$. The corresponding input polynomials are

$$u^{(1)}(D) = 1 + D^2$$
$$u^{(2)}(D) = 1 + D$$
$$u^{(3)}(D) = D + D^2.$$

Using eqn 8.23 and the generator polynomials $g^{(i,j)}(D)$ already derived, the first output polynomial is

$$\begin{aligned} v^{(1)}(D) &= u^{(1)}(D)g^{(1,1)}(D) + u^{(2)}(D)g^{(2,1)}(D) + u^{(3)}(D)g^{(3,1)}(D) \\ &= (1 + D^2)1 + (1 + D)0 + (D + D^2)0 \\ &= 1 + D^2 \end{aligned}$$

and the second

$$\begin{aligned} v^{(2)}(D) &= u^{(1)}(D)g^{(1,2)}(D) + u^{(2)}(D)g^{(2,2)}(D) + u^{(3)}(D)g^{(3,2)}(D) \\ &= (1 + D^2)1 + (1 + D)(1 + D) + (D + D^2)D \\ &= D^2 + D^3. \end{aligned}$$

Likewise $v^{(3)}(D)$ and $v^{(4)}(D)$ can be determined so giving

$$v^{(1)}(D) = 1 + D^2$$
$$v^{(2)}(D) = D^2 + D^3$$
$$v^{(3)}(D) = 1 + D^2 + D^3 + D^4$$
$$v^{(4)}(D) = D^3 + D^4.$$

□

The output polynomial $v(D)$ has contributions from $v^{(1)}(D), v^{(2)}(D), \ldots, v^{(n)}(D)$ suitably delayed to take into account the order in which bits leave the encoder and it can be shown that

$$v(D) = \sum_{j=1}^{n} D^{j-1} v^{(j)}(D^n). \tag{8.24}$$

The components of $\boldsymbol{v}$ are then given by the coefficients of $v(D)$.

Example 8.8
The output polynomial $v(D)$ for the (4, 3, 2) code, given by eqn 8.24 is

$$v(D) = v^{(1)}(D^4) + Dv^{(2)}(D^4) + D^2v^{(3)}(D^4) + D^3v^{(4)}(D^4)$$

and using $v^{(1)}(D)$, $v^{(2)}(D)$, $v^{(3)}(D)$, and $v^{(4)}(D)$ obtained in Example 8.7 gives

$$\begin{aligned} v(D) &= (1 + D^8) + D(D^8 + D^{12}) + D^2(1 + D^8 + D^{12} + D^{16}) + D^3(D^{12} + D^{16}) \\ &= 1 + D^2 + D^8 + D^9 + D^{10} + D^{13} + D^{14} + D^{15} + D^{18} + D^{19}. \end{aligned}$$

Expressing this as the vector

$$v = (v_0\ v_1\ v_2\ v_3,\ v_4\ v_5\ v_6\ v_7,\ v_8\ v_9\ v_{10}\ v_{11},\ v_{12}\ v_{13}\ v_{14}\ v_{15},\ v_{16}\ v_{17}\ v_{18}\ v_{19})$$

gives

$$v = (1\ 0\ 1\ 0,\ 0\ 0\ 0\ 0,\ 1\ 1\ 1\ 0,\ 0\ 1\ 1\ 1,\ 0\ 0\ 1\ 1).$$

This therefore is the output given the inputs $u^{(1)} = (1\ 0\ 1)$, $u^{(2)} = (1\ 1\ 0)$, and $u^{(3)} = (0\ 1\ 1)$ and is the same as that obtained before, using convolution and using matrices. □

The output polynomial $v(D)$ can be expressed directly in terms of the input polynomials, substituting eqn 8.23 into eqn 8.24 gives

$$v(D) = \sum_{i=1}^{k} u^{(i)}(D^n)g^{(i)}(D) \tag{8.25}$$

where

$$g^{(i)}(D) = \sum_{j=1}^{n} D^{j-1}g^{(i,j)}(D^n) \tag{8.26}$$

is the generator polynomial of the ith input summed over all n outputs.

Example 8.9
The (4, 3, 2) code has generator polynomials $g^{(1)}(D)$, $g^{(2)}(D)$, and $g^{(3)}(D)$ representing the individual contributions of the three inputs to the total output from the encoder. From eqn 8.26 we get

$$\begin{aligned} g^{(1)}(D) &= g^{(1,1)}(D^4) + Dg^{(1,2)}(D^4) + D^2g^{(1,3)}(D^4) + D^3g^{(1,4)}(D^4) \\ g^{(2)}(D) &= g^{(2,1)}(D^4) + Dg^{(2,2)}(D^4) + D^2g^{(2,3)}(D^4) + D^3g^{(2,4)}(D^4) \\ g^{(3)}(D) &= g^{(3,1)}(D^4) + Dg^{(3,2)}(D^4) + D^2g^{(3,3)}(D^4) + D^3g^{(3,4)}(D^4) \end{aligned}$$

and substituting the generator polynomials gives

$$\begin{aligned} g^{(1)}(D) &= 1 + D + D^2 + D^3 \\ g^{(2)}(D) &= D + D^3 + D^5 + D^6 \\ g^{(3)}(D) &= D^2 + D^3 + D^5 + D^{10} + D^{11}. \end{aligned}$$

If we once again consider the inputs $\boldsymbol{u}^{(1)} = (1\ 0\ 1)$, $\boldsymbol{u}^{(2)} = (1\ 1\ 0)$, and $\boldsymbol{u}^{(3)} = (0\ 1\ 1)$, so that $u^{(1)}(D) = 1 + D^2$, $u^{(2)}(D) = 1 + D$ and $u^{(3)}(D) = D + D^2$, then using eqn 8.25 gives

$$\begin{aligned} v(D) &= u^{(1)}(D^4)g^{(1)}(D) + u^{(2)}(D^4)g^{(2)}(D) + u^3(D^4)g^{(3)}(D) \\ &= (1 + D^8)(1 + D + D^2 + D^3) + (1 + D^4)(D + D^3 + D^5 + D^6) \\ &\quad + (D^4 + D^8)(D^2 + D^3 + D^5 + D^{10} + D^{11}) \\ &= 1 + D^2 + D^8 + D^9 + D^{10} + D^{13} + D^{14} + D^{15} + D^{18} + D^{19} \end{aligned}$$

as obtained in Example 8.8. □

8.5 Graphical representation of convolutional codes

Convolutional codes can be represented graphically using tree, trellis and state diagrams, all of which show the state of a register and the encoder output for all possible inputs. The *state* of a register is defined as the contents of the stages at a given point during encoding. The diagrams provide an interesting way of looking at the structure and encoding of convolutional codes, and furthermore, the trellis diagram plays an important role when decoding using the Viterbi algorithm (considered in Section 8.6).

We first consider *state diagrams.* Figure 8.6 shows an encoder for a (2, 1, 2) convolutional code. At any point during encoding, the encoder can be in one of the 4 states 00, 01, 10 and 11, referred to as states S_0, S_1, S_2, and S_3 respectively. The states are shown in Fig. 8.7 as *nodes* (filled circles) connected together by *branches* (solid or dashed lines) that represent transitions from one state to another, depending upon whether the input is 1 (solid line) or 0 (dashed line). The output occurring with each transition is shown next to the relevant branch. The state diagram gives a complete description of encoding in that for a given input the output and the next state of the encoder can be determined. To arrive at the state diagram we need to consider the operation of the encoder, taking in to account all possible transitions.

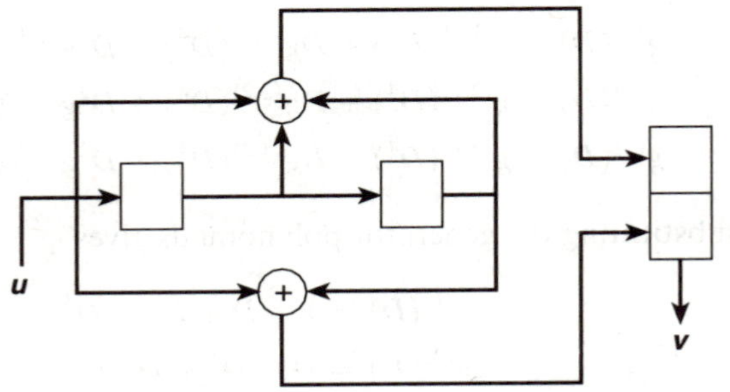

Fig. 8.6 A (2, 1, 2) convolutional encoder.

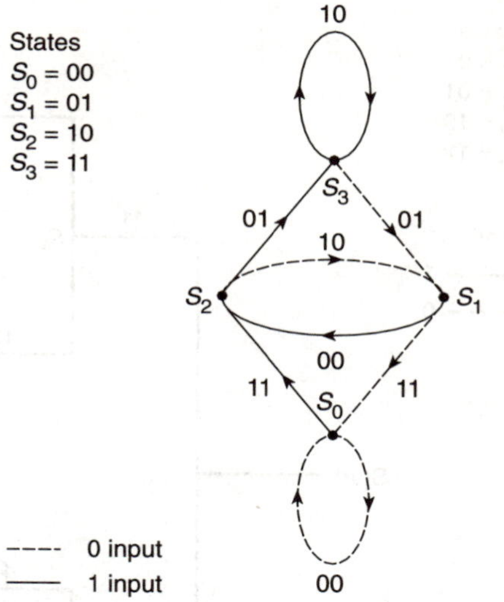

Fig. 8.7 State diagram for the (2, 1, 2) convolutional code.

Referring to Fig. 8.7 we assume that encoding starts at S_0, that is at the zero state with the stages at 00. Whilst the input stays at 0, transitions occur to the same state S_0 and the output is 00. At the first nonzero input, the stage contents change to 10 and the transition is to S_2 giving an output of 11. If the next input is also a 1, then the stage contents change to 11, the transition is to S_3 and the output is 01. Otherwise a 0 input gives a transition to S_1 and an output of 10. The state diagram is completed by considering transitions from S_3 and S_1 for inputs 0 and 1. Given an arbitrary input $\boldsymbol{u} = (u_0\, u_1\, u_2 \ldots)$ the output can be determined by following the transitions through the state diagram.

Example 8.10
Consider the input sequence $\boldsymbol{u} = (1\ 1\ 0\ 1\ 0\ 0)$. Referring to the state diagram in Fig. 8.7, and starting from the state S_0 we get

Input 1, transition to S_2 giving an output 11
Input 1, transition to S_3 giving an output 01
Input 0, transition to S_1 giving an output 01
Input 1, transition to S_2 giving an output 00
Input 0, transition to S_1 giving an output 10
Input 0, transition to S_0 giving an output 11

The output sequence is therefore $\boldsymbol{v} = (11, 01, 01, 00, 10, 11)$ □

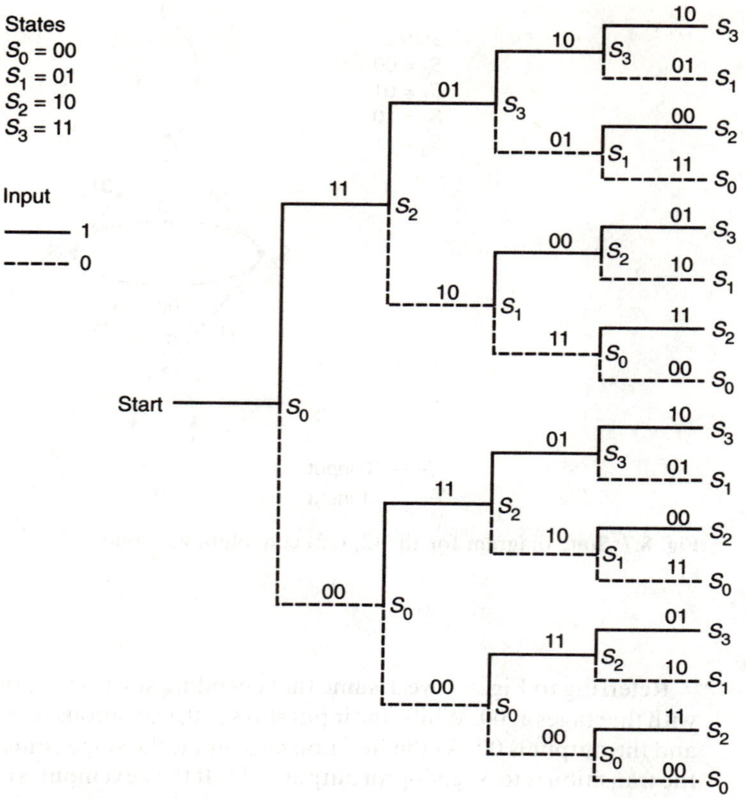

Fig. 8.8 Tree diagram for the (2, 1, 2) convolutional code.

Figure 8.8 shows a *tree diagram* for the (2, 1, 2) code. This again shows transitions from state to state, with corresponding outputs, for all inputs. The state and tree diagrams differ in that the latter shows the evolution of encoding with time. In the state diagram there is no way of knowing how a state was arrived at, from any specified state it is only possible to determine the next state and output for a given input. In the tree diagram moving from left to right represents the flow of time as the input sequence enters the encoder. At any node transitions can only occur to one of the two nodes to the right of the node. The output at each transition is shown above the horizontal portion of each branch. Encoding starts at the far left at S_0 and progresses from left to right. The tree diagram shown in Fig. 8.8 has been truncated after the 4th input, but theoretically the tree extends to the right infinitely.

A tree diagram gives an interesting graphical view of a convolutional code but is of limited practical use due to its size. However tree diagrams have a repetitive structure that allow the diagrams to be simplified whilst maintaining the basic feature of showing how encoding progresses with time. The resulting diagrams are known as *trellis diagrams* and Fig. 8.9 shows the trellis diagram for the (2, 1, 2) code. The trellis

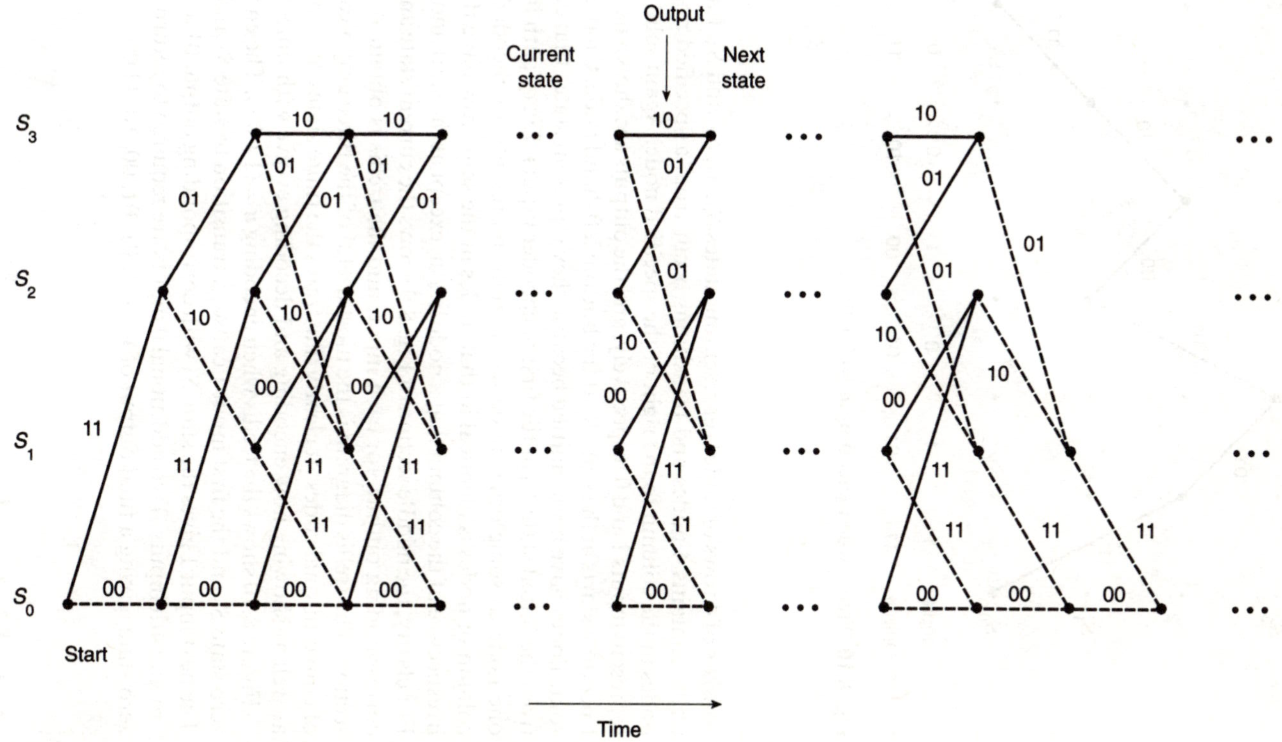

Fig. 8.9 Trellis diagram for the (2, 1, 2) convolutional code.

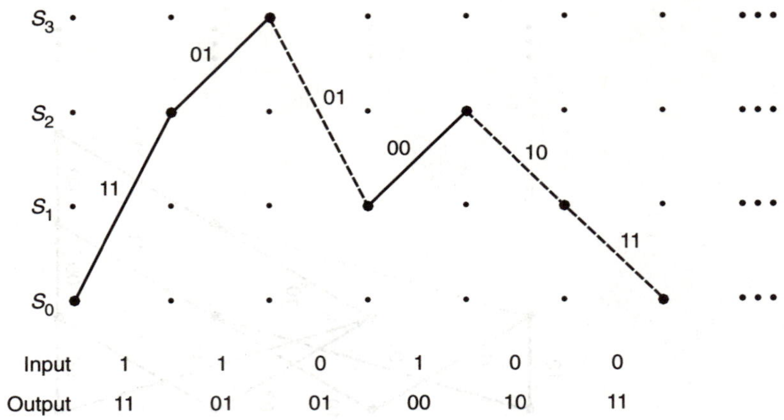

Fig. 8.10 Trellis when encoding $\boldsymbol{u} = (1\ 1\ 0\ 1)$.

consists of 4 rows of nodes representing states S_0, S_1, S_2, and S_3. Each column in the trellis contains 4 nodes and transitions occur from a specified node to one of two nodes in the column to the right of the specified node. Again solid and dashed lines represent inputs 1 and 0 respectively, and the outputs are shown next to each branch. Encoding starts at the state S_0 at the bottom left-hand node. On the left of the trellis, some nodes have been omitted because they represent states that cannot be reached from the initial state S_0 in the first m encoder inputs. Every path in the trellis, from one node to some other node to the right represents an encoded sequence. Each column of nodes is identical in that nodes in the same row have the same branches irrespective of the column that a node lies in, except at the start and end of the trellis. For clarity, the middle section of Fig. 8.9 shows the current state and next state, with corresponding output, for each input, and the trellis is obtained by repeating this section. The trellis diagram is the most useful of the three graphical representations of convolutional codes. It clearly illustrates the finite states of a convolutional code as well as showing how encoding and decoding evolve with time.

Figure 8.10 shows the trellis when encoding $\boldsymbol{u} = (1\ 1\ 0\ 1)$. The encoder starts at the zero state S_0 and the first input 1 causes a transition to state S_2 and an output of 11. The next input 1 gives the state S_3 with corresponding output 01 and so forth for the 3rd and 4th inputs. Two additional 0 inputs are required to return the encoder to its zero state, giving a final output of $\boldsymbol{v} = (11, 01, 01, 00, 10, 11)$.

8.6 The Viterbi decoder

A decoder for a convolutional code has to establish the path through the code trellis that corresponds to the decoder input $\boldsymbol{w}$. The Viterbi algorithm is a maximum-likelihood decoder that determines the path that $\boldsymbol{w}$ most likely corresponds to. This path is then taken to be the encoder output $\boldsymbol{v}$, from which the encoder input $\boldsymbol{u}$ can be derived. If no errors occur then $\boldsymbol{w}=\boldsymbol{v}$ and the path found by the decoder will be $\boldsymbol{v}$, so giving the correct $\boldsymbol{u}$. Recall that every path through the trellis is an encoded sequence, but the decoder input $\boldsymbol{w}$ may not be a valid path through the trellis.

To achieve maximum-likelihood decoding each branch in the trellis is assigned a *metric* which, for a binary-symmetric channel, is the Hamming distance between the branch output and the corresponding decoder input. Each path in the trellis is also assigned a metric that is the sum of the branch metrics forming the path. The aim of the decoder is to find the path that is the least distance away from the decoder input $\boldsymbol{w}$. Figure 8.11 shows the metrics for the (2, 1, 2) code with input (00, 11, 01, 10, 10). The branch metrics are shown in parenthesis alongside each branch. At each node two paths arrive and the smallest path metric is shown next to the node. The branch of the path with the larger metric is marked with an **X** to show that it has been deleted and is no longer of interest. The other branch and the path, with the lower metric, are referred to as the *survivor* branch and path respectively. If the path metrics are the same then either path can be taken as the survivor. The node is then labelled with a **T** to show that the path metrics are tied. Note that metric values depend on the decoder input and are not fixed attributes of a trellis.

As decoding evolves the only paths that survive are those with the smallest metrics and these are stored by the decoder. On completion of the decoder input, the path starting at S_0 and ending at S_0 going backwards through the trellis, from right to left, is the required maximum-likelihood path. This path can be uniquely followed when going backwards through the trellis because there is only one branch, the survivor, feeding forward into each node. Recall that when encoding the encoder starts and ends at the state S_0 and the decoder must likewise do so.

Figure 8.12 shows the trellis resulting from a Viterbi decoder with input $\boldsymbol{w}=(11, 10, 00, 01, 10, 01, 00, 10, 11)$ for the (2, 1, 2) code. Here $\boldsymbol{w}$ is the decoder input for an encoder output $\boldsymbol{v}$ that has incurred no errors, so $\boldsymbol{v}=\boldsymbol{w}$. Furthermore $\boldsymbol{v}$ is the encoded sequence for $\boldsymbol{u}=(1\,0\,1\,1\,1\,0\,1)$. As in Fig. 8.11 the path and branch metrics are shown along with the deleted branches, for each decoder input. At the end of the input the decoder starts at the state S_0 and follows the path backwards along the surviving branches. The dotted path shows the resulting maximum-likelihood path. Note that at each node in the dotted path the path metric is zero, this is to be expected as $\boldsymbol{w}$ contains no errors. To determine the decoder output, that the constructed path corresponds to, we need to know the decoder output at each branch, for clarity this is not shown in Fig. 8.12. However, referring to Fig. 8.9 we find that the path in Fig. 8.12 gives $\boldsymbol{v}=(11, 10, 00, 01, 10, 01, 00, 10, 11)$ and $\boldsymbol{u}=(1\,0\,1\,1\,1\,0\,1)$, where the last two 0s of $\boldsymbol{u}$ (for returning the encoder to its zero state) have been excluded. Decoding has therefore been successful.

Figure 8.13 shows the trellis when $\boldsymbol{v}=(11, 10, 00, 01, 10, 01, 00, 10, 11)$ incurs the 2-bit error $\boldsymbol{e}=(00, 00, 01, 00, 00, 10, 00, 00, 00)$ so that $\boldsymbol{w}=\boldsymbol{v}+\boldsymbol{e}=(11, 10, 01, 01, 10, 11, 00, 10, 11)$. The maximum-likelihood path is again shown by a dotted line. Note that for the first two inputs the branch and path metrics are 0, but on the third input (which contains the first error) they rise to 1. The path metric stays at 1 for the next two inputs (which are error free) but rises to 2 at the next input (which contains the second error). As there are no more errors the final path metric is 2, which means that all other paths through the trellis are at a distance of 2 or greater away from $\boldsymbol{w}$. The dotted path shown in Fig. 8.13 is the correct path for $\boldsymbol{v}$ (see Fig. 8.12) and therefore decoding has been successful, despite the incurred errors.

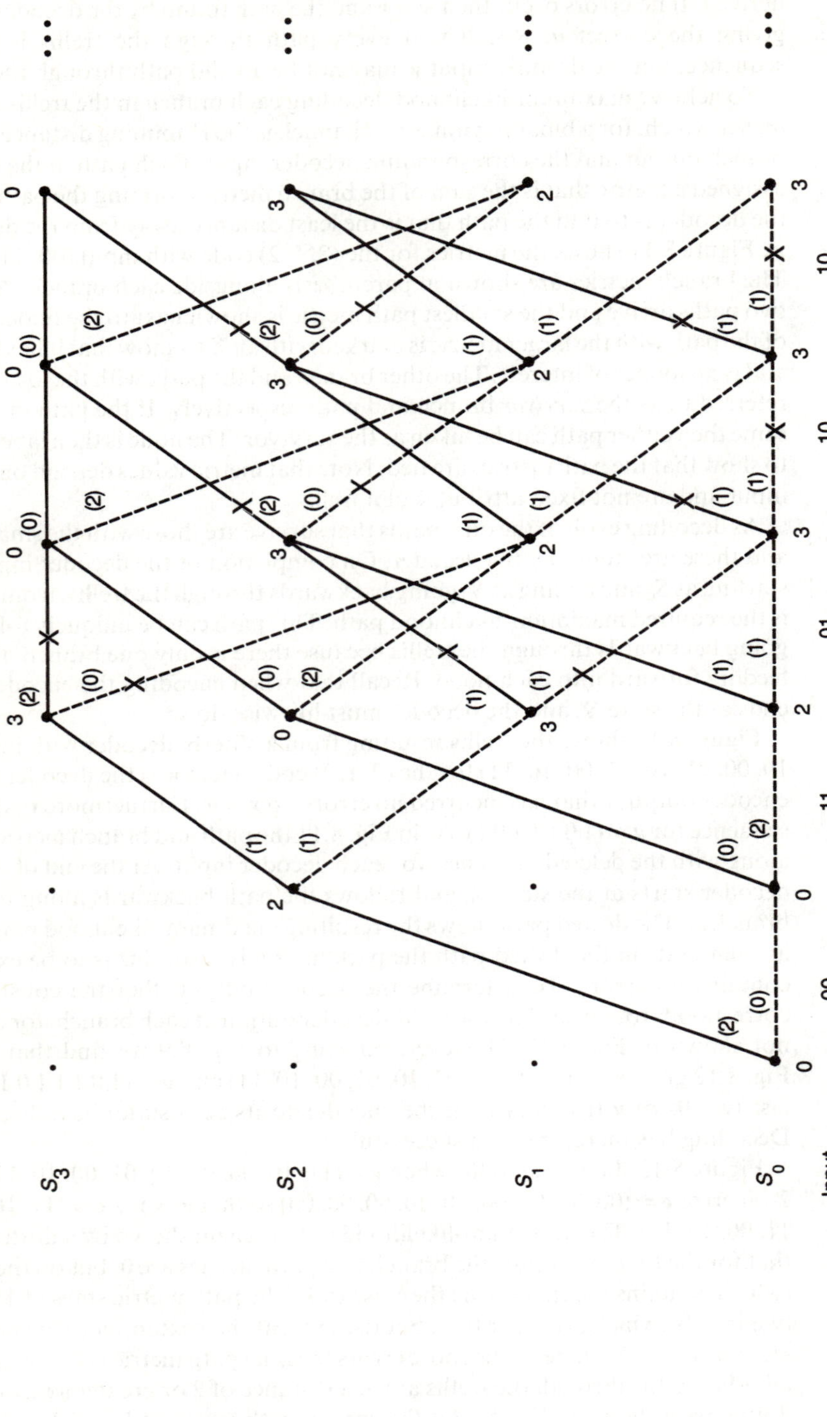

Fig. 8.11 Metrics and survivors in a trellis.

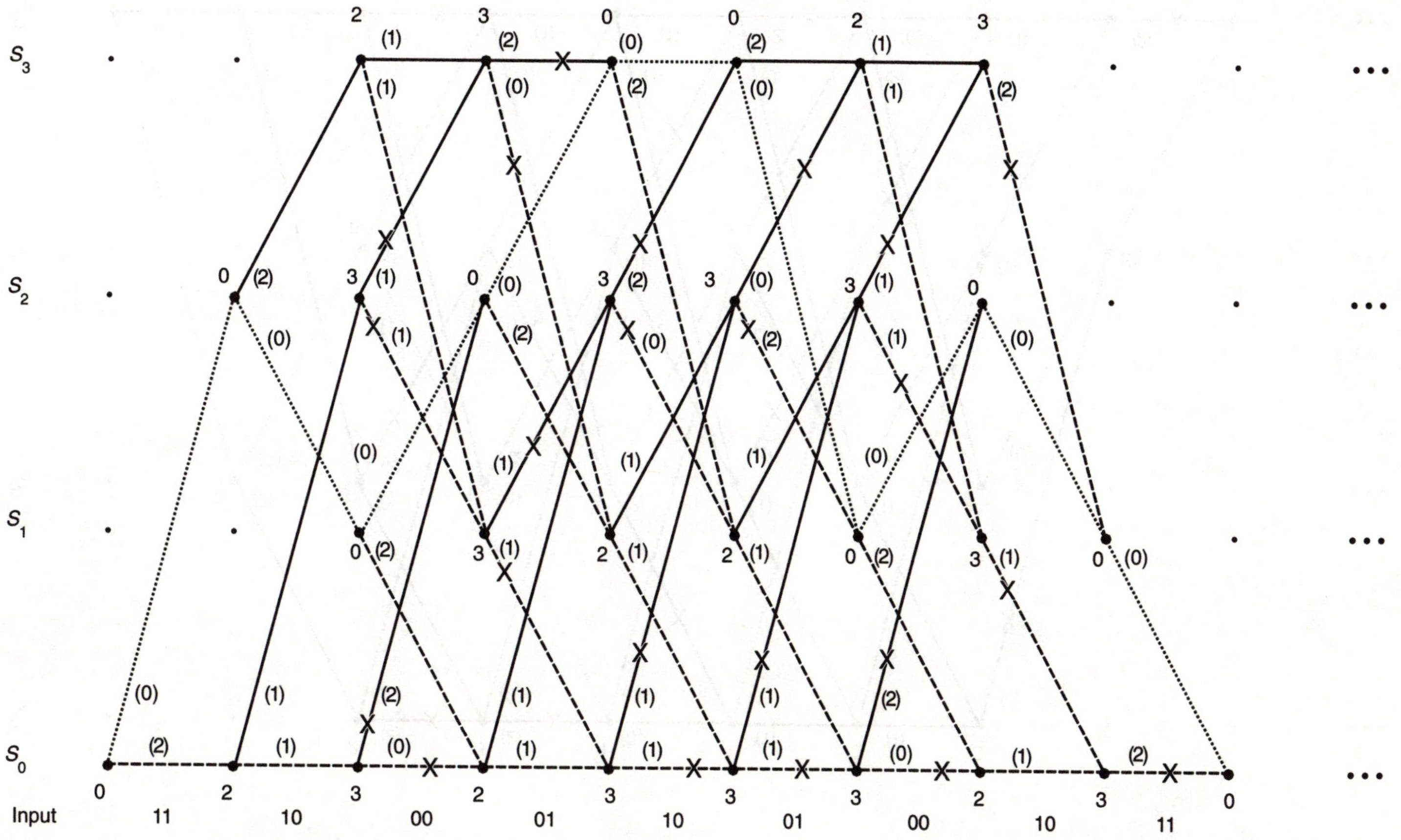

Fig. 8.12 Trellis for decoder input with no errors.

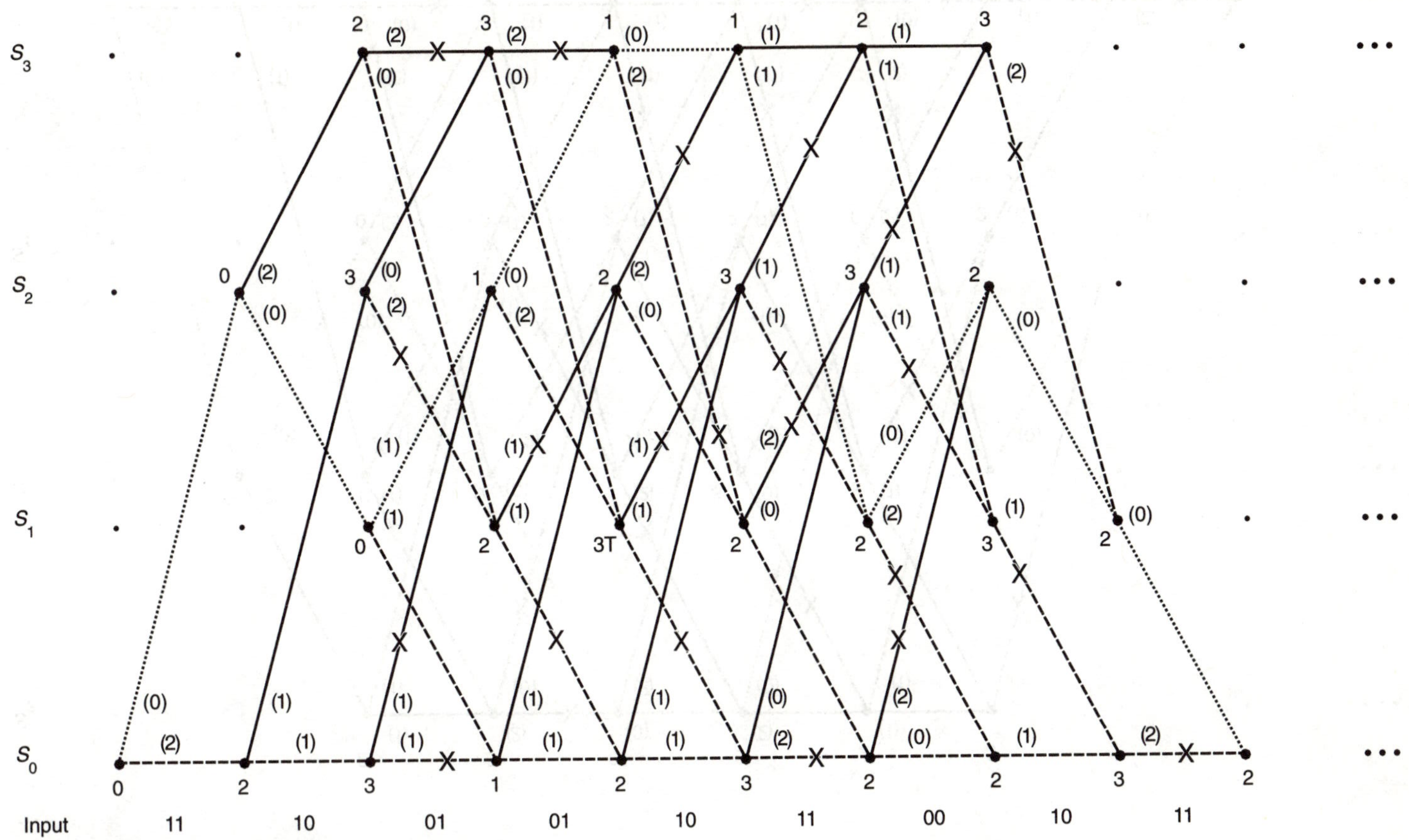

Fig. 8.13 Trellis for decoder input with 2 errors.

Problems

8.1 Determine the generator sequence of the shift register shown in Fig. 8.2. Hence, using convolution, determine the output sequences when the input sequences are
(i) (1 1 0 1 ...);
(ii) (1 1 0 1 1 1).

8.2 Determine the generator sequences for the two outputs of the shift register shown in Fig. 8.3.

8.3 Show that the generator sequences connecting the input to the output sequences of the (4, 3, 2) convolutional code, shown in Fig. 8.5, are given by eqns 8.8. Hence, using eqns 8.10 and 8.12, show that rth components of the output sequences are given by eqns 8.13.

8.4 Given the inputs $\boldsymbol{u}^{(1)} = (0\ 1\ 1\ 0)$, $\boldsymbol{u}^{(2)} = (1\ 0\ 0\ 1)$, and $\boldsymbol{u}^{(3)} = (0\ 1\ 0\ 1)$ to the (4, 3, 2) code use eqns 8.13 to determine the output sequence $\boldsymbol{v}$.

8.5 Draw an encoder for the (2, 1, 3) convolutional code with generator sequences $\boldsymbol{g}^{(1)} = (1\ 0\ 1\ 1)$ and $\boldsymbol{g}^{(2)} = (1\ 1\ 1\ 1)$.

8.6 Determine a generator matrix for the (2, 1, 3) code described in Problem 8.5. Hence find the output $\boldsymbol{v}$ when $\boldsymbol{u} = (0\ 1\ 0\ 1)$.

8.7 Using the generator matrix for the (4, 3, 2) code find $\boldsymbol{v}$ when $\boldsymbol{u}^{(1)} = (1\ 0\ 0\ 1)$, $\boldsymbol{u}^{(2)} = (1\ 1\ 0\ 0)$, and $\boldsymbol{u}^{(3)} = (0\ 1\ 0\ 1)$.

8.8. Determine the generator polynomials corresponding to the 12 generator sequences connecting the input sequences to the output sequences of the (4, 3, 2) code. Therefore determine the generator polynomials $g^{(1)}(D)$, $g^{(2)}(D)$ and $g^{(3)}(D)$, and find the output $\boldsymbol{v}$ when $\boldsymbol{u}^{(1)} = (1\ 0\ 0\ 1)$, $\boldsymbol{u}^{(2)} = (1\ 1\ 0\ 0)$, and $\boldsymbol{u}^{(3)} = (0\ 1\ 0\ 1)$.

Index